# Lecture Notes on Data Engineering and Communications Technologies

Volume 82



**Series Editor**

Fatos Xhafa, Technical University of Catalonia, Barcelona, Spain

The aim of the book series is to present cutting edge engineering approaches to data technologies and communications. It will publish latest advances on the engineering task of building and deploying distributed, scalable and reliable data infrastructures and communication systems.

The series will have a prominent applied focus on data technologies and communications with aim to promote the bridging from fundamental research on data science and networking to data engineering and communications that lead to industry products, business knowledge and standardisation.

Indexed by SCOPUS, INSPEC, EI Compendex.

All books published in the series are submitted for consideration in Web of Science.

More information about this series at http://www.springer.com/series/15362

Zhengbing Hu · Qingying Zhang ·
Sergey Petoukhov · Matthew He
Editors

# Advances in Artificial Systems for Logistics Engineering

Springer

*Editors*
Zhengbing Hu
School of Educational Information Technology
Central China Normal University
Wuhan, Hubei, China

Qingying Zhang
Department of Logistics Management
Wuhan Institute of Technology
Wuhan, China

Sergey Petoukhov
Mechanical Engineering Research Institute of the Russian Academy of Sciences
Moscow, Russia

Matthew He
Halmos College of Natural Sciences
Nova Southeastern University
Plantation, FL, USA

ISSN 2367-4512 ISSN 2367-4520 (electronic)
Lecture Notes on Data Engineering and Communications Technologies
ISBN 978-3-030-80474-9 ISBN 978-3-030-80475-6 (eBook)
https://doi.org/10.1007/978-3-030-80475-6



This Springer imprint is published by the registered company Springer Nature Switzerland AG
The registered company address is: Gewerbestrasse 11, 6330 Cham, Switzerland

# Contents

**Mathematical Advances in Schemes for Logistics Engineering**

**Deep Learning for Grasp-and-Lift Movement Forecasting Based on Electroencephalography by Brain-Computer Interface** . . . . . . . . . . . 3
Yuri Gordienko, Kostiantyn Kostiukevych, Nikita Gordienko, Oleksandr Rokovyi, Oleg Alienin, and Sergii Stirenko

**Inland Waterborne Commerce Study Based on Variance Decomposition and Cross-Spectral Analysis** . . . . . . . . . . . . . . . . . . . . . . 13
Yumeng Xie, Peilin Zhang, and Yong Wang

**RBF-Based Input Doubling Method for Small Medical Data Processing** . . . . . . . . . . . . . . . . . . . . . . . . . . . . . . . . . . . . 23
Ivan Izonin, Roman Tkachenko, Solomia Fedushko, Dmytro Koziy, Khrystyna Zub, and Olena Vovk

**Vehicles and Cargos Two-Sided Matching Based on Similarity Calculation** . . . . . . . . . . . . . . . . . . . . . . . . . . . . . . . . . . . . . . . 32
Meng Wang, Qing Xia, and Xuejiang Wei

**Improving Pedestrian Detection Methods by Architecture and Hyperparameter Modification of Deep Neural Networks** . . . . . . . . 44
Roman Statkevych, Yuri Gordienko, and Sergii Stirenko

**Warehousing Process Mining Research Based on Petri Net** . . . . . . . . . . 54
Fei Yin, Jingjing Cao, Jie Chu, and Shurchkova Olga

**Mathematical Methods of Algebraic Biology** . . . . . . . . . . . . . . . . . . . . 66
Georgy K. Tolokonnikov and Sergey V. Petoukhov

**Forecast and Analysis of Domestic Iron Ore Shipping Pricing Based on BP Neural Network** . . . . . . . . . . . . . . . . . . . . . . . . . . . . . . . . 76
Lu Cao and Meng Wang

**On Trilateral Evolutionary Game and Simulation Analysis of College-Enterprise-Government Collaborative Digital Talent Cultivation in the Context of Digital Economy** . . . . . . . . . . . . . . . . . . . . 85
Ping Liu, Wei Ma, and Yi Zhang

**Design and Development of Mobile Learning Platform Based on WeChat Mini Programs** . . . . . . . . . . . . . . . . . . . . . . . . . . . . . . . . 96
Kai Zeng, Yuyan Zhang, Yong Liu, and Fuan Wen

**A New Fog Enabled Sensor Cloud Platform for Smart Logistics Park** . . . . . . . . . . . . . . . . . . . . . . . . . . . . . . . . . . . . . . . . 107
XueJiang Wei and Meng Wang

**The Location of Company's Logistics Hub Based on Principal Component Analysis** . . . . . . . . . . . . . . . . . . . . . . . . . . . . . . . . . . . . 118
Zhuoni Zhang and Zhengguo Wang

**Advances in Technological and Educational Approaches**

**Deep Learning for Melanoma Detection with Testing Time Data Augmentation** . . . . . . . . . . . . . . . . . . . . . . . . . . . . . . . . . . . . 131
Volodymyr Doms, Yuri Gordienko, Yuriy Kochura, Oleksandr Rokovyi, Oleg Alienin, and Sergii Stirenko

**Research and Development of Interactive Virtual Disassembly Teaching System Considering Difficulty Gradient** . . . . . . . . . . . . . . . . 141
Qi Wang, YanFang Yang, Qiang Bai, XiaoShu Zhang, YaHui Xing, and ZhiPing Liu

**Teaching Reform and Empirical Research of Logistics Professional English Based on Super Star Learning APP Platform** . . . . . . . . . . . . . . 152
Jinling Xia and Jingjing Wu

**Integrated Operation Mode of Warehousing and Distribution in the O2O Environment** . . . . . . . . . . . . . . . . . . . . . . . . . . . . . . . . . 161
Huiling Jian, Chuanlei Wang, and Xiaocheng Jiang

**A Comparison of Machine Learning Algorithms for Prediction Higher Education Institution's Entrants Admissions** . . . . . . . . . . . . . . . . . . . . 171
Pavlo Zhezhnych, Khrystyna Zub, Oleksandr Berezko, and Anna Shilinh

**Ocean Carrier Price Competition with Bulk Cargo Transportation and Empty Container Repositioning** . . . . . . . . . . . . . . . . . . . . . . . . . . 180
Mingzhu Yu and Hanzhe Liu

**Practical Personnel Training Mode of Logistics Engineering Specialty Based on CDIO Concept Under the Background of Emerging Engineering Education** . . . . . . . . . . . . . . . . . . . . . . . . . . . . . . . . . . . . . 190
Yanhui Liu, Xiaoguang Zhou, Ruiceng Meng, and Liang Fang

**Teaching Reform of Sports Dance Course in Ethnic Areas Under the Background of "Intelligent Teaching"** . . . . . . . . . . . . . . . . . . . . . . 200
Qunqun Sun and Qi Shan

**Design and Implementation of "Task Driven" Teaching Mode Based on Flipped Classroom in Logistics Management Course** . . . . . . . . . . . . 209
Qinxian Chen and Saipeng Xing

**The Operation Mode and Vehicle Scheduling Problem of Urban Co-distribution** . . . . . . . . . . . . . . . . . . . . . . . . . . . . . . . . . . . . . 219
Mingwang Dong, Yi Qiu, and Yong Gu

**Interdisciplinary Training Models for Shipping Talents in the Universities** . . . . . . . . . . . . . . . . . . . . . . . . . . . . . . . . . . 228
Ying Zhang, Haojie Zhu, Dexin Tao, and Zhiping Liu

**The Application of Data-Driven Technologies to Enhance Supply Chain Resilience in the Context of COVID-19** . . . . . . . . . . . . . . . . . . . 238
Zhong Zheng, Yun Lin, Liwei Li, Liuqing Lu, and Yanfang Pan

**Design and Application of Tag Identification APP for Warehouse Sorting System** . . . . . . . . . . . . . . . . . . . . . . . . . . . . . . . . . . . . 254
Hao Li, Yanjia Li, Zanxing Yang, Haiyuan Wang, and Gang'ao Chen

**Modeling in Logistics Engineering**

**Game Analysis of Military-Civil Integration in Emergency Logistics Distribution Under the Context of Pandemic** . . . . . . . . . . . . . . . . . . . 265
Bingbing Li, Lulu Wang, and Zili Liu

**Pricing Game Model of Pallet Sharing Platform Under Bidding** . . . . . . 277
Yu Li, Yong Gu, and Yangyang Zheng

**Oblique Projection Operators in Matrix Representations of Genetic Coding Systems** . . . . . . . . . . . . . . . . . . . . . . . . . . . . . . . . . . . . . 288
Sergey V. Petoukhov

**Simulation of AGV Dynamic Grouping Scheduling Considering Real-Time Operation of Quay Crane** . . . . . . . . . . . . . . . . . . . . . . . . 298
Meng Yu, Ruiyu Wang, and Zhiling He

**Literature Review of Cold Chain Logistics Research in China Based on CiteSpace** . . . . . . . . . . . . . . . . . . . . . . . . . . . . . . . . . . . . . 307
Di Kang, MengMeng Guo, and Gang Chen

**Channel Strategy of Supply Chain Extended Warranty Service Considering Retailers' Fairness** . . . . . . . . . . . . . . . . . . . . . . . . . . . . 317
Jiali Pan, Xingjian Zhou, and Yong Wang

**Strategic Choice of Smart Logistics Business Model Based on Supply Chain Perspective** . . . . . . . . . . 327
Na Liu and Xiaojun Zhang

**Investment Income Analysis of Cold Chain Transportation Platform Based on the Construction of Third Party Logistics** . . . . . . . . . . 336
Xudong Deng, Ning Gao, and Ao Liu

**Simulation and Optimization of M-commerce Small Commodities Logistics System Based on Arena** . . . . . . . . . . 345
Xin Li, Yong Wang, Yuan Chen, Hong Jiang, and Xuanzhi Ye

**Role of Government and Regulatory System of Medical Waste Recycling and Transportation Based on Game Theory** . . . . . . . . . . 355
Hengyu Zhang and Qiguang Du

**Simulation and Analysis of Handling Technology Equipped with Single Box Gauge Crane** . . . . . . . . . . 365
Jingjing Zheng, Qiang Zhou, and Ting Wang

**Author Index** . . . . . . . . . . 379

# Mathematical Advances in Schemes for Logistics Engineering

# Deep Learning for Grasp-and-Lift Movement Forecasting Based on Electroencephalography by Brain-Computer Interface

Yuri Gordienko, Kostiantyn Kostiukevych(✉), Nikita Gordienko, Oleksandr Rokovyi, Oleg Alienin, and Sergii Stirenko

National Technical University of Ukraine "Igor Sikorsky Kyiv Polytechnic Institute", Kyiv, Ukraine

**Abstract.** Several deep neural networks (DNNs) were compared to classify some basic hand movements (actions) on the grasp-and-lift (GAL) dataset by analysis of prior-action, in-action, and posterior-action segments of the correspondent brain activity. These results demonstrate possibility to classify GAL-like hand movements reliably by simple and small CNNs even with AUCs > 0.92 with in-process labeling that is important for porting DNN models to Edge Computing layer devices with the limited computational power. Moreover, the possibility to "forecast" intentions up to 0.6 s before movement (i.e. to classify the mental precursors of the intended physical actions) was observed by the more sophisticated labeling.

**Keywords:** Deep learning · Convolutional neural network · Fully-connected neural network · Brain-computer interface · EEG · Grasp-and-lift

## 1 Introduction

Deep learning (DL) methods on the basis of deep neural networks (DNNs) become popular for processing various objects, including structured table data and unstructured information like images, sounds, video, etc. [1]. These works become especially intensive for processing various multimodal time sequences measured and gathered in health care and elderly care from wearable electronics or by external sensors [2–8]. Recently, they were successfully applied for analysis of electroencephalography (EEG) signals collected by brain-computer interface (BCI) for different applications [9–11]. This work is dedicated to investigation of various DNNs applied to EEG obtained by BCI with a purpose to recognize (classify) some basic hand movements (actions) by analysis of prior-action, in-action, and posterior-action segments of the correspondent brain activity. The structure of this paper is as follows: Sect. 2 *Background and Related Work* contains a brief review of some similar approaches to characterize EEG by DNNs, Sect. 3 *Methodology* describes the dataset, hand actions, DNNs, and metrics used, Sect. 4 *Experimental* gives the results obtained by DNNs, Sect. 5 *Discussion* summarizes all results with analysis of advantages and disadvantages of the methods used, and Sect. 6 Conclusions and Future Work outlines the findings and the possible ways for their improvements.


Z. Hu et al. (Eds.): ICAILE 2021, LNDECT 82, pp. 3–12, 2021.
https://doi.org/10.1007/978-3-030-80475-6_1

## 2 Background and Related Work

Various DNNs have been investigated and proposed for a wide range of potential applications in medicine, sport, psychology, military, transport, entertainment, and general purpose supportive applications. For example, EEGNet DNN, a compact convolutional neural network (CNN) for EEG-based BCIs, uses depthwise and separable convolutions to construct an EEG-specific model which encapsulates well-known EEG feature extraction concepts for BCI with the relatively high performance in comparison to other similar models [11]. 3D CNN was proposed for EEG signal classification method based on multi-dimensional feature fusion, which achieves improvement in the average classification accuracy of the brain's sensorimotor region activated tasks [12]. CNNs and recurrent neural networks (RNN) were proposed for alleviation of the influence of the translation problem in convolution-based models [13]. Additional CNN-related researches were related to driving fatigue detection, sleep stage classification, stress recognition, motor imagery classification and emotion recognition and classification [14]. Some interesting use cases for CNN application were demonstrated also, for example, for prediction of hand movement speed and force from single-trial EEG with the overall accuracy of 84% [15] and to predict the presence of conflicts in a given moment with the average prediction accuracy ~60% [16]. Electrocorticography (EcoG) Annotated Joints in Long-term ECoG (AJILE) dataset and the multimodal model that combines CNN with long short-term memory (LSTM) blocks, leveraging both ECoG and video modalities, was used for detection and prediction hand movement up to 800 ms before movement initiation [17]. Taking into account these results, this work was aimed on investigation of dependence between various DNN architectures and their performance in classification tasks for some basic hand movements (actions). The special attention was paid to analysis of the prior-action, in-action, and posterior-action segments of the correspondent brain activity to foresee the actions before they were started even.

## 3 Methodology

Below the dataset with action types, some models, metrics, and workflow are described that were used to pursue the main aim, namely to investigate the various DNNs applied to EEG signals obtained by BCI with a purpose to recognize (classify) some basic hand movements by analysis of the pre-process, in-process, and post-process segments of the correspondent brain activity.

**Dataset.** In this paper, the grasp-and-lift (GAL) dataset was used that contains 3,936 grasp and lift trials collected and recorded by 32 channels of EEG (sampling rate of 500 Hz) from 12 participants who perform the following movements [18]:

1) HandStart - the beginning of the movement of the hand to the object,
2) FirstDigitTouch - the researched person has to touch the object,
3) BothStartLoadPhase - the subject should squeeze an object,
4) LiftOff - the subject lifts an object into the air with two fingers,
5) Replace - the subject rotates the object with two fingers back,

6) BothReleased - the subject releases two fingers.

**Models.** In this paper, the following DNNs were used to:

1) Fully connected neural network (FCN) is the version of multi-layered perceptron [19] with 3 fully connected dense layers (128 units and ReLU activation function), and dropout (0.4) layer compiled with Adam optimizer;
2) Lenet-like DNN is the variant of the classical CNN [20] with 3 1D convolutional layers (6, 16, 120 channels and tanh activation function), 2 average pooling layers with a stride 2 × 1, and 1 fully connected dense layer (84 units and tanh activation function) compiled with Adam optimizer;
3) Alexnet-like DNN is the variant of the other classical CNN [21] with 5 1D convolutional layers (96, 256, 384, 384, 256 channels with kernel sizes from 11 × 1 to 3 × 1, and ReLU activation function), 4 average pooling layers with a stride 2 × 1, and 1 fully connected dense layer (4096 units and ReLU activation function), and dropout (0.4) layer compiled with Adam optimizer;
4) VGG-like DNN is the variant of the other classical image classification CNN [22] with the simplified architecture: 4 contiguous blocks of 13 convolution layers (2 × 64, 2 × 128, 3 × 256, 3 × 512 to 3 × 512 channels with kernel size 3 × 1 and ReLU activation function) followed by average pooling with a stride 2 × 1, 2 fully connected dense layers with dropout (0.4) layer compiled with Adam optimizer;
5) "vanilla" is the simplified pyramid-like DNN proposed by us for comparison: 3 convolution layers (64, 128, 256 channels with kernel sizes 3 × 1, 5 × 1, 7 × 1 and ReLU activation function) followed by average pooling with a stride 2 × 1, flattening layer with dropout layer compiled with Adam optimizer.

**Metrics.** The following standard metrics were used: accuracy, loss, and their statistical parameters like minimal value (for loss), maximal value (for accuracy); area under curve (AUC) for receiver operating characteristic (ROC), micro-AUC and macro-AUC, and their statistical parameters like mean and standard deviation.

**Workflow.** The experimental workflow included Exploratory Data Analysis (EDA), training, validation and testing phases for all above mentioned DNNs. The single epoch was used to provide the minimal training time (because the highest accuracy was not the aim of this research). After each Iterative training, the best models (with the minimal validation loss and maximal validation accuracy) were saved as checkpoints for the further testing phase. The number of samples ($N$) in the input sequence (IS) was varied from 100 up to 2000 (and up to 10000 for the vanilla DNN) and the input sequences were selected randomly from the whole timeline of the experimental data. The three variants of action labeling of each IS were used which were set by ground truth (GT) values at the first, medium, and last moment inside IS with regard to the offset value defined at the data pre-processing stage, namely: if offset = 0, then label(IS) = GT(first); if offset = 0.5, then label(IS) = GT(medium); if offset = 1, then label(IS) = GT(last). The training, validation, and testing phases were performed iteratively for all 20 values of $N$ separately with the increase step of 100, where $N$ was in the range of: 100, 200, 300, ..., 1900, 2000 (9900 for the vanilla DNN) that lead to 20 iterations of training/validation/testing stages. Per each iteration it allowed to get 20 models (one per each input sequence of length $N$), calculate 20 sets of metrics, namely, AUC, micro-AUC, and macro-AUC, and construct their plots as a function of $N$ (see below in the next sections).

# 4 Experimental Analysis

## 4.1 Exploratory Data Analysis (EDA)

The GAL-dataset [11] was analyzed with regard of the mutual timeline location of different actions (hand movements) and their timing (Fig. 1a). Actually, all data were collected by the BCI-sensors with their standard spatial location (Fig. 1b).

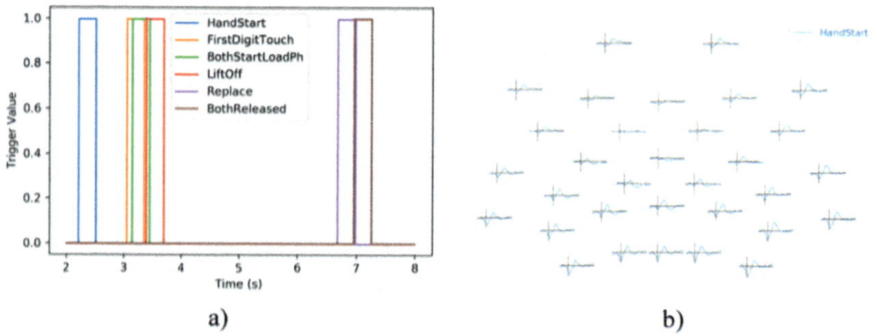


**Fig. 1.** Timeline of grasp-and-lift actions (a), and 2D scalp topography of the evoked responses at sensor locations (EEG channels) for HandStart action

Due to their electromagnetic nature, they interfered with each other (Fig. 2) that allowed for various types of the combination of the correspondent data channels for their further processing by DNNs: from FCN to CNN, and to other DNN types.

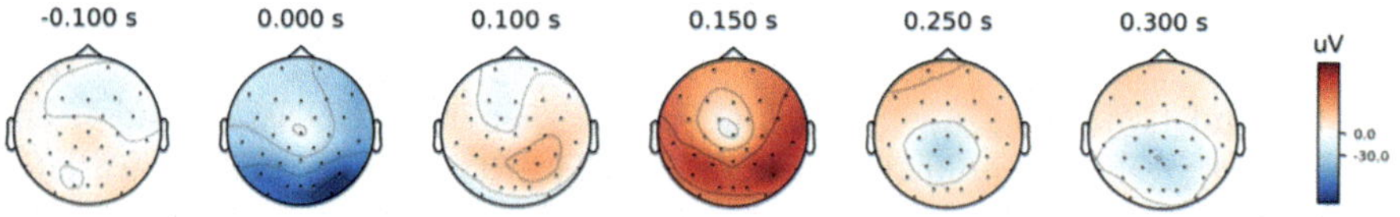


**Fig. 2.** 2D scalp topography of the power spectral density (PSD) for the evoked responses at sensor locations (EEG channels) for HandStart action

EDA of the pre-processed EEG signals for all actions allow to plot the time sequences for the evoked responses for all 32 EEG channels (Fig. 3). For HandStart action is clear that some brain activity was started some time before the movement itself, but it is dubious to make the same statement about other actions. The idea behind this work was motivated by the assumption that DNNs can catch the distinctive EEG patterns not only for the classification of the undergoing actions (like it was done before in the similar research [9, 10]), but also for "forecasting" non-started actions, i.e. before the start of the correspondent physical activity.

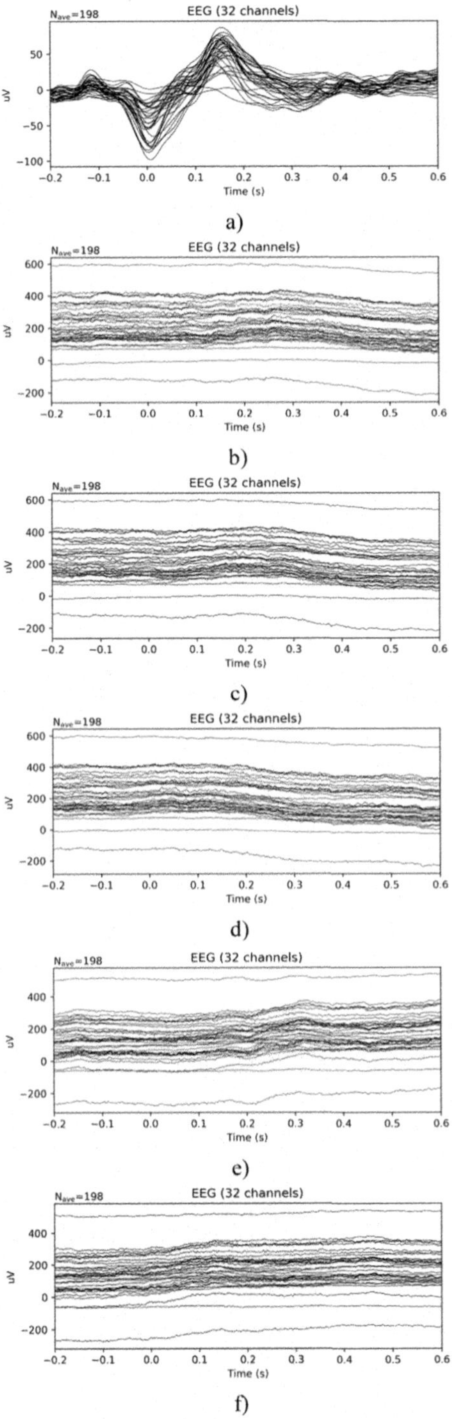


**Fig. 3.** The time sequences for the responses by 32 EEG channels for actions: a) HandStart, b) FirstDigitTouch, c) BothReleased, d) LiftOff, e) Replace, f) BothReleased actions

## 4.2 DNN Training/Validation/Testing Phases

**FCN.** After training FCN-like DNN during the single epoch only, the AUC values were calculated at testing stage and plotted as a function of the number of samples in the input sequence ($N$) with the various offset values (Fig. 4).

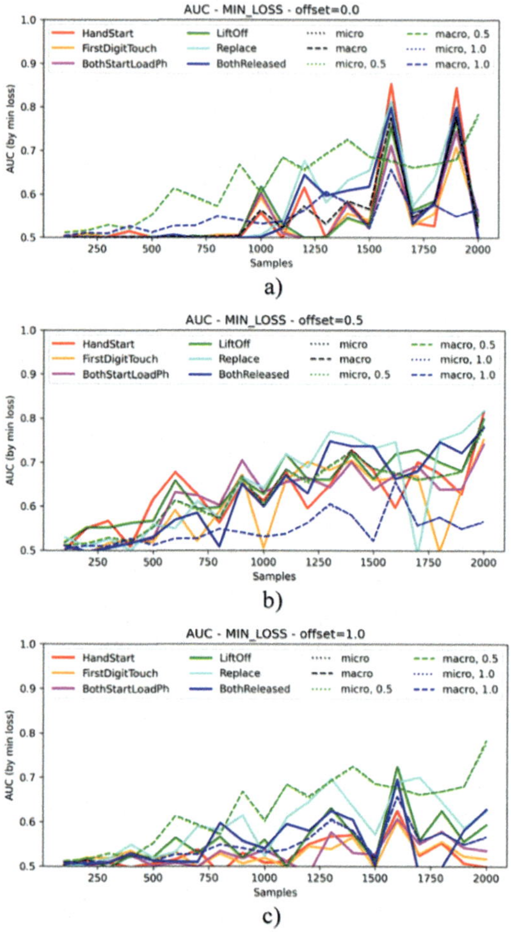


**Fig. 4.** AUC as a function of the number of samples in the input sequence ($N$) for FCN-like DNN with the various offset values: a) 0, b) 0.5, and c) 1.0

**Lenet.** After training Lenet-like DNN during the single epoch only, the AUC values were calculated at testing stage and plotted as a function of the number of samples in the input sequence with the various offset values (Fig. 5).

**Other DNNs.** In addition to FCN and Lenet-like DNNs the similar training/testing experiments were carried out with the other before mentioned DNNs (Alexnet, VGG, vanilla). The detailed AUC vs. the number of samples in the input sequence ($N$) plots are not shown here because of the lack of the space, but they are summarized below. The values of the metrics obtained for the other DNNs are shown in Fig. 6.

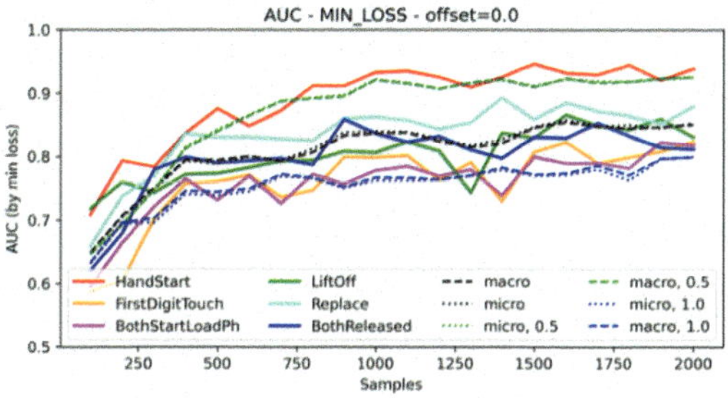


a)

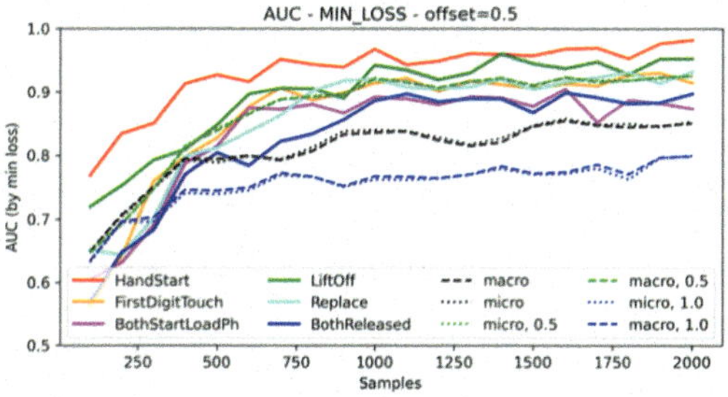


b)

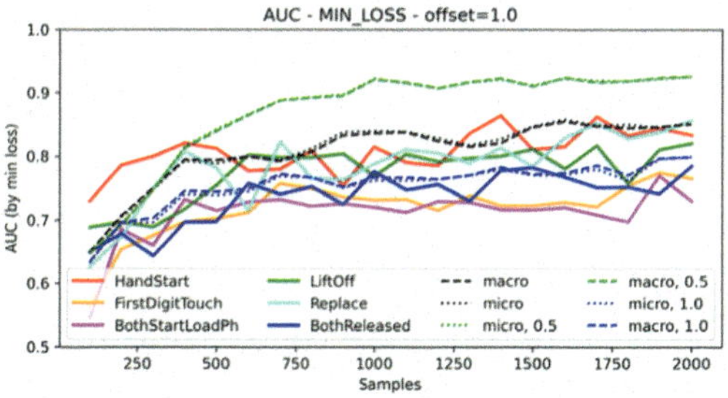


c)

**Fig. 5.** AUC as a function of the number of samples in the input sequence ($N$) for Lenet DNN with the various offset values: a) 0, b) 0.5, and c) 1.0

One can see the quite different sensitivity of the performance to DNN architecture, where FCN gives the worst AUCs (Fig. 6a), and CNNs (Fig. 6b–e) – the best ones up to ~0.92 which is close to the previously published values by other authors in similar experiments [9–11]) despite only 1 training epoch and the very simple architectures like Lenet-like (Fig. 6b) and Alexnet-like (Fig. 6c). It can be explained by the lower sensitivity of FCNs to the correlated patterns in comparison to CNNs (especially after only 1 training epoch) with taking into account spatial (here time) local correlations.

## 5 Discussion

The growth of AUC with the number of samples ($N$) in IS is observed up to the different saturation levels of AUC for different DNN architectures, but saturation is faster for the more complicated CNNs like VGG than for the simpler CNNs like Lenet or Alexnet (or FCN). The most important observation is based on the possibility to make better classification of GAL-actions (with the various accuracy) for offset = 0 (red lines in Fig. 6) with the higher contribution of the prior-action segment of the correspondent brain activity than for offset = 1 (blue lines in Fig. 6) with the higher contribution of posterior-action segment.

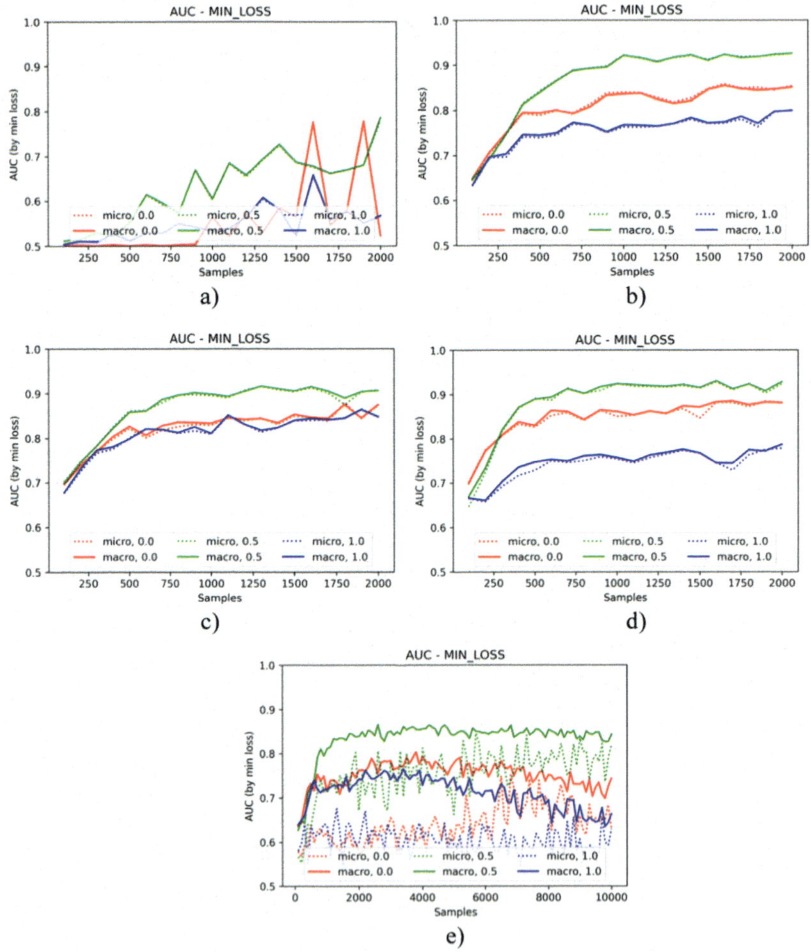


**Fig. 6.** AUC as a function of the number of samples ($N$) in the input sequence with the various offset values for: a) FCN, b) Lenet, c) Alexnet, d) VGG, and e) vanilla DNNs

It is intriguing to observe that for the most complicated CNN like VGG the runs for offset = 0 with the dominance of prior-action segments (red line in Fig. 6d) gives the higher micro-AUC and macro-AUC values than for offset = 1 with the dominance of posterior-action segments (blue line in Fig. 6d), and, especially, for offset = 0.5 with the dominance of in-action segments (green line in Fig. 6d) even for $N$ up to ~300 (i.e. time $< 0.6$ s). It allows making assumption about possibility to detect and "forecast" mental intentions for non-started actions even, i.e. before the start of the correspondent physical activity that supports the previous similar findings [17].

## 6 Conclusions and Future Work

These results demonstrate the possibility to classify GAL-like hand movements reliably by simple CNN (with AUCs $> 0.92$ after 1 training epoch) with in-process labeling that is very important for porting DNN models to Edge Computing layer devices with the limited computational power [23]. Moreover, the possibility to classify the mental precursors of the intended physical actions by the more sophisticated labeling was observed. But due to shortage of the training time these research should be continued for the longer training time and with inclusion of RNN-like and RNN-CNN hybrid DNN architectures that are carried out right now. That is why the future work will be targeted on real-time monitoring the EEG signals by BCI in mobile installment [24] to provide the immediate surveillance and processing. It could be potentially useful for foreseeing GAL-like and any other movements with the physical feedback like fulfillment of some physical actions which were started by brain activity of people, but cannot be finished due to their physical disabilities, and in other various applications of DNNs [25, 26].

**Acknowledgment.** The work was partially supported by the National Research Foundation of Ukraine by the grant 2020.01/0490 "Artificial Intelligence Platform for Distant Computer-Aided Detection (CADe) and Computer-Aided Diagnosis (CADx) of Human Diseases".

## References

1. LeCun, Y., Bengio, Y., Hinton, G.: Deep learning. Nature **521**, 436–444 (2015)
2. Chen, Y.W., Jain, L.C.: Deep Learning in Healthcare. Springer, Cham (2020). https://doi.org/10.1007/978-3-030-3206-7
3. Esteva, A., et al.: A guide to deep learning in healthcare. Nat. Med. **25**(1), 24–29 (2019)
4. Gordienko, N., Lodygensky, O., Fedak, G., Gordienko, Y.: Synergy of volunteer measurements and volunteer computing for effective data collecting, processing, simulating and analyzing on a worldwide scale. In: Proceedings of the 38th International Convention on Information and Communication Technology, Electronics and Microelectronics, pp. 193–198 (2015)
5. Gordienko, Y., Stirenko, S., Kochura, Y., Alienin, O., Novotarskiy, M., Gordienko, N.: Deep learning for fatigue estimation on the basis of multimodal human-machine interactions. ArXiv (2017)
6. Stirenko, S., et al.: Parallel statistical and machine learning methods for estimation of physical load. In: Vaidya, J., Li, J. (eds.) ICA3PP 2018. LNCS, vol. 11334, pp. 483–497. Springer, Cham (2018). https://doi.org/10.1007/978-3-030-05051-1_33

7. Gang, P.: Prediction of physical load level by machine learning analysis of heart activity after exercises. In: IEEE Symposium Series on Computational Intelligence (SSCI), pp. 557–562 (2019)
8. Gang, P., et al.: User-driven intelligent interface on the basis of multimodal augmented reality and brain-computer interaction for people with functional disabilities. In: Arai, K., Kapoor, S., Bhatia, R. (eds.) FICC 2018. AISC, vol. 886, pp. 612–631. Springer, Cham (2019). https://doi.org/10.1007/978-3-030-03402-3_43
9. Roy, Y., Banville, H., Albuquerque, I., Gramfort, A., Falk, T.H., Faubert, J.: Deep learning-based electroencephalography analysis: a systematic review. J. Neural Eng. **16**(5), 051001 (2019)
10. Kotowski, K., Stapor, K., Ochab, J.: Deep learning methods in EEG. Mach. Learn. Paradigms **18**, 191–212 (2020)
11. Lawhern, V.J., Solon, A.J., Waytowich, N.R., Gordon, S.M., Hung, C.P., Lance, B.J.: EEGNet: a compact convolutional neural network for EEG-based brain–computer interfaces. J. Neural Eng. **15**(5), 056013 (2018)
12. Wei, M., Lin, F.: A novel multi-dimensional features fusion algorithm for the EEG signal recognition of brain's sensorimotor region activated tasks. Int. J. Intell. Comput. Cybern. **13**(2), 239–260 (2020)
13. Lin, B., Deng, S., Gao, H., Yin, J.: A multi-scale activity transition network for data translation in EEG signals decoding. IEEE/ACM Trans. Comput. Biol. Bioinform. (2020)
14. Gu, X., et al.: EEG-based brain-computer interfaces (BCIs): a survey of recent studies on signal sensing technologies and computational intelligence approaches and their applications. IEEE/ACM Trans. Comput. Biol. Bioinform. (2020)
15. Gatti, R., Atum, Y., Schiaffino, L., Jochumsen, M., Manresa, J.B.: Convolutional neural networks improve the prediction of hand movement speed and force from single-trial EEG. BioRxiv (2019)
16. Vahid, A., Mückschel, M., Stober, S., Stock, A.K., Beste, C.: Applying deep learning to single-trial EEG data provides evidence for complementary theories on action control. Commun. Biol. **3**(1), 1–11 (2020)
17. Wang, N.X.R., Farhadi, A., Rao, R., Brunton, B.: AJILE movement prediction: multimodal deep learning for natural human neural recordings and video. ArXiv (2017)
18. Luciw, M.D., Jarocka, E., Edin, B.B.: Multi-channel EEG recordings during 3,936 grasp and lift trials with varying weight and friction. Sci. Data **1**(1), 1–11 (2014)
19. Rosenblatt, F.: The perceptron: a probabilistic model for information storage and organization in the brain. Psychol. Rev. **65**(6), 386 (1958)
20. LeCun, Y., et al.: Backpropagation applied to handwritten zip code recognition. Neural Comput. **1**(4), 541–551 (1989)
21. Hinton, G.E., Krizhevsky, A., Sutskever, I.: Imagenet classification with deep convolutional neural networks. Adv. Neural. Inf. Process. Syst. **25**, 1106–1114 (2012)
22. Simonyan, K., Zisserman, A.: Very deep convolutional networks for large-scale image recognition. ArXiv (2014)
23. Gordienko, Y., et al.: "Last mile" optimization of edge computing ecosystem with deep learning models and specialized tensor processing architectures. Adv. Comput. (2020)
24. Kostiukevych, K., Gordienko, Y., Stirenko, S.: Multimodal data integration and processing method for brain-computer interface research. In: Presented at 6th International Conference High Performance Computing (2020)
25. Shafiee, N.S.M., Mutalib, S.: Prediction of mental health problems among higher education student using machine learning. Int. J. Educ. Manag. Eng. (IJEME) **10**(6), 1–9 (2020)
26. Stepanyan, I., et al.: Neural network modeling and correlation analysis of brain plasticity mechanisms in stroke patients. Int. J. Intell. Syst. Appl. (IJISA) **11**(6), 28–39 (2019)

# Inland Waterborne Commerce Study Based on Variance Decomposition and Cross-Spectral Analysis

Yumeng Xie[1], Peilin Zhang[1](✉), and Yong Wang[1,2,3]

[1] School of Transportation, Wuhan University of Technology, Wuhan 430063, China
[2] School of Logistics, Wuhan Technology and Business University, Wuhan, Hubei, China
[3] Department of Production Engineering, KTH Royal Institute of Technology, Stockholm, Sweden

**Abstract.** Much research has focused on the prediction field; however, few are concerned about the detail between forecasting and influence factors. It is an important area to help decision-makers decide in time. River and relevant factors were studied utilizing variance decomposition and cross-spectral analysis. The results show that there is an inflection point between the shipping volume and the related variables. After passing a high-speed development period, the inland shipping volume will enter a stable period of fluctuation, and the correlation with other variables will gradually decrease. The amplitude spectrum verifies the degree of influence; the cross-spectral analyzed each variable is driving relationship; the phase spectrum shows the shipping volume and the time required for effecting.

**Keywords:** Transportation economy · The inflection point of shipping volume · Cross-spectrum analysis · Shipping volume · Variance decomposition

## 1 Introduction

Most of the Quantitative analysis of waterway transportation research, has focused on time-series analyses and econometric models, and the research has become an interdisciplinary field.

In the inland waterway traffic market, cargo owner enterprise makes their decisions not only based on historical data but also consider other influence factors. Port authorities could use inland river traffic spectrum analysis for port planning; shipping business managers could make business plans, planning equipment levels and hiring operating staff by those analysis; government policymakers may measure the impact of past and future transport policies by this analysis.

Time-series easily ignore the correlation between different influence factors. Regularly the shipping volume appears to be influenced mainly by the GDP, industrialization rate, and urbanization rate. Therefore, we propose a spectrum analysis with influence factors, which provided a comprehensive analysis based on frequency domain information collection, to puzzle out this problem. For extracts information from time series, the Fourier transform is an appropriate method.

Through, time series can transform into frequency-domain data.


Z. Hu et al. (Eds.): ICAILE 2021, LNDECT 82, pp. 13–22, 2021.
https://doi.org/10.1007/978-3-030-80475-6_2

## 2 Early Theories of Emotions

Spectral analysis can provide a complete analysis of multiple variables and establish predictively [1, 2]. Based on the stability of the spectrum analysis, CHEN J. H., et al. deal with the periodicity of it empirically and theoretically for the shipping market of Capsize dry bulk by Fourier transform [3]. Li Yue qi et al., utilize the Logistic curve and co-integration theory, concluded a strong associated relationship between inland waterway shipping and regional economy [4]. Zhang Bing believed that the US had a 5–6-year economic cycle through factor analysis and cross-spectrum analysis [5].

In contrast, through spectrum analysis, Wang Yue thought that the U.S had a 3-year sub-cycle in addition to a 7–8-year economic cycle [6]. According to the theory of structural school, Maya Hua divided the Industrialization of the U.S into several stages [7]. Hein A. M. et al., using the Rhine River Delta's shipping rate data, demonstrated the Rhine River's shipping competitiveness [8]. Rayevnyeva O uses the Fourier analysis to determine the long-term and short-term cycles in the index [9]. Mittal P. et al. proposed a spectrum-aware routing scheme that uses throughput as a performance metric for path selection [10]. Karabutov N. introduces impressive structures that describe a modification of the Lyapunov exponent to identify a spectrum eigenvalue, and then spectrum tentative estimations apply from the secant structures method [11].

With the factor analysis, Zeng W. et al. tested ' the path and degree of various factors' influence through the empirical method with four aspects [13].

Above all research, verify different factor analysis with spectral method could help to have a better understanding.

## 3 Methodology

The spectral analysis involves the wave structure of considered variables of stochastic processes, allowing us to, through the use of trigonometric functions as functions sine and cosine, analyze time series in the frequency domain. The traffic data $y_t$ collected during different time series t, which often exhibits some repetitive behaviors over time. To describe the spectral analysis process, we can define one cycle in one complete period of sine or cosine function, over a unit time interval as Eq. (1).

$$y_t = \sum\nolimits_{j=1}^{m} \left[a_j cos\left(w_j\ t\right) + \delta_j sin\left(w_j\ t\right)\right] \tag{1}$$

Where j represents the index for individual periodic series, m is the total number of periodic elements in the series, $a_j$ and δj are often taken to be zero-mean random variables. The t value is satisfied E($yt$) $= 0$. These undoubtedly indicate that the time series must be static when a spectral analysis is processed, and the mean value and variance of the series remain unchanged in the whole-time range. The time series is likely to be non-static, so the trend of time series must be eliminated.

For the frequency domain analysis, we can utilize the Fourier transform to change the time series into a few different indicators, including coherence, amplitude, and phase position; through those indicators, the relationship with influences factors is apparent. There are three steps to achieve this goal, as follows:

1) Remove all trends from the data.
2) Select any number of points and estimate the size of the spectrum at each point.
3) Calculate the spectrum of each sequence one by one. Used parent weights to smooth the data, and the formula for calculating the spectrum at each point ($\pi_i/m$) is as follows:

$$s_j = \left(\frac{c_0}{2\pi}\right) + \left(\frac{1}{\pi}\right)\sum\nolimits_{k=1}^{m} C_k\lambda_k \, cos\left(\frac{\pi_j}{m}\right) k \tag{2}$$

In the equation, λ is a filter, which is defined as:

$$\lambda_k = 1 - \left(\frac{6k^2}{m^2}\right)\left(1 - \frac{k}{m}\right) \text{ When } 0 < k < (\frac{m}{2}) \tag{3}$$

$$\lambda_k = 2(1 - \frac{k}{m})^3 \text{ When } (\frac{m}{2}) < k < m \tag{4}$$

Cross covariance $C_{xy}$ and $C_{yx}$ equation:

$$C_{yx}(k) = \left(\frac{1}{n}\right)\sum\nolimits_{i=1}^{n-k} (x_t - \bar{x})(y_{t+k} - \bar{x}) \tag{5}$$

$$C_{xy}(k) = \left(\frac{1}{n}\right)\sum\nolimits_{t=1}^{n-k} (y_t - \bar{y})(x_{t+k} - \bar{y}) \tag{6}$$

Co-spectrum and quadrature spectrum equation:

$$\begin{aligned} \text{Cospectrum} &= \left(\frac{1}{4\pi}\right)\left[C_{xy}(0) + C_{yx}(0)\right] + \\ &+\left(\frac{1}{2\pi}\right)\sum\nolimits_{k=2}^{m} \lambda k \left[c_{xy}(k) + c_{yx}(k)\right]\cos\left(\frac{\pi_j}{m}\right)k \end{aligned} \tag{7}$$

$$\text{Quadrature Spectrum} = \left(\frac{1}{2\pi}\right)\sum\nolimits_{k=1}^{m} \lambda k \left[c_{xy}(k) - c_{yx}(k)\right]\sin\left(\frac{\pi_j}{m}\right)k \tag{8}$$

Coherence; amplitude; Phase equation:

$$Coherence = \frac{Amplitude^2}{(S_x \times S_y)} \tag{9}$$

$$\text{Amplitude} = \left(Cospectrum^2 + Quadrature\ Spectrum^2\right)^{\frac{1}{2}} \tag{10}$$

$$\text{Phase} = \arctan\left(\frac{-\text{Quadrature Spectrum}}{\text{Cospectrum}}\right) \tag{11}$$

Where, $S_x$ and $S_y$ are respectively spectra of one sequence, belong to $S_j$.

Cross spectral analysis for cycle measure provides a relatively complete system. From each related series of statistical tests to index sequence correlation between the degree of measurement, after that the lead or lag between sequences is determined.

# 4 Experiment Setup and Case Study

## 4.1 Data Description and Pre-processing

The Mississippi River accounts for more than 60 percent of the nation's inland waterway traffic. It contains many industrial raw materials and commodities along the coast, especially bulk commodities, more than 90 percent of transport through the river. Therefore, the GDP, industrialization rate, and urbanization rate of the United States are selected as the research objects to analyze the interaction between the Mississippi River transportation and these factors. From 1947 to 2016, selected 69 groups of observe data, and the trend curve was as follows:

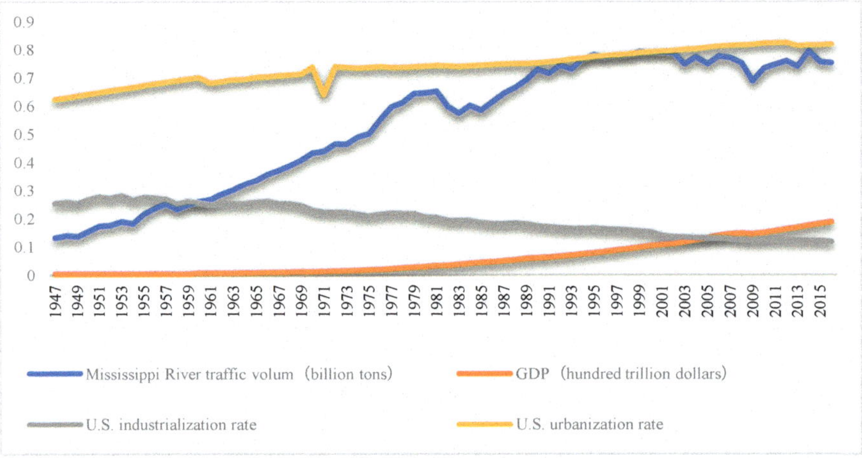


**Fig. 1.** Mississippi River trend chart

In Fig. 1 we can observe that in 2000, the volume of the Mississippi River in the United States reached a periodic peak at the point of 789 million tons; meanwhile, the GDP was 10.28 trillion US dollars, the industrialization rate was 16.4%, and the urbanization rate was 79%.

Before 2000, freight traffic on the Mississippi River grows markable; meanwhile, GDP and urbanization rates also increased.

After 2000, although GDP and urbanization rates are still increasing, however, total freight volume keeps steady at around 750 million tons. (Data from Waterborne Transportation Lines OF the United States 1979, 2005, 2016; Bureau of Economic Analysis Boston University Libraries; United Nations DESA / Population Division).

According to Fig. 1 and Table 1, the year 2000 is a dividing line of correlation, and before 2000 freight volume increase rate is significantly higher than the value after 2000.

The urbanization rate and GDP Compared with the Mississippi River's volume are positively correlated, with correlation coefficients of 0.98 and 0.92. Furthermore, there was a negative correlation between the industrialization rate and the Mississippi River volume, with a correlation coefficient of -0.88. After 2000, those factors are uncorrelation.

**Table 1.** Correlation between the inland waterway and different factors

| Mississippi River traffic volume (year) | 1947–2000 | 2000–2016 |
|---|---|---|
| America's GDP | 0.98 | −0.22 |
| US industrialization rate | −0.88 | 0.42 |
| US urbanization rate | 0.92 | −0.46 |

The main Rhine river is 1230 km long, passing through Germany and many Euro countries. Along the Rhine river, industrial output accounts for 50 percent of the entire German economy. Therefore, we select the Rhine river trunk stream volume, Germany's overall GDP, and Germany's industrialization rate to observe. In Fig. 2, time-series data have 66 observe objectives from 1950 to 2016, and the time tendency as follows:

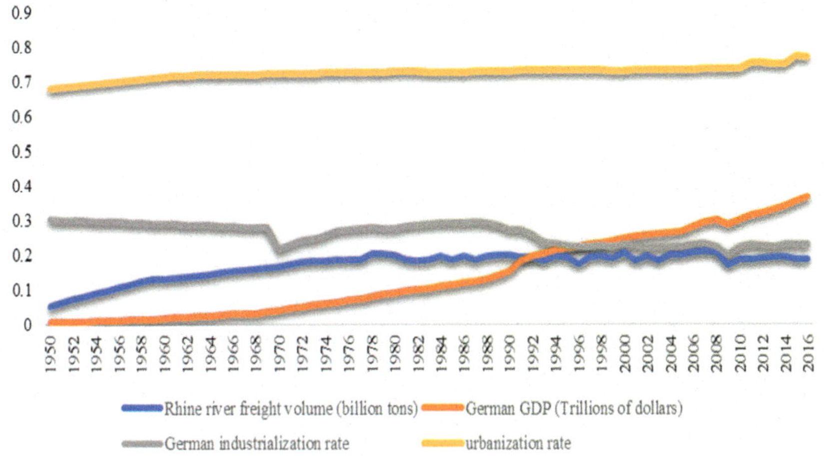


**Fig. 2.** Rhine river trend chart

According to Fig. 2, the Rhine river volume has a relative peak of 199 million tons in 1980. At that time point, the GDP was 14,900 dollars, the urbanization rate was 72.8%, and the industrialization rate was 28%. After 1980, the freight volume stopped growing and showed no downward trend; it remained stable at 2 billion tons. (Data from Inland Navigation in Europe Market Observation 1997–2016; Statistisches; United Nations DESA/Population Division).

By observing Fig. 2 and Table 2, the year 1980 recognizes as the dividing line. The correlation between 1950 and 1980 was significantly higher than that after 1980. Compared with the Rhine River volume, Germany's GDP, and urbanization rate, they have a significant positive correlation, with a correlation coefficient of 1. On the other hand, the industrialization rate negatively correlated with the volume of the Rhine river; the correlation coefficient between them is -0.99. After 1980, they are non-correlation.

**Table 2.** Correlation coefficient

| Rhine river freight volume (year) | 1950–1980 | 1980–2016 |
|---|---|---|
| German GDP | 1 | 0.04 |
| German industrialization rate | −0.99 | 0.14 |
| German urbanization rate | 1 | −0.05 |

## 4.2 Variance Decomposition

According to the analysis of historical data, the variance decomposition can obtain the endogenous variable contribution rate. The contribution rate varies over time (period number). Follow the regular change of observed contribution rate, the observed data divided into 10 phases, and the results are shown in Table 3.

**Table 3.** Variance decomposition of Mississippi River system freight volume, US GDP, US industrialization rate, US urbanization rate

| Periods | Error standard deviation | Traffic contribution rate | GDP contribution | Industrialization contribution rate | Urbanization contribution rate |
|---|---|---|---|---|---|
| 1 | 0.22 | 100 | 0 | 0 | 0 |
| 2 | 0.23 | 97.22 | 0.57 | 1.11 | 1.11 |
| 3 | 0.23 | 93.46 | 4.04 | 1.05 | 1.45 |
| 4 | 0.24 | 92.02 | 5.18 | 1.15 | 1.66 |
| 5 | 0.24 | 90.10 | 7.06 | 1.14 | 1.70 |
| 6 | 0.24 | 89.25 | 8.00 | 1.16 | 1.69 |
| 7 | 0.24 | 88.43 | 8.75 | 1.15 | 1.68 |
| 8 | 0.24 | 87.86 | 9.32 | 1.15 | 1.66 |
| 9 | 0.24 | 87.37 | 9.84 | 1.14 | 1.65 |
| 10 | 0.24 | 86.99 | 10.22 | 1.14 | 1.65 |

Table 3 shows the variance decomposition analysis of The Mississippi River and its corresponding variables. The primary contribution rate is the volume of the Mississippi River itself; the proportion is 90% and decreases year by year. The contribution of GDP is significantly higher than industrialization and urbanization, while urbanization is higher than industrialization.

Before the 5th phase, the contribution of industrialization rate and urbanization rate gradually increased, while after the 5th phase, the contribution of industrialization rate and urbanization rate decreased.

When the volume of inland river traffic reaches a particular stage, the contribution between urbanization, industrialization, and GDP becomes smaller.

**Table 4.** Variance decomposition of Rhine river trunk stream freight volume, German GDP, German industrialization rate, German urbanization rate

| Periods | Error standard deviation | Traffic contribution rate | GDP contribution | Industrialization contribution rate | Urbanization contribution rate |
|---|---|---|---|---|---|
| 1 | 0.01 | 100 | 0 | 0 | 0 |
| 2 | 0.01 | 98.95 | 0.37 | 0.10 | 0.59 |
| 3 | 0.01 | 81.21 | 12.09 | 0.21 | 6.48 |
| 4 | 0.01 | 79.96 | 11.91 | 0.57 | 7.56 |
| 5 | 0.01 | 78.35 | 13.64 | 0.59 | 7.41 |
| 6 | 0.01 | 77.99 | 13.79 | 0.64 | 7.58 |
| 7 | 0.01 | 77.90 | 13.77 | 0.67 | 7.66 |
| 8 | 0.01 | 77.85 | 13.82 | 0.67 | 7.66 |
| 9 | 0.01 | 77.73 | 13.95 | 0.67 | 7.65 |
| 10 | 0.01 | 77.72 | 13.95 | 0.68 | 7.65 |

Table 4 gives the variance decomposition analysis of the Rhine ri.

Like the Mississippi River's situation, the Rhine River is the primary contributing factor. After the 5th phase, when freight volume arrived at 77.7%, GDP was about 13.8%, industrialization was 0.66%, and urbanization was 7.66%. The contribution of GDP is significantly higher than industrialization and urbanization, while the contribution of urbanization is higher than industrialization.

When the volume of inland river traffic reaches a particular stage, the contribution between urbanization, industrialization, and GDP becomes s smaller.

## 4.3 Spectrum Analysis

Through the Fourier transform, the time series have changed to Spectrum set, which focuses on the time lag between different factors. Four indexes need consideration to choose one credible time lag: phase, coherence, and amplitude. Phase presents the time difference between the leading indicator and the synchronous indicator. Coherence usually determines the degree of correlation between the sequences of variables constituting the entire period. Amplitude is usually used to measure the degree of influence between the different factors. Therefore firstly, we chose the high coherence value then consider the phase and amplitude value.

The highest consistent index in the fluctuation cycle for the cross-spectrum analysis between freight volume, GDP, industrialization rate, and urbanization rate are 2.16 years, 8.61 years, and 2.44 years.

Considering the consistency and amplitude of the two indicators, the GDP and industrialization rates of cross-spectrum are 2.16 years and 8.61 years, two cycles, which are the highest consistency index (0.87 and 0.68, respectively). In the amplitude, set chose

**Table 5.** Cross spectral analysis between Mississippi River system freight volume and related factors

| | Period | Frequency | Coherence | Amplitude | Phase | Time Lag (year) |
|---|---|---|---|---|---|---|
| The cross-spectrum between freight volume and GDP | 4.61 | 0.22 | 0.49 | 394.51 | 0.48 | 0.35 |
| | 4.31 | 0.23 | 0.74 | 560.34 | 0.90 | 0.62 |
| | 2.23 | 0.45 | 0.74 | 453.73 | −0.28 | −0.1 |
| | 2.16 | 0.46 | 0.87 | 714.32 | −0.18 | −0.06 |
| Freight volume and industrialization rate are cross-spectrum | 8.61 | 0.12 | 0.68 | 0.03 | −0.86 | −1.18 |
| | 6.99 | 0.14 | 0.51 | 0.01 | −0.32 | −0.35 |
| | 3.89 | 0.26 | 0.53 | 0.02 | 0.09 | 0.05 |
| | 2.22 | 0.45 | 0.54 | 0.03 | 0.24 | 0.08 |
| The freight volume and urbanization rate are cross-spectrum | 11.47 | 0.09 | 0.23 | 0.02 | −2.17 | −3.95 |
| | 3.85 | 0.26 | 0.20 | 0.04 | −0.47 | −0.29 |
| | 2.44 | 0.41 | 0.30 | 0.06 | 1.2 | 0.47 |
| | 2.3 | 0.44 | 0.29 | 0.08 | 0.14 | 0.05 |

the two highest values. Their correspondent lag period is 0.06 years and 1.18 years, respectively.

With careful consideration, the highest consistency representative is more reliable. Therefore, the cycle of 11.47 years has high consistency, so it is considered that the cycle is more consistent with the actual situation. The freight volume lags behind the phase spectrum's urbanization rate, and the lag period is nearly four years.

The cross-spectral cycle lag of Rhine freight volume with GDP, industrialization rate, and urbanization rate are 2.87 years, 3 years, and 6.67 years respectively.

Over the three cycles, GDP's variation reflects in the Rhine river traffic volume after 0.11 years. The change in the industrialization rate will affect the river volume in nearly 0.09 years. The cross-spectrum between the urbanization rate and traffic volume is a long cycle. Generally, the growth of the urbanization rate needs 2.1 years to drive the growth of traffic volume.

From Table 5 and Table 6, the amplitude spectrum value of freight volume and GDP is the highest, followed by the urbanization rate and the industrialization rate.

From the spectral analysis perspective, freight volume is the most sensitive to the change of GDP; the variance decomposition contribution rate also proves that.

**Table 6.** Cross spectral analysis between Rhine river freight volume and related factors

| | Period | Frequency | Coherence | Amplitude | Phase | Time Lag (year) |
|---|---|---|---|---|---|---|
| The cross spectrum between freight volume and GDP | 10.85 | 0.09 | 0.52 | 5.17 | 2.19 | 3.78 |
| | 3.85 | 0.26 | 0.54 | 5.17 | 0.86 | 0.53 |
| | 3.65 | 0.27 | 0.61 | 7.30 | 0.70 | 0.41 |
| | 2.87 | 0.35 | 0.59 | 9.32 | −0.25 | −0.11 |
| Freight volume and industrialization rate are cross - spectrum | 3.88 | 0.26 | 0.31 | 7.60 | 0.19 | 0.12 |
| | 3.65 | 0.27 | 0.45 | 0.01 | 0.42 | 0.25 |
| | 3.19 | 0.31 | 0.46 | 0.01 | 0.01 | 0.01 |
| | 3.00 | 0.33 | 0.35 | 0.01 | −0.18 | −0.09 |
| The freight volume and urbanization rate are cross-spectrum | 6.67 | 0.15 | 0.57 | 2.83 | −1.98 | −2.1 |
| | 3.88 | 0.26 | 0.48 | 4.57 | 0.77 | 0.47 |
| | 3.66 | 0.27 | 0.52 | 6.1 | 1.27 | 0.74 |
| | 3.48 | 0.29 | 0.47 | 6.23 | 1.69 | 0.94 |

## 5 Conclusion

After spectral analysis, the following conclusions were drawn:

1) According to the correlation analysis and variance decomposition with the observation on Figs. 1 and 2, they provide an inflection point in the volume of inland river traffic. Before the inflection point, the volume of inland river traffic shows a trend of rapid growth, which is strongly correlated with GDP, industrialization rate, and urbanization rate. After the inflection point, the volume of the inland waterway grows Slowly and steadily. The industrialization rate and urbanization rate are also obviously uncorrelated.

2) The influences of various factors on the volume of the inland river are also different. GDP is the most significant factor through variance decomposition and amplitude spectrum analysis, followed by urban and industrial rates.

3) From the frequency domain perspective, it displays a particular time difference in the influence of relevant factors on the volume of the inland river. Based on the different national-territorial areas, the USA usually spends more time reacting; also, consider the financial field in the USA, GDP is exceptive. It takes one and a half months for GDP to drive the growth of inland river volume in the United States, otherwise three months in Germany. The urbanization rate on inland river volume takes an extended period, almost four years in the United States; however, two years in Germany. The industrialization rate change takes one year in the United States and only one month in Germany.

## References


1. Grzesica, D., Wicek, P.: Advanced forecasting methods based on spectral analysis. Procedia Eng. **161**, 253–258 (2016)

2. Zhang, Y., Zhang, Y., Haghani, A.: A hybrid short-term traffic flow forecasting method based on spectral analysis and statistical volatility model. Transp. Res. Part C Emerg. Technol. **43**, 65–78 (2014)
3. Chen, J., Zeng, X., Zhen, H., et al.: Spectral analysis model for cyclical characteristics of capesize dry bulk shipping market. J. Transp. Syst. Eng. Inf. Technol. **16**(06), 204–209+235 (2016)
4. Yueqi, L., Yin, W., Xin, Z., et al.: Correlation between inland waterway transport and local economy. J. Traffic Transp. Eng. **9**(06), 97–101 (2009)
5. Bing, Z.: The comparison between the supply and demand factors' influences on the business cycle in the USA: empirical test based on factor analysis and cross spectrum analysis. Nankai Econ. Stud. **04**, 3–17 (2012)
6. Yue, W.: Spectral analysis method and its application in the study of business cycle—a case study of spectral analysis on the business cyclein the USA (1930–2009). Finance Econ. **11**, 34–43 (2011)
7. Ma, Y.H.: Historical assessment of the US industrialization phase. World Region. Stud. **19**(3), 81–87 (2010)
8. Klemann, H.A.M., Schenk, J.: Competition in the Rhine delta: waterways, railways and ports, 1870–1913. Econ. Hist. Rev. **66**(3), 826–847 (2013)
9. Rayevnyeva, O., Stryzhychenko, K.: Investigation of the wave nature of the ukrainian stock market. Int. J. Intell. Syst. Appl. **4**(1), 585 (2012)
10. Karabutov, N.: Identification of quality indicators dynamic system on basis of analysis data "Input-Output." Int. J. Intell. Syst. Appl. **09**(9), 1–11 (2014)
11. Wuling, Z., Keyi, W., Yingchao, J.: An empirical study on determinants of sustainable development of coastal eco-tourism. Int. J. Educ. Manage. Eng. **2**(6), 34–40 (2012)

# RBF-Based Input Doubling Method for Small Medical Data Processing

Ivan Izonin(✉), Roman Tkachenko, Solomia Fedushko, Dmytro Koziy, Khrystyna Zub, and Olena Vovk

Lviv Polytechnic National University, S. Bandera str., 12, Lviv 79013, Ukraine

**Abstract.** Many studies in the medical area require large training datasets for intellectual analysis. The process of collecting such information is time consuming or costly, but in some situations it is simply not possible to collect a large dataset. In such cases, there is a task of developing new soft computing methods for efficient processing and analysis of limited datasets. The paper proposes a new small data approach to solving this task. The authors design a new input doubling method based on the use of the classical iterative RBF neural network. This nonlinear artificial neural network provides the ability for modeling complex response surfaces that are difficult to represent by combinations of sigmoid slopes. The algorithmic implementation of the developed method in both modes: training and application is described. The optimal parameters of the method are selected. The training of the developed method is performed using the MSE loss function. The simulation was performed using a medical dataset of 79 instances. The accuracy of the developed method is compared with a number of existing modifications of the RBF neural networks (iterative and non-iterative). The highest accuracy of the proposed method based on MAE and RMSE was established experimentally.

**Keywords:** Small datasets · Nonlinear ANN · RBF · Medical data · Limited data · Small data approach

## 1 Introduction

Modern development of medicine is increasingly accompanied by the use of soft computing [1]. The variety of medical data [2], the large number of diagnostic, prevention and treatment tasks [3] and their high complexity [4] necessitate the use of computational intelligence, which must ensure high accuracy and speed [5]. However, the existing tools do not always provide the expected result. It largely depends on the amount of input data. Insufficient number of them leads to the fact that a number of problems, which reduces the efficiency of intelligent data processing, and characterizes ML-based regression methods, artificial neural networks and other computational intelligence tools.

The task of short datasets processing is not new. However, there are currently no effective methods and algorithms for solving it. In recent years, interest in this topic has grown significantly. Many scientific papers have been published with attempts to effectively solve this task. In particular in [6] the authors developed a multiple runs method


Z. Hu et al. (Eds.): ICAILE 2021, LNDECT 82, pp. 23–31, 2021.
https://doi.org/10.1007/978-3-030-80475-6_3

using artificial neural networks. It provides high accuracy of processing short data sets. However, the ensemble approach, which essentially underlies the work of the method, requires many resources for its implementation. A number of other ensemble methods are also can be used to solve the stated task [7, 8]. However, there may be problems with over-fitting, which greatly affects the result of the ensemble. In [9] the authors describe the use of feed-forward neural networks with back-propagation algorithm to solve the short data processing task. The authors used different ANN models to analyze nonlinearities in small datasets. The main purpose is to find causal links between the data. In particular, the author used a single-layer perception, explaining this by the fact that only one hidden layer is sufficient to approximate any continuous function. However, this type of ANN does not always provide the highest results according to the accuracy, as well as quite time-consuming training.

The main contribution of this work is the development of a new input doubling method based on the use of classical RBF neural network, which improves the accuracy of solving the short medical data processing task.

## 2 Materials and Methods

### 2.1 Dataset Description

The solution of many real diagnosis, prevention or treatment tasks using soft computing methods, in particular neural networks, largely depends on the amount of the training data. In some cases, they are too expensive or impossible to obtain. This paper uses the Urine Analysis Dataset collected in [10]. This dataset contains only 79 instances with 7 attributes. The actual dataset can be downloaded from [11]. Table 1 summarizes the attributes of this data set, their maximum, minimum and average values.

**Table 1.** Main dataset characteristics

| Attribute's title | MIN value | MAX value | MEAN value |
|---|---|---|---|
| Indicator of the presence of calcium oxalate crystals | 0 | 1 | 0.436 |
| The specific gravity of the urine | 1.005 | 1.04 | 1.018 |
| The pH reading of the urine | 4.76 | 7.94 | 6.042 |
| The osmolarity of the urine | 187 | 1236 | 613.61 |
| The conductivity of the urine | 5.1 | 38 | 20.901 |
| The urea concentration in millimoles per litre | 10 | 620 | 264.141 |
| The calcium concentration in millimoles per litre | 0.17 | 14.34 | 4.161 |

The first six attributes are used as independent variables; the seventh attribute should be predicted. A description of data collection procedures and more detailed characteristics are given in [12].

## 2.2 RBF Neural Network

RBF neural networks are widely used to solve approximation of multidimensional dependences tasks, including classification, regression, prediction. The main advantage of RBF neural networks in comparison with MLP, is the ability for modeling the complex response surfaces that are difficult to represent by combinations of sigmoid slopes. At the same time, almost complete inability of RBF networks to extrapolate, increased requirements for the size of the training sample limit the possibility of their use in many intelligent data processing tasks [13].

RBF neural network in the most cases has three layers: input layer, hidden and output layers (Fig. 1). The mapping from the input layer to the hidden layer is nonlinear, and the mapping from the hidden layer to the output layer is linear. In most cases neurons number in the hidden layer is greater than neurons number in the input layer. The main idea of this ANN type is to transform linear problem in some non-linear way in the space of increased dimension. It provides a higher probability to obtain precisely results. This is confirmed by well-known Cover's theorem [13].

Figure 1 shows the topology of the RBF neural network. Detailed mathematical calculations of the procedure of it training are given in [13].

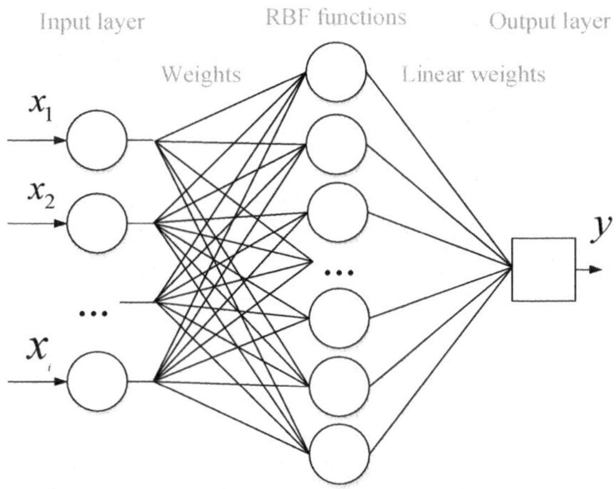


**Fig. 1.** RBF network topology

Among the main problems during the practical implementation of neural networks of this type are: the choice of the method of forming the coordinates of the RBF-centers and the type of basic functions; the setting the number of RBF centers and smooth factor value; the training the output layer of the network.

This paper uses the k-means algorithm for the formation of RBF centers, and Means Square Error loss function during the implementation of the neural network training procedure. Smooth factor value was selected empirically. It should be noted that the same parameters were used for the developed input doubling method based on the classical RBF neural network.

### 2.3 RBF-Based Input Doubling Methods

RBF networks provide possibility to simulate response surfaces of arbitrary complexity. This property became the basis for the development of the input doubling method for efficient processing and analysis of short data samples.

The developed method provides two modes of operation: training and application.

- **The Main Steps of the Training Mode:**

1) To form all possible pairs of double vectors using all available $N$-vectors $\overline{X_i}$ from the training data sample, in the next form: $\overline{X_k}, \overline{X_i}$, where $i$ indicate the index of current vector and $k$ indicate all other evaluable vectors, $i = k = \overline{1, N}$.
2) To form the corresponding output $Z_{k,i}$ for each new (extended) vector from item 1 using the following notation:

$$Z_{k,i} = (y_k - y_i) \tag{1}$$

where $y_i$ is the output of *i-th* vector from training dataset and $y_k$ is the output of k-*th* vector from the training dataset.
3) To set the number of radial inputs in the RBF neural network.
4) To train the RBF neural network using inputs and outputs formed on items 1 and 2 of the training procedure.

- **The Main Steps of the Application Mode:**

1) To form a set of extended vectors for the current $u$-th vector with unknown output $\overline{X_u}$ in the form: $\overline{X_u}, \overline{X_i}$, where $\overline{X_i}$ - known vector from the training sample. The number of such vectors will be equal to the number of vectors of the training sample.
2) To obtain the predicted values $Z_{u,i}^{pred}$ for each extended vector from the set formed in item 1 of the application mode using a pre-trained RBF neural network;
3) To calculate the result of the developed method for the *u-th* input vector using the following equation:

$$y_u \cong \frac{\sum_{i=1}^{N} y_i + \sum_{i=1}^{N} Z_{u,i}^{pred}}{N} \tag{2}$$

where $y_i$ are the actual outputs for all observation in the training dataset, $N$ is the number of these vectors in training mode.

### 2.4 Performance Evaluation

To assess the accuracy of the developed method, as well as to compare its work with existing ones, the following regression metrics were used [14]: Root mean square error

(3) and Mean absolute error (4):

$$RMSE = \sqrt{\frac{\sum_{i=1}^{N} \left(y_i^p - y_i^{real}\right)^2}{N}} \tag{3}$$

$$MAE = \frac{1}{N} \sum_{i=1}^{N} \left| y_i^p - y_i^{real} \right| \tag{4}$$

where: $y_i^{real}$ are the real values of the independed attribute; $y_i^p$ are the predicted values obtained by current regression method; $N$ is the number of vectors $i = 1, N$.

## 3 Modeling and Results

The modeling of the developed method was performed using a real data set [11], which contained only 79 observations. Two vectors of this data set contained missing values. Therefore, they were removed from the dataset used for modeling. For experimental evaluation of the developed method, this short data set was divided into three parts: training, validation and test samples in the ratio of 70%, 10% and 20%, respectively.

The main parameters of the method as well as the results of its work based on regression metrics (3)–(4) are summarized in Table 2.

**Table 2.** Parameters of the proposed method and its results

| Parameter/indicator | Value | Training mode: MAE, RMSE | | Test mode: MAE, RMSE | |
|---|---|---|---|---|---|
| Neurons number in the input layer | 14 | 2.7522 | 3.6774 | 2.7703 | 3.7056 |
| Neurons number in the output layer | 1 | | | | |
| Epoch number | 25 | | | | |
| Loss function | MSE | | | | |
| Clustering method | k-means | | | | |
| Number of clusters | 5 | | | | |
| Iteration number for clustering algorithm | 100 | | | | |
| Smooth factor | 0.8 | | | | |

Figure 2 shows the MSE-based loss function graph for 25 epoch of the proposed method training procedure.

As can be seen from Fig. 2, a significant increase in the dimension of the training data sample (from 61 to 3721), as well as doubling the attributes of each input vector is not accompanied with the overfitting for the developed method. However, it should be noted that the increase in the number of epochs, in particular to 500, is accompanied by a high probability of overfitting of the developed method.

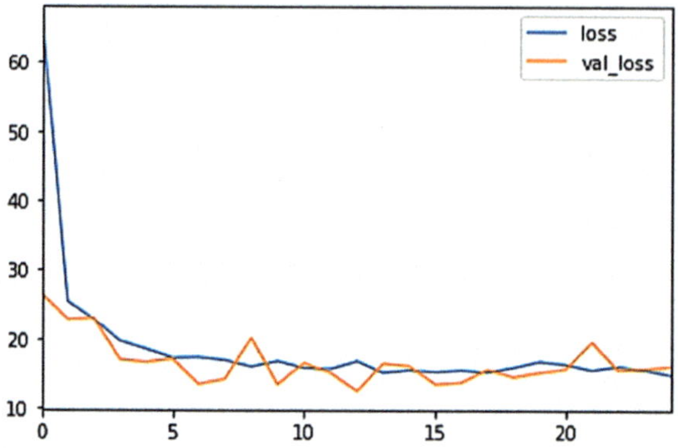


**Fig. 2.** Graph of MSE loss function (*oy* axis) vs. number of epoch of the training procedure for the proposed method (*ox* axis)

Therefore, in order to reduce the duration of the training procedure, as well as to avoid the possibility of overfitting of the developed method, it is recommended to choose a small number of epochs for its training.

## 4 Comparison and Discussion

The results of the developed method were compared with the results of the known neural network methods based on RBF. In particular, the authors used:

1) classical neural network of this type, which is the basis of the developed method;
2) two modifications of the RBF network of non-iterative type [15]:

   – the first of them, SGTM neural-like structure with RBF kernel is based on replacement of initial inputs of the task on the transformed with use of rbf-functions with the subsequent application of greedy algorithm of training of SGTM neural-like structure [16];
   – the second non-iterative version of the RBF network, Combined RBF-SGTM neural-like structure, combines the initial inputs with their extension based on *rbf*-functions [17].

It should be noted that the last two neural networks do not require a learning procedure.

The results of experimental studies comparing all the studied methods based on regression metrics (3)–(4), in the both modes: training (if required) and application are shown in Figs. 3 and 4.

The parameters of the classical RBF neural network are the same as for the proposed method (Table 2) and the main difference is only the number of neurons in the input layer, which is equal to 7.

The parameters of SGTM neural-like structure with RBF kernel are as follows: the number of neurons in the input layer is 15, the smooth factor is 0.8. The same value of the last indicator is for Combined RBF-SGTM neural-like structure, but the neuron number of the input layer is equal 22 (15 *rbf*-inputs and 7 initial inputs). Both methods, due to the peculiarities of SGTM neural-like structure, do not require adjustment of other additional parameters.

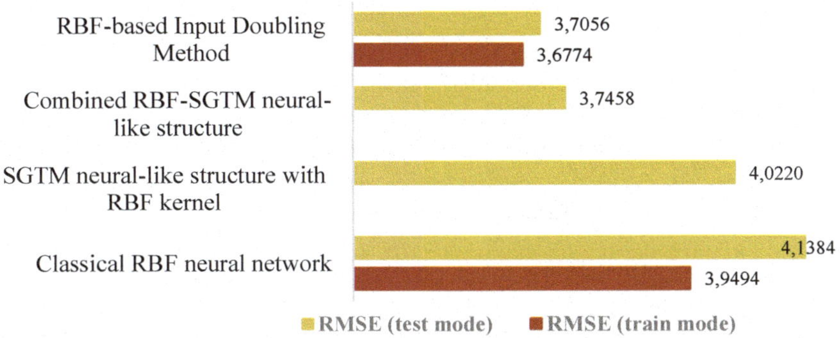


**Fig. 3.** Comparison of the results for all investigated methods based on RMSE

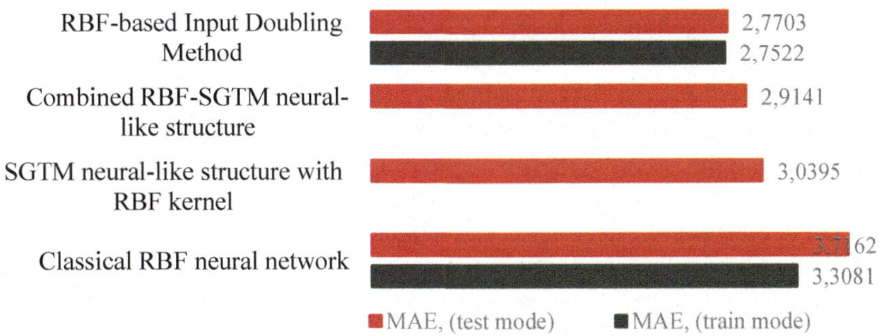


**Fig. 4.** Comparison of the results for all investigated methods based on MAE

As can be seen from both figures, the classical RBF neural network shows the least accuracy. Both of its non-iterative modifications show greater accuracy, in particular due to the peculiarities of construction and training of SGTM neural-like structure, which is their basis. In addition, due to the combination of *rbf*- and the initial inputs, the Combined RBF-SGTM neural-like structure demonstrates the least error among non-iterative methods when solving the problem. However, its disadvantage is the increase in time delays of the application procedure by processing more inputs (*rbf*- and the initial inputs).

The developed method demonstrates the highest accuracy in both training and application modes on both regression metrics among the considered methods. This creates the preconditions for its practical use to solve applied tasks of medicine in the context of processing limited data samples. Among the disadvantages of the method should be noted a significant increase in the duration of its training procedure due to the quadratic increase of the dimension of the initial sample. However, in the conditions of the need to obtain the highest possible accuracy, which is typical for the tasks in medicine domain, where it is about human life and health, sometimes this shortcoming can be missed.

## 5 Summary and Conclusion

The paper presents a new approach to the effective processing of small medical data. The authors developed a input doubling method based on the use of the classical RBF neural network. The main steps of algorithmic implementation of the method in the both modes: training and application are given. The simulation was performed using a short medical dataset. The optimal parameters of the developed method are selected.

The method was trained using the MSE-based loss function. An experimental comparison of the accuracy of its operation based on MAE and RMSE measures with different modifications of the RBF neural network (iterative and non-iterative) are performed. The highest accuracy of the developed method based of both accuracy indicators among all investigated methods is established. It can be used in different areas [18].

Among the disadvantages of the developed method based on the classical iterative neural network RBF should be noted significant time delays in training mode due to the quadratic increase in the number of vectors of the training sample.

Further research will be conducted in the following areas:

1) an application of optimization methods for selection of optimal parameters of the developed method (reduce the time for this procedures) [19];
2) an increasing the accuracy of the method by improving the procedure for forming the result (2) to compensate for errors of various signs of the predicted values (1);
3) the use of other nonlinear computational intelligence tools [20, 21] as the basis of the method, in particular non-iterative analogues of the RBF network [15], to improve the accuracy of the prediction and reduce time delays of the developed input doubling method.

**Acknowledgment.** This research is supported by National Research Foundation of Ukraine within the project "Methods of managing the web community in terms of psychological, social and economic influences on society during the COVID-19 pandemic", grant number 94/01-2020.

## References

1. Boyko, N., Kuba, M., Mochurad, L., Montenegro, S.: Fractal distribution of medical data in neural network. CEUR-WS **2488**, 307–318 (2019)

2. Petrovska, B., Stojanovic, I., Atanasova-pacemska, T.: Classification of small sets of images with pre-trained neural networks. IJEM **8**, 40–55 (2018)
3. Umezuruike, C., Nwankwo, W., Okolie, S.O., Adebayo, A.O., Jonah, J.V., Ngugi, H.: Health informatics system for screening arboviral infections in adults. IJITCS **11**, 10–22 (2019)
4. Bodyanskiy, Y., et al.: Hybrid clustering-classification neural network in the medical diagnostics of the reactive arthritis. IJISA **8**, 1–9 (2016)
5. Melnykova, N., Melnykov, V., Vasilevskis, E.: The personalized approach to the processing and analysis of patients' medical data. CEUR-WS **2255**, 103–112 (2018)
6. Shaikhina, T., Khovanova, N.A.: Handling limited datasets with neural networks in medical applications: a small-data approach. Artif. Intell. Med. **75**, 51–63 (2017)
7. Bodyanskiy, Y.V., Tyshchenko, O.K.: A hybrid cascade neural network with ensembles of extended neo-fuzzy neurons and its deep learning. In: Kulczycki, P., Kacprzyk, J., Kóczy, L.T., Mesiar, R., Wisniewski, R. (eds.) ITSRCP 2018. AISC, vol. 945, pp. 164–174. Springer, Cham (2020). https://doi.org/10.1007/978-3-030-18058-4_13
8. Hu, Z., Bodyanskiy, Y., Tyshchenko, O.K.: Self-learning procedures for a kernel fuzzy clustering system. Adv. Intell. Syst. Comput. **754**, 487–497 (2019)
9. Pasini, A.: Artificial neural networks for small dataset analysis. J. Thorac. Dis. **7**(5), 953–960 (2015)
10. Andrews, D.F., Herzberg, A.M.: Data: A Collection of Problems from Many Fields for the Student and Research Worker. Springer, New York (1985). https://doi.org/10.1007/978-1-4612-5098-2
11. R: Urine Analysis Data (n.d.). https://vincentarelbundock.github.io/Rdatasets/doc/boot/urine.html. Accessed 12 Dec 2020
12. Davison, A.C., Hinkley, D.V.: Bootstrap Methods and Their Application. Cambridge University Press, Cambridge (1997)
13. Medykovskvi, M., Pavliuk, O., Sydorenko, R.: Use of machine learning technologies for the electric consumption forecast. In: Proceedings of the CSIT2018, pp. 432–435 (2018)
14. Fedushko, S., Ustyianovych, T.: Predicting pupil's successfulness factors using machine learning algorithms and mathematical modelling methods. In: Hu, Z., Petoukhov, S., Dychka, I., He, M. (eds.) ICCSEEA 2019. AISC, vol. 938, pp. 625–636. Springer, Cham (2020). https://doi.org/10.1007/978-3-030-16621-2_58
15. Tkachenko, R., et al.: Non-iterative neural-like predictor for solar energy in libya. CEUR-WS **2105**, 35–45 (2018)
16. Izonin, I., et al.: Committee of SGTM neural-like structures with RBF kernel for insurance cost prediction task. In: Proceedings of the UKRCON, pp. 1037–1040 (2019)
17. Tkachenko, R., et al.: Committee of the combined RBF-SGTM neural-like structures for prediction tasks. In: Awan, I., Younas, M., Ünal, P., Aleksy, M. (eds.) MobiWIS 2019. LNCS, vol. 11673, pp. 267–277. Springer, Cham (2019). https://doi.org/10.1007/978-3-030-27192-3_21
18. Sheremeta, B., et al.: Marketing tools as the competitiveness enhancer of the Ukrainian film distribution market entities. Innov. Mark. **15**(4), 88–101 (2019)
19. Rzheuskiy, A., et al.: The information support of virtual research teams by means of cloud managers. IJISA **10**(2), 37–46 (2018)
20. Shakhovska, N., Montenegro, S., Kryvenchuk, Y., Zakharchuk, M.: The neurocontroller for satellite rotation. IJISA **11**(3), 1–10 (2019)
21. Leoshchenko, S., et al.: Parallel method of neural network synthesis based on a modified genetic algorithm application. In: MoMLeT 2019, vol. 2386, pp. 11–23 (2019)

# Vehicles and Cargos Two-Sided Matching Based on Similarity Calculation

Meng Wang[1], Qing Xia[2](✉), and Xuejiang Wei[1]

[1] School of Logistics, Wuhan Technology and Business University, Wuhan 430065, China
[2] School of Economics, Wuhan Donghu University, Wuhan 430212, China

**Abstract.** Vehicles and cargos matching is an important issue in the interconnection of logistics parks, which affects the efficiency and benefit of logistics. In this paper, a vehicles and cargos two-sided matching model based on similarity calculation is proposed, which engages to achieve stable two-sided matching between vehicles and cargos under optimal overall benefit. Firstly, the information similarity between candidate resources and demands is calculated, including status information, ability information and level information. Then, the comprehensive similarity is calculated to screen and sort. Finally, considering satisfaction of both subjects, a two-sided integrated matching model is proposed, and the overall matching scheme with best comprehensive benefit is obtained. Through the example verification, the model and algorithm have high recall, accuracy and matching efficiency, and have good applicability.

**Keywords:** Similarity calculation · Two-sided matching · Vehicles and cargos matching

## 1 Introduction

To promote the construction of modern circulation system, it is required to realize the interconnection of logistics parks and smooth flow of logistics elements. The vehicles and cargos matching is one of the important problems, which affect the efficiency and benefit of parks interconnection and elements flow.

At present, most of information platforms of logistics parks provide customers with information release, query and simple matching functions as a 3rd party. The matching function can only provide a list of available resources for vehicle owners or cargo owners. With the deepening of interconnection, more and more logistics parks have changed their roles from a 3rd party to an integrated operator, which require the parks to provide overall matching scheme of vehicles and cargos. The goal of this matching problem is to achieve a stable two-sided matching between vehicles and cargos.

This problem has the following characteristics:

1) It is a supply-demand matching problem under big data and cloud platform, and the matching algorithm must be more efficient.
2) Hard constraints and soft constraints need to be considered.


Z. Hu et al. (Eds.): ICAILE 2021, LNDECT 82, pp. 32–43, 2021.
https://doi.org/10.1007/978-3-030-80475-6_4

3) The principle of two-sided equality should be obeyed as far as possible, so as to promote stable cooperation and form best overall benefit.

In the existing literature, Vijay Verma (2019) proposed a new similarity measure based on simple matching coefficient [1]. Hu (2018) proposed the idea of matching and scheduling vehicles and cargos based on the credit evaluation system, and used TS algorithm to solve the problem [2]. Li (2015) constructed a post transaction reputation evaluation model and a multi-objective vehicles and cargos matching ranking model [3]. But they are mainly used to solve matching problem of one side. Mahantesh G (2020) brought a resource matching scheme for vehicular cloud networks based on cost optimization [4]. Fan (2014), Li (2013) constructed a multi-objective two-sided matching optimization model based on the order value information of bilateral subjects' preference [5, 6]. Alessandro S (2019) designed an efficient resource sharing method in intelligent cloud manufacturing platform [7]. Zhao (2017) based on prospect theory, determine the preference order of matching subjects, and then propose a bilateral stable matching scheme. Jia (2020) proposes a resource-demand efficient matching strategy of cloud manufacturing platform [8, 9]; Bi (2019) proposes a resource multi-level intelligent rent-seeking and matching strategy for the problem of huge scale of resources [10].

The above researches proposed two-sided matching methods, but for the vehicles and cargos matching problem in the case of parks intercommunication, there is a lack of targeted solutions, and it is necessary to further put forward applicable solutions, models and algorithms.

## 2 Matching Process

The description of resources (vehicles) and demands (cargos) information mainly involves three dimensions: status information ($S_{status}$), ability information ($S_{ability}$) and level information ($S_{level}$). Based on these dimensions, similarity calculation and comprehensive matching are carried out.

The matching process consists of five steps, as shown in Fig. 1.

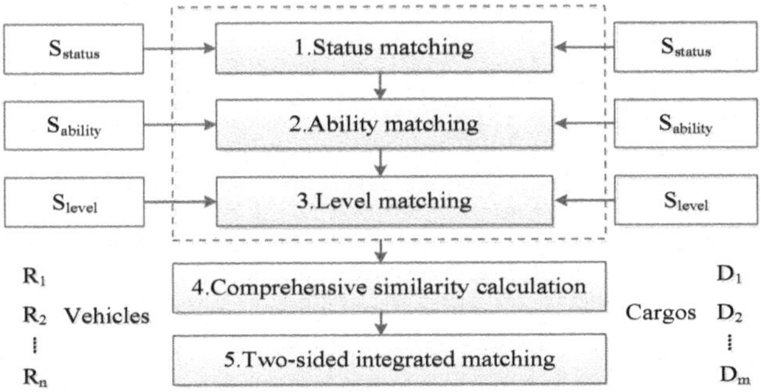


**Fig. 1.** Matching process

# 3 Matching Model and Algorithm

## 3.1 Status Matching

Match the candidate resource status information *(RS)* with the candidate demand status information *(DS)*.

RS mainly includes four statuses: idle, under loaded, full loaded and invalid.

$$RS = \begin{cases} 0 \; when\, the\, status\, is\, full\, load\, and\, invalid \\ 1 \; when\, the\, status\, is\, idle \\ 2 \; when\, the\, status\, is\, under\, loaded \end{cases} \tag{1}$$

DS mainly includes 4 statuses: unexecuted, in execution, completed and invalid.

$$DS = \begin{cases} 0 \; when\, the\, status\, is\, completed\, and\, invalid \\ 1 \; when\, the\, status\, is\, unexecuted \\ 2 \; when\, the\, status\, is\, in\, execution \end{cases} \tag{2}$$

The matching process is as follows:

1) Confirm input. Candidate resources set: $D_R = \{d_{1R}, d_{2R}, \cdots, d_{kR}, \cdots, d_{nR}\}$; candidate demands set: $D_D = \{d_{1D}, d_{2D}, \cdots, d_{kD}, \cdots, d_{nD}\}$.
2) Confirm output. The set of status information matching between resource and demand sides $D_{SR}, D_{SD}$.
3) Determine initial state. $D_{SR}, D_{SD} = \emptyset, k = 1$, where $k$ is the sequence number of the candidate resource or demand.
4) If $RS \neq 0$, then $D_{SR} = D \cup d_{kR}$; if $DS \neq 0$, then $D_{SD} = D \cup d_{kD}$.
5) In order to meet the demand in execution or the under full load resource, the higher *RS* and *DS* values are, the higher the selection priority is.
6) Do $k = k + 1$, if $k \leq n$, go to *4)*, otherwise go to *7)*.
7) The algorithm ends and outputs $D_{SR}, D_{SD}$.

## 3.2 Ability Matching

In terms of matching index system, it can be divided into hard matching indexes and soft matching indexes:

1) Hard matching indexes mainly include *vehicle type matching* ($H_1$)*, weight matching* ($H_2$)*, vehicle length matching* ($H_3$)*, volume matching* ($H_4$)*, time matching of arrival at application point* ($H_5$)*, route matching* ($H_6$)*.*
2) Soft matching indexes mainly include *load ratio* ($S_1$)*, vehicle length compliance* ($S_2$)*, volume compliance* ($S_3$)*.*

The above matching indexes can be divided into two categories.

1) One is to use the phrase concept to describe (such as $H_1, H_5, H_6$), and for this kind of attribute matching, the concept semantic similarity algorithm can be used.
2) The other is described by numerical parameters (such as $H_2, H_3, H_4, S_1, S_2, S_3$). For this kind of attribute matching, numerical parameter similarity algorithm is used.

### 3.2.1 Hard Matching Indexes Calculation

For $H_1, H_5, H_6$, using the concept semantic similarity algorithm, considering the semantic distance and semantic coincidence.

$$sim_C(RF_w, DF_w) = \eta_1 \frac{1}{1 + dist(RF_w, DF_w)^2} + \eta_2 OL(RF_w, DF_w) \quad (3)$$

$$OL(RF_w, DF_w) = \frac{|Path(RF_w) \cap Path(DF_w)|}{|Path(RF_w) \cup Path(DF_w)|} \quad (4)$$

Where, $\mathrm{sim}_C(RF_w, DF_w)$ is the concept semantic similarity of the $w$-th concept parameter $(RF_w)$ in resource ability information and the $w$-th concept parameter $(DF_w)$ in demand ability information; $dist(RF_w, DF_w)$ is the semantic distance; $OL(RF_w, DF_w)$ is the semantic coincidence degree; $\eta_1$ and $\eta_2$ are the influence weights of semantic distance and semantic coincidence degree; *Path ()* is the number of upper concept nodes from the resource concept parameter to the root node of the resource ontology tree.

Set $\varepsilon_{CH}$ (close to 1), when $sim_C(RF_w, DF_w) \geq \varepsilon_{CH}$, keep it in the candidate set; otherwise reject it.

For $H_2, H_3, H_4$, numerical parameter similarity algorithm is used.

$$\text{For the resource side,} \quad sim_N(R_h, D_h) = \begin{cases} 1 \; R_h \cap D_h = D_h \\ 0 \; R_h \cap D_h < D_h \end{cases} \quad (5)$$

$$\text{For the demand side,} \quad sim_N(R_h, D_h) = \begin{cases} 1 \; R_h \cap D_h = R_h \\ 0 \; R_h \cap D_h < R_h \end{cases} \quad (6)$$

Where, $R_h$ is the $h$-th numerical parameter information of resource side; $D_h$ is the $h$-th numerical parameter information of demand side; $\mathrm{sim}_N(R_h, D_h)$ is the numerical parameter similarity of $R_h, D_h$. When the similarity is *1*, keep it in the candidate set; otherwise reject it.

### 3.2.2 Soft Matching Indexes Calculation

For $S_1, S_2, S_3$, the similarity algorithm of numerical parameters is used.

$$\text{For the resource side,} \quad sim_S(R_h, D_h) = \begin{cases} 1 & R_h = D_h \\ \frac{|D_h|}{|R_h|} & R_h \cap D_h = D_h \\ 0 & R_h \cap D_h < D_h \end{cases} \quad (7)$$

$$\text{For the demand side,} \quad sim_S(R_h, D_h) = \begin{cases} 1 & R_h = D_h \\ \frac{|R_h|}{|D_h|} & R_h \cap D_h = R_h \\ 0 & R_h \cap D_h < R_h \end{cases} \quad (8)$$

Where, $|R_h|$, $|D_h|$ are the length of the numerical interval of $R_h, D_h$.

If there are $n$ soft matching index parameters, the soft matching index similarity can be expressed as follows:

$$\mathrm{sim}_G(RF, DF) = \frac{1}{n} \sum\nolimits_{h=1}^{n} \mathrm{sim}_S(R_h, D_h) \quad (9)$$

If carpooling is allowed, $H_2, H_3, H_4$ can participate in the matching again after removing occupied value until there is no matching object. In this case, the soft matching indexes can be considered after all the hard matching is completed, or not considered.

### 3.2.3 Ability Matching Process

1) Confirm input. The resource set matched by status information $D_{SR} = \{d_{SR1}, d_{SR2}, \cdots, d_{SRk}, \cdots, d_{SRn}\}$; the demand set matched by status information $D_{SD} = \{d_{SD1}, d_{SD2}, \cdots, d_{SDk}, \cdots, d_{SDn}\}$; $\varepsilon_F$ is the threshold of ability information matching.
2) Confirm output. Establish the set of ability information matching $D_{FR}, D_{FD}$.
3) Determine initial state. $D_{FR}, D_{FD} = \phi, k = 1$.
4) According to *Formulas* (3)–(6), calculate $sim_C(RF_k, DF_k)$, if $sim_C(RF_k, DF_k) \geq \varepsilon_{CH}, sim_N(R_k, D_k) = 1, sim_G(d_{SRk} \cdot RF, DF) \geq \varepsilon_F$, then $D_{FR} = D_{FR} \cup d_{SRk}, D_{FD} = D_{FD} \cup d_{SDk}$.
5) Do $k = k + 1$, if $k \leq n$, go to *4)*, otherwise go to *6)*.
6) The algorithm ends and outputs $D_{FR}, D_{FD}$.

## 3.3 Level Matching

For both sides, the required "level" connotation is not consistent.

- The level requirements of "demand side to resource side" mainly include: price level, service level (punctual service rate, accurate service rate, damage compensation rate, accident rate, etc.), qualification level, reputation level, technical level, etc.
- The level requirements of "resource side to demand side" mainly include: price level, payment level, reputation level, etc.

Because of the uncertainty and complexity of above indexes, fuzzy evaluation method can be used to calculate the similarity. Triangular fuzzy number is a common method of fuzzy evaluation, which is represented by five granularity language set. It is shown as Table 1.

**Table 1.** Triangular fuzzy number corresponding to five granularity language set

| Language | Triangular fuzzy number |
|---|---|
| Low \| weak | (0, 0, 0.25) |
| Lower than medium \| weaker than medium | (0, 0.25, 0.5) |
| Medium | (0.25, 0.5, 0.75) |
| Higher than medium \| stronger than medium | (0.5, 0.75, 1) |
| High \| strong | (0.75, 1, 1) |

By measuring the above matching indexes, two groups of triangular fuzzy numbers are obtained, which are $\tilde{a}_{ki} = \left(a^l_{ki}, a^m_{ki}, a^u_{ki}\right), \tilde{a}'_{ki} = \left(a^{l\prime}_{ki}, a^{m\prime}_{ki}, a^{u\prime}_{ki}\right)$ and $\tilde{b}_{kj} =$

$\left(b_{kj}^{l}, b_{kj}^{m}, b_{kj}^{u}\right), \tilde{b}_{kj}' = \left(b_{kj}^{l\prime}, b_{kj}^{m\prime}, b_{kj}^{u\prime}\right)$. Where, $a_{ki}^{l}, a_{ki}^{m}, a_{ki}^{u}$ corresponds to lower, median and upper value of triangular fuzzy number respectively; $\tilde{a}_{ki}$ represents the real value of resource $R_i$ under index $O_k$, $\tilde{a}_{ki}'$ represents the expected value of resource $R_i$ under index $O_k$; $\tilde{b}_{kj}$ represents the real value of resource $D_j$ under index $O_k$, $\tilde{b}_{kj}'$ represents the expected value of resource $D_j$ under index $O_k$.

The level information similarity of resource side $R_i$ and demand side can be expressed as follows:

$$\mathrm{sim_L}(RL_i, DL_i) = \begin{cases} \frac{\sum_{k=1}^{e}[(a_{ki}^{l})^2+(a_{ki}^{m})^2+(a_{ki}^{u})^2]w_k}{\sum_{k=1}^{e}(a_{ki}^{l}a_{ki}^{l\prime}+a_{ki}^{m}a_{ki}^{m\prime}+a_{ki}^{u}a_{ki}^{u\prime})w_k} & a_{ki}^{m\prime} \geq a_{ki}^{m} \\ 1 & a_{ki}^{m\prime} < a_{ki}^{m} \end{cases} \tag{10}$$

Where, $w_k$ is the proportion of index $O_k$ in level matching indexes; $e$ is the number of level matching indexes.

In the same way, the level information similarity of demand side $D_j$ and resource side can be expressed as follows:

$$\mathrm{sim}_L(RL_j, DL_j) = \begin{cases} \frac{\sum_{k=1}^{f}[(b_{kj}^{l})^2+\left(b_{kj}^{m}\right)^2+\left(b_{kj}^{u}\right)^2]w_k}{\sum_{k=1}^{f}(b_{kj}^{l}b_{kj}^{l\prime}+b_{kj}^{m}b_{kj}^{m\prime}+b_{kj}^{u}b_{kj}^{u\prime})w_k} & b_{kj}^{m\prime} \geq b_{kj}^{m} \\ 1 & b_{kj}^{m\prime} < b_{kj}^{m} \end{cases} \tag{11}$$

Where, $f$ is the number of level matching indexes.

The level information matching process is as follows:

1) Confirm input. The resource set matched by ability information $D_{FR} = \{d_{FR1}, d_{FR2}, \cdots, d_{FRk}, \cdots, d_{FRn}\}$; the demand set matched by ability information $D_{FD} = \{d_{FD1}, d_{FD2}, \cdots, d_{FDk}, \cdots, d_{FDn}\}$; $\varepsilon_L$ is the threshold of level information matching.
2) Confirm output. Establish the set of level information matching $D_R, D_D$.
3) Determine initial state. $D_R, D_D = \phi$, $k = 1$.
4) Normalize the attribute values of level information in $D_{FR}, D_{FD}$.
5) Calculate $sim_L(RL_k, DL_k)$, if $sim_L(RL_k, DL_k) \geq \varepsilon_L$, then $D_R = D_R \cup d_{FRk}$, $D_D = D_D \cup d_{FRk}$.
6) Do $k = k + 1$, if $k \leq nF$, go to 5), otherwise go to 7).
7) The algorithm ends and outputs $D_R, D_D$.

### 3.4 Comprehensive Similarity Calculation

After the above status matching, ability matching and level matching, the resources/demands in set $D_R, D_D$ can meet the needs of both sides. In order to determine clear matching value and ranking, comprehensive similarity calculation is needed.

Since the hard matching indexes in status matching and ability matching are both necessary conditions, similarity ranking is no longer needed. Therefore, the calculation is mainly based on $\mathrm{sim}_G(RF, DF)$ and, $\mathrm{sim}_L(RL, DL)$. Under the same conditions, and priority is given to the object whose status value is “2”.

Assuming that the proportion of $sim_G(RF, DF)$ and $sim_L(RL, DL)$ is α, β, then the comprehensive similarity can be expressed as follows:

$$\mathrm{sim(R, D)} = \alpha \cdot \mathrm{sim}_G(RF, DF) + \beta \cdot \mathrm{sim}_L(RL, DL) \tag{12}$$

According to the comprehensive similarity value, the resources and demands can be sorted, which can provide the basis for the selection of resource side and demand side.

### 3.5 Two-Sided Integrated Matching

The above results mainly provide matching scheme with maximum satisfaction of one side (demand/resource). If the satisfaction of two sides is considered, further two-sided integrated matching is needed.

Assuming that there are $m$ resources and $n$ demands in $D_R$, $D_D$, the demand must be satisfied, under the condition of stable matching, with the objective of maximizing overall satisfaction of both sides, the following multi-objective optimization model can be established [11]:

$$maxZ = w_R \sum_{i=1}^{m} \sum_{j=1}^{n} x_{ij} \underset{R \to D}{sim} (R_i, D_j) + w_D \sum_{i=1}^{m} \sum_{j=1}^{n} x_{ij} \underset{D \to R}{sim} (R_i, D_j) \tag{13}$$

$$s.t. \sum_{j=1}^{n} x_{ij} = 1 \; i = 1, 2, \cdots, m$$

$$\sum_{i=1}^{m} x_{ij} \leq 1 \; j = 1, 2, \cdots, n$$

$$x_{ij} = 0 \; or \; 1 \; i = 1, 2, \cdots, m; j = 1, 2, \cdots, n$$

Where, $\underset{R \to D}{sim} (R_i, \mathrm{D}_j)$ is matching similarity of resource $i$ to demand $j$, $w_R$ is weight of resource side; $\underset{D \to R}{sim} (R_i, \mathrm{D}_j)$ is matching similarity of demand $j$ to resource $i$, $w_D$ is weight of resource side.

$w_R$, $w_D$ reflect the importance of resource/demand in actual decision-making. If the fairness of the two parties is taken into account, then $w_R = w_D$.

The objective function and constraint conditions in *model* (13) are linear, and can be solved by integer programming method. Since the two-sided matching problem in the set has been filtered many times, the scale is not very large. For the model with a small number of variables, MATLAB integer programming function can be used to solve the model. If the scale of bilateral matching problem is large, heuristic method or genetic algorithm can be designed to solve the problem [12].

## 4 Case Analysis

Taking a logistics park interconnection freight platform (involving freight transportation in Wuhan, Xi'an, Qingdao, etc.), some resource information and demand information as the case, shown in Table 2 and Table 3.

**Table 2.** Resource information of the logistics park

| Serial number | Business | Category | RS | RF | | | | | | RL | | | |
|---|---|---|---|---|---|---|---|---|---|---|---|---|---|
| | | | Status | Type | Load | Length | Volume | Source | Destination | Price | Service | Qualification | Reputation |
| R1 | FLT | Heavy cargo | Idle | Flatbed truck | 40 | 17.5 | 65 | Wuhan | — | M | $M^{+}$ | H | $M^{+}$ |
| R2 | FLT | Heavy cargo | Full load | Flatbed truck | 50 | 19.5 | 85 | Wuhan | Xi'an | M | $M^{+}$ | H | M |
| R3 | FLT | Heavy cargo | Idle | Flatbed truck | 50 | 19.5 | 85 | Wuhan | — | M | $M^{+}$ | H | H |
| R4 | FLT | Heavy cargo | Idle | Flatbed truck | 50 | 19.5 | 85 | Wuhan | — | M | H | M | $M^{+}$ |
| R5 | FLT | Heavy cargo | Idle | Flatbed truck | 50 | 19.5 | 85 | Xi'an | Wuhan | M | H | M | $M^{+}$ |
| R6 | FLT | Heavy cargo | Idle | Flatbed truck | 40 | 17.5 | 65 | Xi'an | — | M | H | M | $M^{+}$ |
| R7 | FLT | Heavy cargo | Idle | Flatbed truck | 40 | 17.5 | 65 | Wuhan | — | M | $M^{+}$ | H | $M^{+}$ |
| R8 | FLT | Heavy cargo | Idle | Flatbed truck | 40 | 17.5 | 65 | Xi'an | — | M | H | M | $M^{+}$ |
| R9 | FLT | Heavy cargo | Idle | Flatbed truck | 50 | 19.5 | 85 | Wuhan | — | M | $M^{+}$ | H | M |
| R10 | FLT | Heavy cargo | Idle | Flatbed truck | 50 | 19.5 | 85 | Wuhan | — | $M^{+}$ | M | M | $M^{+}$ |

Note: L-low; $M^{-}$-lower than medium; M-medium; $M^{+}$-higher than medium; H-high.

**Table 3.** Demand information of the logistics park

| Serial number | Business | Category | RS | RF | | | | | | RL | | | |
|---|---|---|---|---|---|---|---|---|---|---|---|---|---|
| | | | Status | Type demand | Cargo weight | Length demand | Cargo volume | Source | Destination | Price | Payment | Qualification | Reputation |
| D1 | FLT | Heavy cargo | Unexecuted | Flatbed truck | 45 | 15 | 50 | Wuhan | Xi'an | M | H | M | $M^+$ |
| D2 | FLT | Heavy cargo | Unexecuted | Flatbed truck | 30 | 15 | 40 | Xi'an | Wuhan | M | H | M | $M^+$ |
| D3 | FLT | Heavy cargo | Unexecuted | Flatbed truck | 39 | 16 | 50 | Wuhan | Xi'an | M | $M^+$ | H | $M^+$ |
| D4 | FLT | Heavy cargo | Unexecuted | Flatbed truck | 45 | 15 | 50 | Wuhan | Qingdao | M | $M^+$ | H | M |
| D5 | FLT | Heavy cargo | Completed | Flatbed truck | 40 | 17 | 70 | Xi'an | Wuhan | M | H | M | $M^+$ |
| D6 | FLT | Heavy cargo | Unexecuted | Flatbed truck | 45 | 18 | 80 | Wuhan | Xi'an | M | H | M | $M^+$ |
| D7 | FLT | Heavy cargo | Unexecuted | Flatbed truck | 40 | 10 | 20 | Wuhan | Xi'an | M | H | M | $M^+$ |

Note: L-low; $M^-$-lower than medium; M-medium; $M^+$-higher than medium; H-high.

**Step 1: Status Information Matching**
From resource set and demand set, remove the full and invalid resources ($R_2$), and completed and invalid demands ($D_5$), the matching set is as follows:

$$D_{SR} = \{R_1, R_3, R_4, R_5, R_6, R_7, R_8, R_9, R_{10}\}; \quad D_{SD} = \{D_1, D_2, D_3, D_4, D_5, D_6\}$$

**Step 2: Ability Information Matching**
Taking the resource set meeting demand $D_1$ as an example, according to the matching results of hard matching indexes such as vehicle type, load, vehicle length, volume, time and route, $R_1, R_5, R_6, R_7, R_8$ are eliminated.
According to the soft matching indexes such as load rate, length compliance and volume compliance, $\text{sim}_G(RF, DF)$ is calculated according to *Formula* (7) (9).
It can be obtained by calculation that $\text{sim}_G(RF_i, DF_1)$ of $R_3, R_4, R_9, R_{10}$ to $D_1$ are all 0.86, which is higher than $\varepsilon_F$ (set as 0.6), so they are retained in $D_{FR}$. Therefore, we can get $D_{FR} = \{R_3, R_4, R_9, R_{10}\}$. The same method is used to calculate remaining matching sets.
**Step 3: Level Information Matching**
According to *Formula* (10), $sim_L(RL_i, DL_i)$ of resources in $D_{FR}$ relative to $D_1$ can be calculated:

$$sim_L(RL_3, DL_1) = 0.95; sim_L(RL_4, DL_1) = 0.91; sim_L(RL_9, DL_1) = 0.90; sim_L(RL_{10}, DL_1) = 0.84.$$

Set $\varepsilon_L = 0.9$, because $\text{sim}_L(RL_{10}, DL_1) < \varepsilon_L$, so reject $R_{10}$. We can get the matching result $D_R = \{R_3, R_4, R_9\}$. The same method is used to calculate remaining matching sets.
**Step 4: Comprehensive Similarity Calculation**
Set $\alpha = 0.3,\ \beta = 0.7$, according to *Formula* (12), calculate comprehensive similarity of each resource in $D_R$ relative to $D_1$, that is:

$$\text{sim}(R_3, D_1) = 0.923; \text{sim}(R_4, D_1) = 0.895; \text{sim}(R_9, D_1) = 0.888.$$

So the output matching result is $D_R = \{R_3, R_4, R_9\}$, and $R_3$ is the best candidate resource. The remaining matching sets can be calculated in the same way.
**Step 5: Two-sided Integrate Matching**
Calculate all of the above $\underset{R \to D}{\text{sim}}(R_i, D_j)$, $\underset{D \to R}{\text{sim}}(R_i, D_j)$, based on *model* (13), and use MATLAB "intlinprog" function to solve the programming problem.

When $x_{31}, x_{62}, x_{13}, x_{94}, x_{10,6}, x_{77} = 1$, $Z$ takes maximum value. So, the matching scheme is $R_3 \leftrightarrow D_1$, $R_6 \leftrightarrow D_2$, $R_1 \leftrightarrow D_3$, $R_9 \leftrightarrow D_4$, $R_{10} \leftrightarrow D_6$, $R_7 \leftrightarrow D_7$. At this time, stable matching is achieved, and the overall satisfaction is the highest.

## 5 Conclusion

Vehicles and cargos matching is an important issue in logistics parks interconnection. This matching problem is based on the satisfaction of two-sided parties. The goal is to achieve a stable two-sided matching between vehicle side and cargo side under optimal overall benefit.

Therefore, this paper designs a vehicles and cargos two-sided matching model based on similarity calculation. Through the example verification, the model and algorithm have high recall, accuracy and matching efficiency, and provide a new idea for the matching of vehicles and cargos in logistics information service platform. Compared with traditional matching scheme, this matching scheme solves the problem of vehicles and cargos comprehensive matching based on two-sided stable matching, and has good applicability.

This paper mainly focuses on the matching problem of same type of resources and demands data information under a certain scale. For the complex data information in big data environment, the applicability and efficiency of the model and algorithm need to be verified. It needs to further optimize the model, and then match on the basis of cluster analysis to improve the accuracy and efficiency of matching. This will be the focus of next research.

**Acknowledgment.** This project is supported by: Doctoral research fund subsidized project of WTBU (D2018007); Natural Science Foundation of Hubei Province (2019CFC930); Distinguished Young and Middle-aged Team Program for Scientific and Technological Innovation in Higher Education of Hubei (T201938).

## References

1. Verma, V., Aggarwal, R.K.: A new similarity measure based on simple matching coefficient for improving the accuracy of collaborative recommendations. Int. J. Inf. Technol. Comput. Sci. (IJITCS) **11**(6), 37–49 (2019)
2. Hu, J., Bing, C., Han, S.: Research on vehicle cargo matching of highway trunk freight platform based on TS algorithm. J. Zhejiang Univ. Sci. Technol. (Soc. Sci. Ed.) **40**(05), 478–486 (2018). (in Chinese)
3. Li, H.: Research on the Matching of Vehicle and Cargo Supply and Demand in the Loading Logistics Information Service Platform. Beijing Jiaotong University, Beijing (2015). (in Chinese)
4. Kambalimath, M.G., Kakkasageri, M.S.: Cost optimization based resource allocation scheme for vehicular cloud networks. Int. J. Comput. Netw. Inf. Secur. (IJCNIS) **12**(2), 22–31 (2020)
5. Fan, Z., Li, M., Le, Q.: A bilaterally satisfactory matching decision-making method considering stable matching conditions. Chin. Manag. Sci. **22**(04), 112–118 (2014). (in Chinese)
6. Li, M., Fan, Z., Li, Q.: 1-n bilateral matching decision making method considering stable matching conditions. J. Syst. Eng. **28**(04), 454–463 (2013). (in Chinese)
7. Zhao, D., Li, R.: Two-sided matching mechanism of cloud manufacturing resources considering subject psychological expectation. Control Decis. Making **32**(05), 871–878 (2017). (in Chinese)
8. Alessandro, S., Alessandra, C., Lev, B., et al.: Intelligent cloud manufacturing platform for efficient resource sharing in smart manufacturing networks. Procedia CIRP **79**, 233–238 (2019)
9. Jia, H., Zhang, J., Qi, Y., Deng, S.: Research on efficient matching strategy of resource demand of cloud manufacturing platform. Mod. Manuf. Eng. (04), 36–43 (2020). (in Chinese)

10. Bi, Y., Yuan, H., Chu, Y., et al.: Resource multi-level intelligent rent-seeking and matching strategy and value creation based on public service platform in big data environment. Comput. Sci. **46**(02), 42–49 (2019). (in Chinese)
11. Zhao, D., Ding, L.: Bilateral matching mechanism and stability of cloud manufacturing platform resources. Syst. Eng. **35**(2), 109–115 (2017). (in Chinese)
12. Kumar, M., Kashyap, N.: QoS aware logical channel prioritization under burst resource allocation for uplink in LTE. Int. J. Inf. Technol. Comput. Sci. (IJITCS) **9**(8), 14–21 (2017)

# Improving Pedestrian Detection Methods by Architecture and Hyperparameter Modification of Deep Neural Networks

Roman Statkevych(✉), Yuri Gordienko, and Sergii Stirenko

National Technical University of Ukraine "Igor Sikorsky Kyiv Polytechnic University", Kyiv, Ukraine

**Abstract.** Now analysis of real-time data streams from vehicle's sensors and cameras by deep neural networks (DNNs) is a very big challenge, especially in the context autonomous driving, unobtrusive health and elderly care for monitoring camera-to-person and inter-person distances. On the basis of the current pedestrian localization methods the several minimal improvements were proposed and investigated by usage of various activation functions (like Swish and GELU) and changes of the underlying neural network (the number of layers and channels in them). The most effective combination of the mentioned measures was found that leads to the smaller DNNs without detriment to their performance. These proposed improvements are considered as very promising for the potential usage of the investigated DNNs in Edge Computing applications on the computing units with the limited resources.

**Keywords:** Neural networks · Deep learning · Pedestrian detection · Pose estimation · Autonomous driving · Health care · Elderly care

## 1 Introduction

Modern vehicles are equipped with a lot of different equipment and sensors, which gather information about the surrounding environment and objects in it. Most commonly, this equipment includes Light Detection and Ranging (LiDAR) system, which performs wide scanning of the environment to detect distances to objects. Despite the fact LiDAR provides comprehensive information about distances, matching it with visual data retrieved from frontal cameras makes environment perception more complete. Also, LiDAR technologies are relatively expensive, so detecting distances using visual information can be a cheap alternative.

Most common approach to detect pedestrians on the image is using a deep neural network (DNN) to estimate distances to them. The numerous research attempts are carried out now for general purpose tasks (which detect all traffic participants, including vehicles) [1] and pedestrian-specific ones [2, 3]. For developing new approaches, researchers are using out-of-box datasets that are publicly available for scientific and research purposes, for example, KITTI [4] and nuScenes [5]. These datasets consist of


Z. Hu et al. (Eds.): ICAILE 2021, LNDECT 82, pp. 44–53, 2021.
https://doi.org/10.1007/978-3-030-80475-6_5

images, taken with the vehicle's camera, and ground-truth files based on LiDAR data, where each object on the image is labeled, as well as its location and its distance to the camera. This data is sufficient to train DNNs and get the reliable estimation of the real locations of objects with regard to the camera.

While training a DNN model is a well-known approach, a choice of input layer plays a crucial role in determining a model's efficiency. Usually, objects on an image are marked with so-called bounding-boxes. While a bounding-box approach works good for rigid-body objects (objects that have fixed form) like vehicles, for pedestrian detection problems it might not be the best solution because of specifics of the human body dynamics which can behave differently in different situations. For example, some researchers used a method based on human pose estimation to estimate a pedestrian position [2]. It has shown better performance compared to bounding-box-based methods.

**Uncertainty Estimation.** While detecting an exact position of an object from the image is an important task, software errors and mispredictions are very critical when talking about autonomous driving. Even a mistake which occurs once in a million times may cost human lives. That's why researchers are trying to define a confidence interval for evaluating prediction results. These confidence intervals are a spacial characteristics, which implies it guarantees that object is located within it. Prediction of the confidence interval depends on the uncertainty factors, that takes effect due to training dataset's specifics or intrinsic uncertainty of the trained model. There are two types of uncertainties: aleatoric and epistemic [6, 7]. Usually, they are "epistemic", if the modeler sees a possibility to reduce them by gathering more data or by refining models, and they are "aleatory" if the modeler does not foresee the possibility of reducing them [7]. Using both types of uncertainties makes predictions more accurate in terms of localizing them within some boundaries.

**Pedestrian Localization.** As mentioned above, a bounding-box detection approach is a default solution for solid and monolithic objects. But the human body is very different because it often changes shapes as humans walk or move their limbs like head, hands, legs, etc. It is possible that poorly trained DNN models can erroneously predict a human position in some specific configurations. That's why it was supposed to use human pose estimation to detect human positions [2, 3]. This method usually use off the shelve pose estimation methods (for example [8, 29]). This approach has several advantages: while human joints might have different configurations, they are connected with each other in a specific order, so, if using a human pose to train a model, this data gives both a good probability distribution and robustness in detecting objects as pedestrians. One more advantage is that using human poses provides a possibility to estimate the position of pedestrians behind obstacles. While not all joints might be visible, it is still possible to estimate a human position, with a higher uncertainty factor, compared to a pedestrian, whose all joints are visible.

## 2 Methodology

While existing methods give a reasonable performance and efficiency in terms of execution time and prediction confidence, there are still ways to improve. Here the method is considered which is based on MonoLoco system which code is publicly available and ready to use [2]. It utilizes PifPaf pose estimation method to extract human poses from the image, and then analyses it with DNN model [8]. The model (developed using PyTorch 1.1.0) accepts 17 pose joints as an input, and consists of several fully-connected (FC) layers with 256 channels including dropout and batch-normalization layers.

For training the DNN model, the preliminary step of extracting poses from the training dataset is required. For the research purpose, the custom training open dataset was used (consisting of 7600+ .png images) which is publicly available [4]. Poses were extracted using publicly available PifPaf ResNet101 model [8]. This dataset is used to evaluate the current performance compared to the original method implementation.

The main metric for evaluation is average localization error (ALE):

$$e = \frac{d_g - d_p}{n} \tag{1}$$

where $d_g$ is a ground-truth (real) distance of the pedestrian to the camera, $d_p$ is a predicted distance, and $n$ is a number of samples in the dataset.

The evaluation results are represented in the tables below, which show the percentages of all pedestrians divided to 3 groups (ALE is less than 0.5 m, 1 m and 2 m). Also, in the original research 3 categories representing detection difficulty (easy, medium, hard) for a particular pedestrian were introduced [2]. These categories rely on pedestrian visibility and are based on ground-truth data. The following possible ways for improvements were considered:

1) activation function, beside the famous Rectified Linear Units (ReLU) [10] activation function (used in the original system), several other activation functions (like Swish [11], GELU [12]) were tested to get the better performance compared to ReLU;
2) network layer parameters, like the number of channels in internal layers, number of the internal layers, dropout ratio, etc.;
3) combination of the best above mentioned measures: activation functions and network parameters leading to the smaller DNNs without detriment to their performance.

All experiments and evaluations were performed using Google Colaboratory platform (commonly known as Google Colab or just Colab) (https://colab.research.google.com/) equipped with 2-core Intel(R) Xeon(R) 2.30GHz central processing units (CPUs). In the methodological sense, these proposed improvements are motivated by our previous attempts to estimate the potential usage of the investigated DNNs for Edge Computing applications on the computing units with the limited resources [13–15].

## 3 Experiment

### 3.1 Activation Function Influence

ReLU is one of the most commonly used activation functions. However, despite widespread popularity, computational efficiency and simplicity, several alternatives were proposed, like Swish [11], GELU (Gaussian exponential linear units) [12], and an improved

ReLU (such as Parametric ReLU (or PreLU [16]) and LeakyReLU [17]). Currently, Swish has shown the promising results after some benchmark tests [11].

That is why some of these activation functions were used to evaluate their influence on the performance of Monoloco [2] and compare the results obtained with default ReLU-based implementation. The results obtained are given in Table 1, where the nonlinear activation functions (like GELU and Swish) demonstrated the slightly better performance than linear ones. As mentioned before, "easy", "medium", and "hard" columns represent the difficulty of pedestrian detection, based on ground-truth by KITTI dataset [4], and "all" column - for entire validation dataset. GELU has shown better performance by having a higher percentage of pedestrians, classified with ALE metric less than 1 m. This is an important improvement, because it classifies more pedestrians with less localization error. However, all activation functions have shown statistically the same result of 66% pedestrians located within 2 m of localization error. By the average error across the entire dataset, Swish has shown the best result, outperforming original method by roughly 4.5%.

**Table 1.** Results of activation function influence on the ALE values.

| Activation function | 0.5 m | 1 m | 2 m | Easy | Medium | Hard | All |
|---|---|---|---|---|---|---|---|
| ReLU (default) | 26.31 | 47.1 | 66.05 | 0.96 | 1.20 | **1.40** | 1.07 |
| PReLU | 27.67 | 48.55 | 66.62 | 0.94 | 1.18 | 1.45 | 1.05 |
| GELU | **28.64** | **49.56** | 66.35 | 0.94 | **1.09** | 1.44 | 1.03 |
| LeakyReLU | 26.79 | 47.63 | 66.62 | 0.94 | 1.19 | 1.43 | 1.06 |
| Swish | 28.2 | 48.99 | 66.57 | **0.92** | 1.12 | 1.41 | **1.02** |

### 3.2 Network Architecture Influence

Complex and deep DNN models sometimes tend to overfit the training set, so simplifying a model's architecture may also lead to improving its performance [9]. As mentioned before, Monoloco uses 17 joints for the input layer (totally 34 values, because joints consist of $x$ and $y$ values), and consists of several layers with 256 channels (Fig. 1). It was decided to evaluate networks with 128 and 64 channels. Also, there are 3 sequential groups of layers, so it was also decided to try 2 and 1 Linear layer in the groups (for example, like in Fig. 2).

Because of decreasing the number of channels and layers, it was expected that prediction and training speed would increase. As the original method uses off-the-shelf pose-estimation as a preliminary step to the prediction, and it is invariant, only position prediction time from the pre-computed pose was measured.

Results of the evaluation of the DNN with varying internal layer size ($S$) and number of layers ($N$) are displayed in Table 2. The original activation function (ReLU) was used. * - denotes parameters, used in the original research. Time ($t$) represents the duration of predicting position from joints omitting pose estimation in microseconds. According to the results the different methods have shown different performance, but all of them outperform the original method. By ALE metric the best choice is the network with the size of internal layer (*S* in Table 2) equal to 256 and number of layer stages (*N* in Table 2) equal to 1, or a network with 3 stages sized 256 elements, which both have shown 1.02 m of average localization, but 1 stage network gives a faster prediction. Also, the improved network is 1.53 times faster in terms of prediction time, then the original network.

```
LinearModel(
  (w1): Linear(in_features=34, out_features=256, bias=True)
  (batch_norm1): BatchNorm1d(256, eps=1e-05, momentum=0.1, affine=True, track_running_stats=True)
  (linear_stages): ModuleList(
    (0): Linear(
      (relu): ReLU()
      (dropout): Dropout(p=0.2)
      (w1): Linear(in_features=256, out_features=256, bias=True)
      (batch_norm1): BatchNorm1d(256, eps=1e-05, momentum=0.1, affine=True, track_running_stats=True)
      (w2): Linear(in_features=256, out_features=256, bias=True)
      (batch_norm2): BatchNorm1d(256, eps=1e-05, momentum=0.1, affine=True, track_running_stats=True)
    )
    (1): Linear(
      (relu): ReLU()
      (dropout): Dropout(p=0.2)
      (w1): Linear(in_features=256, out_features=256, bias=True)
      (batch_norm1): BatchNorm1d(256, eps=1e-05, momentum=0.1, affine=True, track_running_stats=True)
      (w2): Linear(in_features=256, out_features=256, bias=True)
      (batch_norm2): BatchNorm1d(256, eps=1e-05, momentum=0.1, affine=True, track_running_stats=True)
    )
    (2): Linear(
      (relu): ReLU()
      (dropout): Dropout(p=0.2)
      (w1): Linear(in_features=256, out_features=256, bias=True)
      (batch_norm1): BatchNorm1d(256, eps=1e-05, momentum=0.1, affine=True, track_running_stats=True)
      (w2): Linear(in_features=256, out_features=256, bias=True)
      (batch_norm2): BatchNorm1d(256, eps=1e-05, momentum=0.1, affine=True, track_running_stats=True)
    )
  )
  (w2): Linear(in_features=256, out_features=2, bias=True)
  (relu): ReLU()
  (dropout): Dropout(p=0.2)
)
```

**Fig. 1.** Original monoloco DNN model (by Pytorch representation) with several groups of layers with 256 features in each linear layer.

```
LinearModel(
  (w1): Linear(in_features=34, out_features=128, bias=True)
  (batch_norm1): BatchNorm1d(128, eps=1e-05, momentum=0.1, affine=True, track_running_stats=True)
  (linear_stages): ModuleList(
    (0): Linear(
      (relu): Swish()
      (dropout): Dropout(p=0.2)
      (w1): Linear(in_features=128, out_features=128, bias=True)
      (batch_norm1): BatchNorm1d(128, eps=1e-05, momentum=0.1, affine=True, track_running_stats=True)
      (w2): Linear(in_features=128, out_features=128, bias=True)
      (batch_norm2): BatchNorm1d(128, eps=1e-05, momentum=0.1, affine=True, track_running_stats=True)
    )
  )
  (w2): Linear(in_features=128, out_features=2, bias=True)
  (relu): Swish()
  (dropout): Dropout(p=0.2)
)
```

**Fig. 2.** Modified monoloco DNN model (by Pytorch representation) with one group of layers with 128 features in linear layer.

### 3.3 Combination of Various Activation Functions and Network Changes

On the basis of the results from Tables 1 and 2, it seems to be intuitively wise to consider some combinations of various activation functions and network parameters. It was decided to execute experiments for the best-performing activation functions from Table 1 (Swish and GELU) in combination with the layer sizes *S* (64, 128 and 256) and the number of layer stages *N* (1, 2, 3). In Table 3 some of the parameter combinations with the best performance values are shown for each category. Several methods had demonstrated ALE values equal to 1.02 m, but only one method had shown 1.01, which is the best result for average ALE values observed during these experiments. This best method utilizes Swish activation function with the network of 1 stage of internal layers with 128 elements.

**Table 2.** Influence of changing network architecture on performance of ALE metric.

| Network parameters | | Evaluation categories | | | | | | | |
|---|---|---|---|---|---|---|---|---|---|
| *N* | *S* | 0.5 m | 1 m | 2 m | Easy | Medium | Hard | All | *t*, μs |
| 3* | 256* | 26.31 | 47.1 | 66.05 | 0.96 | 1.20 | 1.40 | 1.07 | 1392 |
| 3 | 128 | 26.84 | **49.38** | 66.88 | 0.94 | 1.19 | 1.35 | 1.04 | 1163 |
| 3 | 64 | 27.23 | 48.85 | 66.71 | **0.92** | **1.14** | 1.41 | **1.02** | 1182 |
| 2 | 256 | 28.02 | 49.29 | 66.62 | 0.93 | 1.18 | 1.40 | 1.04 | 1225 |
| 2 | 128 | 27.41 | 48.55 | 66.49 | 0.95 | 1.15 | 1.31 | 1.04 | 1056 |
| 2 | 64 | **28.64** | 48.02 | 65.61 | 0.93 | 1.21 | 1.51 | 1.06 | 947 |
| 1 | 256 | 28.37 | 48.72 | **67.41** | **0.92** | 1.15 | **1.30** | **1.02** | 908 |
| 1 | 128 | 27.85 | 48.68 | 66.92 | 0.94 | 1.18 | **1.30** | 1.04 | **781** |
| 1 | 64 | 27.01 | 48.42 | 66.49 | 0.97 | 1.16 | 1.35 | 1.06 | **822** |

Considering the results of evaluation, it is shown that the best-performing methods in terms of ALE across the entire evaluation set are also outperforming by estimating an "easy" category of pedestrian, which is the most common case. On the basis of these results, it is worth to invest more efforts and time in investigating the harder cases. Figure 3 demonstrates the comparison analysis of the proposed DNN versions with best performance detected in comparison with the original DNN structure. This chart shows ALE values distributed among clusters of ground-truth pedestrian's positions. It shows that the improved method shows visibly better results for predicting a pedestrian's position within 15 m. The original method still performs better it he ranges of distances like 15–25 m and elimination of this difference should be the subject of our further research. Figures 1 and 2 represent original architecture and proposed one correspondingly. The proposed model is much simpler, and yet gives slightly better results, and it is much faster in terms of training and prediction, as seen from Table 3. Figure 3 shows the comparison of original and "improved" method, which implies the method with Swish activation function, 1 stage and 128 elements layer size. "Task Error" means the metric showing the approximate ALE due to ignoring the gender factor, introduced in the original research [2].

## 4 Discussion

The original Monoloco setup uses 17-joints human pose-skeletons to train the DNN model, because of outputs provided by the PifPaf pose estimation method. One of the additional promising ways of improvement could be increase or decrease the number of joints to define their effect on the ALE results. For this purpose, it is supposed to replace pose estimation method with OpenPose body + foot model, and perform evaluation, using the same approach, as demonstrated above [18].

**Table 3.** The best performing methods in each category, where * - denotes parameters, used in the original research

| Network parameters | | | Evaluation categories | | | | | | | |
|---|---|---|---|---|---|---|---|---|---|---|
| *N* | *S* | AF | 0.5m | 1 m | 2 m | Easy | Med | Hard | All | *t*, ns |
| 3* | 256* | ReLU* | 26.31 | 47.1 | 66.05 | 0.96 | 1.20 | 1.40 | 1.07 | 1392 |
| 1 | 128 | Swish | 28.33 | **50.52** | 66.97 | **0.91** | 1.13 | 1.43 | **1.01** | **902** |
| 1 | 256 | ReLU | 28.37 | 48.72 | **67.41** | 0.92 | 1.15 | 1.43 | 1.02 | 908 |
| 2 | 128 | Swish | **28.81** | 49.42 | 66.66 | 0.93 | 1.12 | **1.38** | 1.03 | 1079 |
| 3 | 256 | GELU | 28.64 | 49.56 | 66.35 | 0.94 | **1.09** | 1.44 | 1.03 | 2574 |

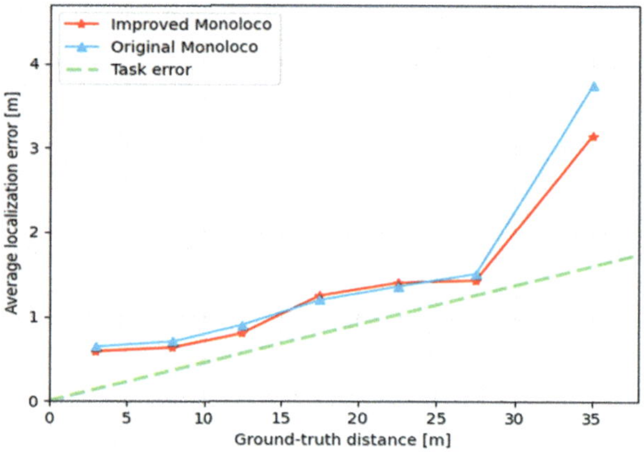


**Fig. 3.** ALE values depending on a pedestrian's ground-truth distance to camera.

Another promising way of improvement is to utilize a gender bias that was noticed in pedestrian predictions due to missing labels that describes genders in the Caltech Pedestrian Benchmark Dataset [19]. This issue is related not only to gender features, but also to age, especially if children will be taken into account.

KITTI dataset does not contain gender features for pedestrians, which makes it not suitable for training a model utilizing gender features. This can be fixed by using different dataset for training [20] or updating their labels by gender information. The possible method's architecture to utilise a gender/age feature s shown in Fig. 4, where the simple schema describes the idea of utilizing gender features with existing methods. For this purpose off-the-shelf age/gender estimation methods can be used, which may provide either numerical or classification values, that opens a possibility for more potential improvements. In the current stage of deep learning research, the generalization problem is very important especially when the number of trainable parameters is far greater than the number of training samples. That is why the current attempts to decrease the number of the trainable parameters could lead to the better generalization also, but this assumption should be proved later in our further research.

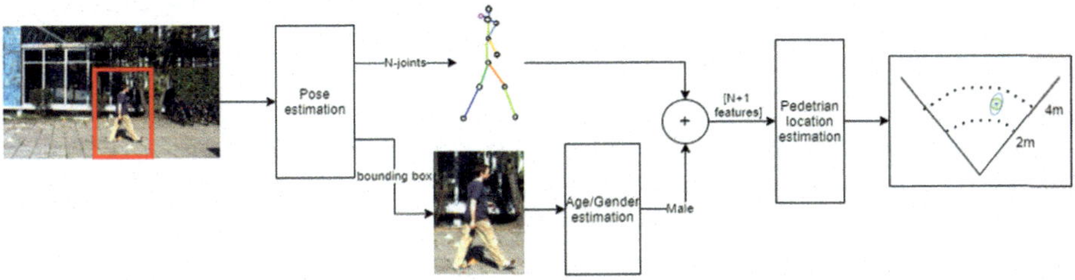


**Fig. 4.** The updated workflow schema describing the idea of utilizing gender features with the existing methods (the pedestrian image was taken from KITTI dataset).

Also, the contextual information may bring additional factors which may affect the performance results. The context-sensitive information from the environment like traffic conditions may help to estimate not only pedestrian positions, but also some related danger levels with regard to the context (such as accident risk due to the dangerous distance between some pedestrians and traffic lanes). The simplest example of the contextual information could be like a pedestrian crossing a road over the crosswalk, or just staying near the road with no intention of crossing it? These factors and other above mentioned aspects will be investigated in a further research.

The additional promising directions of the further research will be oriented to health care applications where the inter-person distances are of great importance for providing safer conditions in the view of infectious diseases (like the current COVID-2019 problem and others). Our previous attempts to provide the patient activity monitoring were mainly based on wearable electronic sensors and actuators [21–24], brain-computer interface [25, 26] which are usually considered as an obtrusive data acquisition method especially by elders or people with limited abilities. That is why these results by unobtrusive computer vision and multimodal data analysis [27] can be used in elderly care applications as an alternative method to estimate the current patient activity and monitor the decrease of activity which is often can be the precursor of the worse health state, diseases and forthcoming fatal events even.

Crowd behavior prediction is one more possible application [28], where distance between people is very critical parameter. Moreover, human pose estimation will allow to predict the current state of the crowd (are all people standing, or running in different directions), so this approach could be beneficial for this problem.

## 5 Conclusions

Despite availability of several modern approaches for pedestrian location, there are still several ways to improve these methods by simple measures. Here some improvements were proposed and investigated by selection of some other activation functions and structural changes of the underlying DNN. The most effective combination of the above mentioned improvement measures was found that leads to the smaller DNNs without detriment to their performance. It was shown that by choosing simpler architecture one can obtain predictions not only faster, but also with lower errors. The proposed improvements are considered as very promising ones for the potential usage of the investigated DNNs in Edge Computing applications on the computing units with the limited resources, for example, for autonomous driving, health care and elderly care applications.

**Acknowledgment.** The work was supported by the National Research Foundation of Ukraine in the grant 2020.01/0490 "Artificial Intelligence Platform for Distant Computer-Aided Detection (CADe) and Computer-Aided Diagnosis (CADx) of Human Diseases".

## References

1. Chen, Y., Tai, L., Sun, K., Li, M.: Monopair: monocular 3D object detection using pairwise spatial relationships. In: Proceedings of the IEEE/CVF Conference on Computer Vision and Pattern Recognition, pp. 12093–12102 (2020)
2. Bertoni, L., Kreiss, S., Alahi, A.: Monoloco: monocular 3D pedestrian localization and uncertainty estimation. In: Proceedings of the IEEE International Conference on Computer Vision, pp. 6861–6871 (2019)
3. Hayakawa, J., Dariush, B.: Recognition and 3D localization of pedestrian actions from monocular video (2020). http://arxiv.org/abs/2008.01162
4. Geiger, A., Lenz, P., Stiller, C., Urtasun, R.: Vision meets robotics: the kitti dataset. Int. J. Robot. Res. **32**(11), 1231–1237 (2013)
5. Caesar, H., et al.: Nuscenes: a multimodal dataset for autonomous driving. In: Proceedings of the IEEE/CVF Conference on Computer Vision and Pattern Recognition, pp. 11621–11631 (2020)
6. Der Kiureghian, A., Ditlevsen, O.: Aleatory or epistemic? Does it matter? Struct. Saf. **31**(2), 105–112 (2009)
7. Hüllermeier, W.W.: Aleatoric and epistemic uncertainty in machine learning: an introduction to concepts and methods (2019). http://arxiv.org/abs/1910.09457
8. Kreiss, S., Bertoni, L., Alahi, A.: Pifpaf: composite fields for human pose estimation. In: Proceedings of the IEEE Conference on Computer Vision and Pattern Recognition, pp. 11977–11986 (2019)
9. He, K., Zhang, X., Ren, S., Sun, J.: Deep residual learning for image recognition. In: Proceedings of the IEEE Conference on Computer Vision and Pattern Recognition, pp. 770–778 (2016)
10. Glorot, X., Bordes, A., Bengio, Y.: Deep sparse rectifier neural networks. In: Proceedings of the Fourteenth International Conference on Artificial Intelligence and Statistics, pp. 315–323 (2011)
11. Ramachandran, P., Zoph, B., Le, Q.V.: Searching for activation functions (2017). http://arxiv.org/abs/1710.05941
12. Hendrycks, D., Gimpel, K.: Gaussian error linear units (GELUs) (2016). http://arxiv.org/abs/1606.08415

13. Kochura, Y., Gordienko, Y., Rokovyi, A., Alienin, O., Stirenko, S.: Batch size influence on performance of graphic and tensor processing units during training and inference phases. In: Proceedings of International Conference on Computer Science, pp. 658–668 (2019)
14. Gordienko, Y., et al.: Scaling analysis of specialized tensor processing architectures for deep learning models. In: Deep Learning: Concepts and Architectures, pp. 65–99 (2020)
15. Gordienko, Y., et al.: "Last mile" optimization of edge computing ecosystem with deep learning models and specialized tensor processing architectures. In: Advances in Computers. Elsevier, Amsterdam (2021). (in print)
16. He, K., Zhang, X., Ren, S., Sun, J.: Delving deep into rectifiers: surpassing human-level performance on imagenet classification. In: Proceedings of IEEE International Conference on Computer Vision (ICCV), pp. 1026–1034 (2015)
17. Maas, A.L., Hannun, A.Y., Ng, A.Y.: Rectifier nonlinearities improve neural network acoustic models. In: Proceedings of International Conference on Machine Learning, vol. 30, p. 3 (2013)
18. Cao, Z., Simon, T., Wei, S.E., Sheikh, Y.: Realtime multi-person 2D pose estimation using part affinity fields. In: Proceedings of the IEEE Conference on Computer Vision and Pattern Recognition, pp. 7291–7299 (2017)
19. Brandao, M.: Age and gender bias in pedestrian detection algorithms (2019). http://arxiv.org/abs/1906.10490
20. Deng, Y., Luo, P., Loy, C.C., Tang, X.: Pedestrian attribute recognition at far distance. In: Proceedings of the 22nd ACM International Conference on Multimedia, pp. 789–792 (2014)
21. Gordienko, N., Lodygensky, O., Fedak, G., Gordienko, Y.: Synergy of volunteer measurements and volunteer computing for effective data collecting, processing, simulating and analyzing on a worldwide scale. In: Proceedings of 38th International Convention on Information and Communication Technology, Electronics and Microelectronics (MIPRO), pp. 193–198. IEEE (2015)
22. Gordienko, Y., et al.: Deep learning for fatigue estimation on the basis of multimodal human-machine interactions (2017). http://arxiv.org/abs/1801.06048
23. Stirenko, S., et al.: Parallel statistical and machine learning methods for estimation of physical load. In: International Conference on Algorithms and Architectures for Parallel Processing, pp. 483–497 (2018)
24. Gang, P., et al.: Prediction of physical load level by machine learning analysis of heart activity after exercises. In: Proceedings of IEEE Symposium Series on Computational Intelligence (SSCI), pp. 557–562 (2019)
25. Gang, P., et al.: User-driven intelligent interface on the basis of multimodal augmented reality and brain-computer interaction for people with functional disabilities. In: Arai, K., Kapoor, S., Bhatia, R. (eds.) FICC 2018. AISC, vol. 886, pp. 612–631. Springer, Cham (2019). https://doi.org/10.1007/978-3-030-03402-3_43
26. Kostiukevych, K., Gordienko, Y., Stirenko, S.: Multimodal data integration and processing method for brain-computer interface research. Presented at 6th International Conference High Performance Computing (HPC-UA 2020) (2020)
27. Shulha, M., Gordienko, Y., Stirenko, S.: Comparative analysis of the datasets with multimodal content. Presented at 6th International Conference High Performance Computing (HPC-UA 2020) 06–07 November 2020, Kyiv/Ukraine (2020)
28. Dey, A., Mohammad, F., Ahmed, S., Sharif, R., Saif, A.F.M.S.: Anomaly detection in crowded Scene by pedestrians behaviour extraction using long short term method: a comprehensive study. Int. J. Educ. Manag. Eng. (IJEME) **9**(1), 51–63 (2019)
29. Kishore, D.R., Mohan, M.C., Rao, A.A.: A novel joint chaining graph model for human pose estimation on 2D action videos and facial pose estimation on 3D images. Int. J. Image Graph. Sig. Process. (IJIGSP) **9**(3), 21–32 (2017)

# Warehousing Process Mining Research Based on Petri Net

Fei Yin, Jingjing Cao(✉), Jie Chu, and Shurchkova Olga

School of Logistics Engineering, Wuhan University of Technology, Wuhan 430077, China
bettycao@whut.edu.cn

**Abstract.** With the advent of the era of big data, enterprises pay increasing attention to the business process information recorded in the business system, which plays an important role in optimizing the business process. The smooth operation of warehousing business process can save operation time and cost for enterprises to a great extent. In this paper, it is proposed a Petri net which reflects the real warehousing and out-storage process of the enterprise. Specifically, the event log stored in the enterprise business system is obtained. As for building the Petri net, Prom software and α algorithm are applied to mine the event log. Then, the real event log is replayed on the Petri net of the theoretical warehousing process for consistency checking. Finally, the problems arise in the consistency checking results are analyzed and solved, and the original Petri net are improved.

**Keywords:** Petri net · Warehousing process mining · Event log · Consistency check · Prom

## 1 Introduction

Process mining has become a hot research topic [1]. It refers to how to mine the valuable data of a certain business process from the large amount of data recorded by the information system for the enterprise to achieve optimization. There are three types of process mining: process discovery, process consistency detection and process improvement [2]. Petri net is one of the modeling methods of process mining [3]. It can not only describe the process graphically, but also model the process by using mathematical model. It provides an effective solution for describing and analyzing the process, and has great advantages in the process of building enterprise process mining modeling [4].

Many scholars have proposed different algorithms for process discovery. The α algorithm is one of the most classical algorithms, which generates process models according to activity dependencies [5]. This method can identify the parallel structure in the model, but it cannot mine the short cycle structure and the non-free choice structure [6]. Sun et al. designed $\alpha^{TR}$ algorithm using association rules to describe the indirect dependencies for indirect dependencies between process, and formalized process model with extended Petri net [7]. Yu proposed based process mining algorithm based on statistical α algorithm, effectively eliminated the influence of noise to the event logs [8]. Xu et al.


Z. Hu et al. (Eds.): ICAILE 2021, LNDECT 82, pp. 54–65, 2021.
https://doi.org/10.1007/978-3-030-80475-6_6

proposed a tracking clustering method to divide the event logs, assign values and make up missing data, to establish a process model [9].

Modeling representations of process mining include Petri net, causality net [6], working flow net, BPMN model, etc. [10, 11]. These modeling methods are respectively proposed for dealing with different problems, so they have advantages and disadvantages in modeling direction, comprehensibility, dynamism, computerization ability, etc. The Petri net modeling method is used in the paper [12]. Petri net has made many improvements in practical applications.

In our work, the warehousing process, and use the process mining technology to mine the warehousing process are analyzed, which includes the preprocessing of event logs, Petri net modeling and consistency checking, so as to improve the quality and efficiency of warehousing services for enterprises and enhance their competitiveness [13].

The remaining part is organized as follows. A brief review of related work is presented in Sect. 2. The proposed process of build a Petri net is introduced in Sect. 3. A description of establishment of the enterprise instance and analysis of the results are reported in Sect. 4. Finally, the work and results are summarized in Sect. 5.

## 2 Related Work

This section is dedicated to the martials needed to express the proposed method. The reminder of the section is organized as follows: firstly, Petri net is discussed, then event log is explained, and final subsection deals with the consistency check.

### 2.1 Petri Net

Petri net, first proposed by Carl Adam Petri of Germany in 1962, is the oldest and most deeply researched process modeling language that supports modeling concurrency semantics.

**Definition 1 (Petri Net).** A Petri net is a triple-group $N = (P, T, F)$. $P$ is the finite set of places. $T$ is the finite set of transition, $P \cap T = \varnothing$. $F \subseteq (P \times T) \cup (T \times P)$ is the set of directed arc, named flow relationship. Place, transition, and flow relationship are represented by $\circ$, $\square$ and $\rightarrow$ in Petri net.

**Definition 2 (Marked Petri Net).** A marked Petri net is a binary-pair $(N, M)$, in which $N = (P, T, F)$ represents the Petri net. $M$ represents the mark, which is a multi-set defined on place $P$, can be represented as: $M \in \boldsymbol{B}(P)$.

**Definition 3 (Firing Rule).** $(N, M)$ is a Petri net, in which $N = (P, T, F)$, $M \in \boldsymbol{B}(P)$. Only when every input place of transition $t \in T$ have tokens, can transition $t$ take place, and use $(N, M)[t >$ to represent. After this, one token is removed from each input place of transition $t$, and one token is added to each output place.

**Definition 4 (Token).** Token is an identifier that flows on Petri net under the control of firing rule, represented by ● in Petri nets. The distribution of the token in the place, also known as the marking of Petri nets, determines the state of the Petri nets.

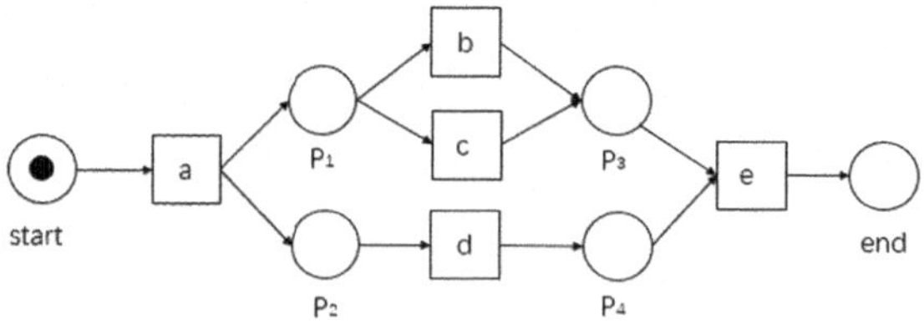


**Fig. 1.** An instance Petri net

A instance Petri net is shown in Fig. 1, place $P = \{\text{start}, P_1, P_2, P_3, P_4, end\}$, set of transitions $T = \{a, b, c, d, e\}$, set of flow relationship $F = \{(start, a), (a, P_1), (a, P_2), (P_1, b), (P_1, c), (P_2, d), (b, P_3), (c, P_3), (d, P_4), (P_3, e), (P_4, e), (e, end)\}$.

### 2.2 Event Log

Event logging can be regarded as the starting point and core element of process mining, and the smooth process mining based on effective event logging.

**Definition 5 (Event Log).** Event log is a record of the business process execution files, represented as L. In general, event logs record events that occur in ascending order of time. Each row in the event log represents an event. Events are described by different attributes, typically activities, timestamps, resources, costs, and so on. The activities in the event log correspond to transitions in Petri nets.

### 2.3 Consistency Check

The core idea of consistency checking is to compare the behavior and process models recorded in the event log to find the differences and common points between them, so as to realize the purpose of improving the process model or optimizing the business process.

In this paper, we use fitting degree to check consistency, denoted as fitness. Fitness refers to the degree of match between the process model and the event log.

Fitting degree of Petri net *N* based on event log *L*:

$$fitness(L, N) = \frac{1}{2}\left(1 - \frac{\sum_i x_i \times m_{N,\sigma_i}}{\sum_i x_i \times c_{N,\sigma_i}}\right) + \frac{1}{2}\left(1 - \frac{\sum_i x_i \times r_{N,\sigma_i}}{\sum_i x_i \times p_{N,\sigma_i}}\right) \quad (1)$$

Where $L$ is the event log, $\sigma_i$ is trajectory of the case in event log $L$, $x_i$ is the number of trajectories $\sigma_i$ in event log $L$. $N$ is Petri net. $p_{N,\sigma_i}$ is the produced token when trajectory $\sigma_i$ replay in the Petri net $N$, $c_{N,\sigma_i}$ is the consumed token when trajectory $\sigma_i$ replay in the Petri net $N$, $m_{N,\sigma_i}$ is the missed token when trajectory $\sigma_i$ replay in the Petri net $N$, $r_{N,\sigma_i}$ is the remained token when trajectory $\sigma_i$ replay in the Petri net $N$.

# 3 Method

## 3.1 Analysis of Warehousing Processing

To build a normal Petri net of the inbound flow, the inbound flow must be analyzed first. Although there are many different types of warehouses, they are similar to each other in the setting of warehouse and out-storage process flow. This paper takes the warehousing processing of raw material warehouse as the research object. The overall process of warehouse is summarized as follows and showed in Fig. 2:

1) Receive documents and prepare for warehousing.
2) Receive reached goods. The warehouse operators receive the incoming goods and complete the unloading operation in the unloading area with the scheduled equipment and personnel.
3) Check goods. The warehouse operators carry out overall inspection and sampling inspection of the discharged goods in the unloading area.
4) Stack goods. Warehouse operators in the operation area of the goods separate and stack plates, aiming at facilitating equipment handling goods.
5) Print and paste barcode label. Warehouse operators print and paste bar code labels with goods information in the operation area.
6) Get ready for warehouse. Warehouse operators get ready for the goods allocation, scheduling equipment and warehousing preparation.
7) Warehouse operators carry out warehousing operation of goods. The goods that fail to be put into storage need to be put into storage again after debugging.
8) Print warehouse list. After warehousing, warehouse operators print the corresponding goods warehousing list, and complete the handover procedures.

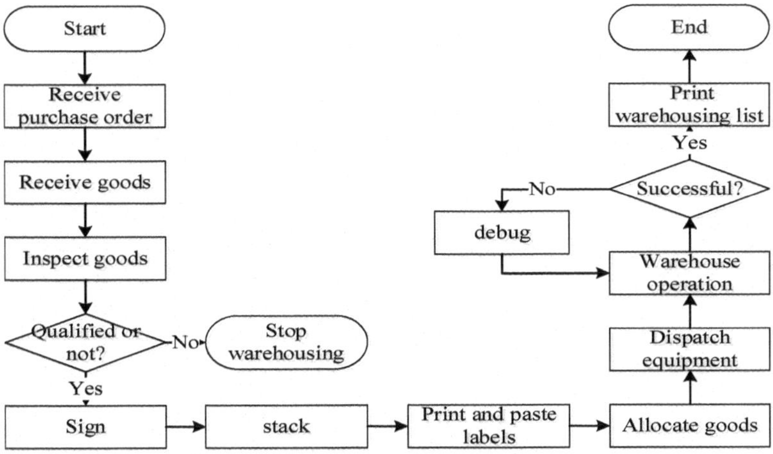


**Fig. 2.** Overall process of warehouse

### 3.2 Process Mining Based on Algorithm α

The α algorithm based on the event log is one of the most classic algorithms used in the process mining, it is a good way to deal with concurrent events in the event log. The input of α algorithm is the event log, the order of the relationship between events are found by scanning the event log in the case, then a Petri net is output.

Input: the event log $L$;

Step 1: find a collection $T_L$ of all activities contained in the event log $L$,$T_L$ is the set for changes of the Petri net;

Step 2: identify the event log $L$ in each case of the first event collection $T_1$;

Step 3: find the event log $L$ in each case the last event of a collection of $T_0$;

Step 4: clear the event log $L$ events in the order of the relationship between, establish a footprint matrix;

Step 5: in the footprint matrix to find a collection of all the causal relationship between $X_L$;

Step 6: in a collection of all the causal relationship between $X_L$ removed not the largest collection of get contains the largest collection of causality $Y_L$;

Step 7: according to the collection $Y_L$, determine the Petri net library collection $P_L$;

Step 8: according to certain library collection $P_L$, determine Petri net flow relationship $F_L$;

Step 9: make sure the final Petri nets, $N = (P_L, T_L, F_L)$;

Output: $N$ Petri net.

### 3.3 Petri Net Modeling of the Warehousing Processing

In this paper, event log processing and Petri net construction are realized by using Prom. Most events of warehousing processing are cascading events, but there also exists parallel, circulation and other special circumstances:

1) When inspect goods in the warehousing processing of the enterprise, the goods that fail to pass the inspection will be intercept and wait for subsequent processing. Therefore, this batch of goods has been stopped from warehousing. Goods that pass the inspection will continue to follow the warehousing process.
2) After the goods are disassembled and stacked, the next steps are to print and paste labels, allocate cargo space and dispatch lifting and handling equipment. These three parts are respectively carried out by different operators. Therefore, there is no strict execution sequence in the aspect of process execution, and it is paralleled.
3) During the warehousing operation of goods, some of the goods need to go through code plate and other operations again when the occasional problems are solved. So there exists a cyclic structure in the process.

Based on the above three characteristics, using a Prom *Alpha Miner* plugin construct Petri net $N_5$ as shown in Fig. 3, and the meanings of the letters are shown in Table 1.

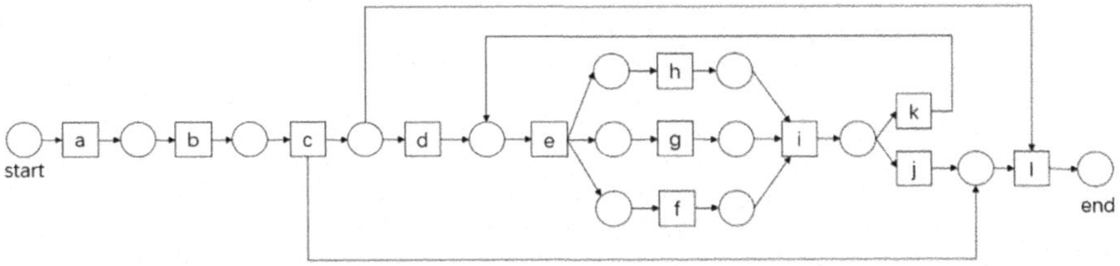


**Fig. 3.** The Petri net of warehouse under normal circumstances

**Table 1.** The meanings of the letters

| Events | Meanings | Events | Meanings |
|---|---|---|---|
| *a* | Receive purchase order | *g* | Allocation goods |
| *b* | Receive goods | *h* | Dispatch equipment |
| *c* | Goods inspection | *i* | Execute warehouse |
| *d* | Sign | *j* | Print entry |
| *e* | Stack | *k* | Debug |
| *f* | Print labels | *l* | End of warehousing, sorting |

# 4 Experiment

In order to apply the process mining technology to the actual process and realize the business process optimization, this paper introduces the actual enterprise warehousing process for analysis.

## 4.1 Event Log Construction and Preprocessing

Event log is a record of the real process, thus when build event log, we should consider many aspects, including warehousing business process analysis of each link, characteristics of predefined Petri net $N_5$ and the circumstance that actual process fail to conform to the predefined one.

The paper builds the event log $L_4$ from May 5, 2019 to May 10, 2019, and assumes it to be the initial event log. The pieces of event log $L_4$ are shown in Table 2.

Where the letters of event name are shown in the Table 1, and the resource of Table 2 refers to the principal.

Preprocessing of the event log in this paper includes converting the event log to a standard XES format and filtering out incomplete cases in the event log. With the help of the *Convert CSV to XES* plug-in on the Prom, after setting up the conversion conditions, the conversion of CSV files to XES files can be realized and the event log in standard format can be obtained. Then with the help of the *Filter Log using Simple Heuristics* plug-in on the Prom, the filter operation is completed. Figure 4 shows the filtered event log $L_5$.

**Table 2.** The pieces of event log $L_4$

| Case ID | Event ID | Time stamp | Event name | Resource |
|---|---|---|---|---|
| 1 | 1101 | 2019/5/5 7:30 | *g* | Jianguo Wu |
| 1 | 1104 | 2019/5/5 8:55 | *l* | Yang Wang |
| 2 | 2101 | 2019/5/5 9:00 | *a* | Li Li |
| 2 | 2104 | 2019/5/5 11:00 | *l* | Yang Wang |
| 3 | 3102 | 2019/5/5 15:00 | *b* | Li Li |
| 3 | 3106 | 2019/5/5 19:00 | *f* | Lei Wang |
| …… | …… | …… | …… | …… |

## 4.2 Warehousing Process Petri Net of Enterprise A

After preprocessing, we build Petri net model based on the event log $L_5$. With the help of the *Alpha Miner* plug-in on the Prom, the corresponding pieces of event log $L_4$ is generated, which is shown in Fig. 5.

**Fig. 4.** The filtered event log $L_5$

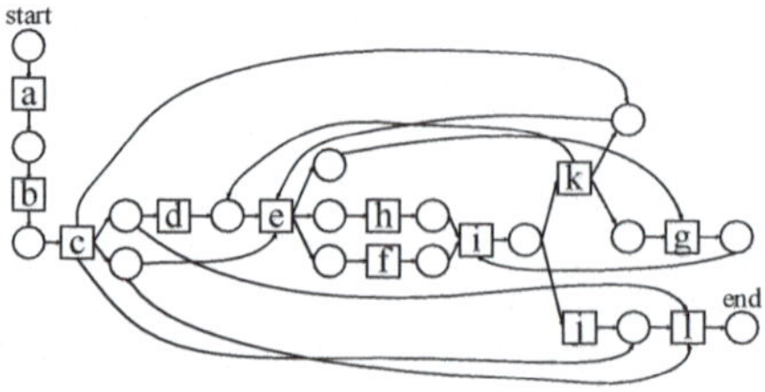


**Fig. 5.** The pieces of event log $L_4$

Compared $N_5$ and $N_6$, it's obvious that there are many differences between the actual business process and the predefined business process, and the actual business process is more flexible and complex. Therefore, it's necessary to do consistency check.

## 4.3 Consistency Check

As one of the most important induces of consistency check, fitness is used in this paper to measure the degree of match between the event log and the model. The event log $L_5$ can be shown as Table 3.

**Table 3.** The event log $\boldsymbol{L_5}$

| Trajectory | Times | Composition |
|---|---|---|
| $\sigma_1$ | 1 | <*a, b, c, l*> |
| $\sigma_2$ | 1 | <*a, b, c, d, e, f, g, h, i, j, l*> |
| $\sigma_3$ | 1 | <*a, b, c, d, e, f, h, g, i, j, l*> |
| $\sigma_4$ | 1 | <*a, b, c, d, e, g, f, h, i, j, l*> |
| $\sigma_5$ | 1 | <*a, b, c, d, e, f, h, g, i, k, e, f, g, h, i, j, l*> |
| $\sigma_6$ | 1 | <*a, b, c, d, e, h, f, g, i, j, l*> |
| $\sigma_7$ | 1 | <*a, b, c, d, e, h, g, f, i, j, l*> |
| $\sigma_8$ | 1 | <*a, b, c, d, e, f, g, h, i, k, g, h, i, j, l*> |
| $\sigma_9$ | 1 | <*a, b, c, e, f, g, h, i, j, l*> |

Take $\sigma_1$ as an example to illustrate the process of track replay in detail. To make it clear, we just use letters to represent $p_{N,\sigma_i}$. Before the replay begins, p = 0, c = 0, m = 0, r = 0 is the initial transition marked *a*. Then add a new token to Petri net $N_5$, p = 1, c = m = r = 0, the process is shown in Fig. 6.

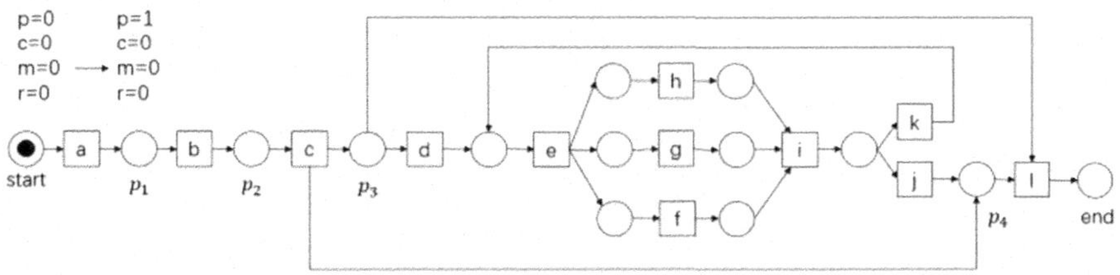


**Fig. 6.** The initial Petri net $\boldsymbol{N_5}$

After transition *a* happens, a token is consumed in place start, and a token is produced in place $p_1$, then p = 2, c = 1, m = r = 0. Figure 7 shows the Petri net $N_5$ after transition *a* happens.

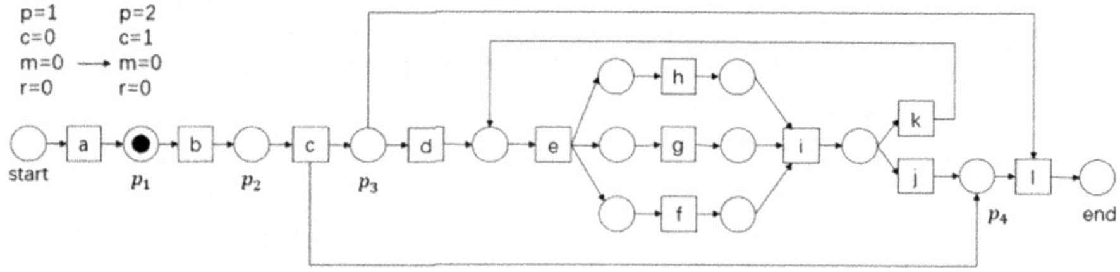


**Fig. 7.** The Petri net $N_5$ after transition a

When the replay of trajectory $\sigma_1$ is over, tokens in place end are consumed by environment, p = c = 6, m = r = 0.

Replay trajectory $\sigma_2$–$\sigma_9$ in accordance with the rules, and ensure corresponding p, c, m, r. Table 4 shows the p, c, m, r value of trajectory $\sigma_1$–$\sigma_9$.

**Table 4.** p, c, m, r value of trajectory $\boldsymbol{\sigma}_1$–$\boldsymbol{\sigma}_9$

| Trajectory | p value | c value | m value | r value |
|---|---|---|---|---|
| $\sigma_1$ | 6 | 6 | 0 | 0 |
| $\sigma_2$ | 15 | 14 | 0 | 1 |
| $\sigma_3$ | 15 | 14 | 0 | 1 |
| $\sigma_4$ | 15 | 14 | 0 | 1 |
| $\sigma_5$ | 23 | 22 | 0 | 1 |
| $\sigma_6$ | 15 | 14 | 0 | 1 |
| $\sigma_7$ | 15 | 14 | 0 | 1 |
| $\sigma_8$ | 19 | 20 | 3 | 2 |
| $\sigma_9$ | 14 | 13 | 1 | 2 |
| Sum | 137 | 131 | 4 | 10 |

According to Table 4, substitute them into the formula (1), fitness between event log $L_4$ and Petri net $N_5$ can be calculated:

$$fitness(L_5, N_5) = \frac{1}{2} \times \left(1 - \frac{4}{131}\right) + \frac{1}{2} \times \left(1 - \frac{10}{137}\right) \approx 0.948 \qquad (2)$$

The obtained fitness value is 0.948, it means that most of the real business process of enterprise A is correspond with the predefined Petri net $N_5$.

### 4.4 Analysis and Optimization of Warehousing Processing

It can be found that when there is no missed or remained token, in the other word, m = r = 0, the replay process finished successfully. However, there exist some problems. The main reasons for the above three problems are analyzed as follows:

1) There is a remained token in trajectory $\sigma_2$–$\sigma_7$. After transitions $c$ and $j$ occur, one token is generated in the input place of transition $l$, and two tokens are generated in total. However, the transition $l$ only consumes one token, which causes the redundant token remained.
2) There are three missed token and two remained tokens in trajectory $\sigma_8$. When the enterprise performs warehousing operation for the first time, there arises a problem, and it needs to remove the fault and re-enter into the warehouse. Moreover, it does not re-stack and print and paste labels when re-entering into the warehouse.
3) There is a missed token and two remained tokens in trajectory $\sigma_9$. Transition $d$ corresponds to the signature confirmation link after goods pass the quality inspection, indicating that there is no signature confirmation after the goods inspection in this case.

The optimization methods are shown as follows:

Improve the selective structure after quality inspection and the cycle structure of secondary storage. In the selection structure, the goods that fail the quality inspection need to trigger the corresponding transition of the ending activity, rather than sharing the same transition of the ending activity with the goods that pass the inspection. The improved cycle structure can trigger different loops for different kinds of warehousing problems. The improved Petri net $N_5'$ of warehousing processing is shown in Fig. 8. The corresponding events and meanings are shown in Table 5.

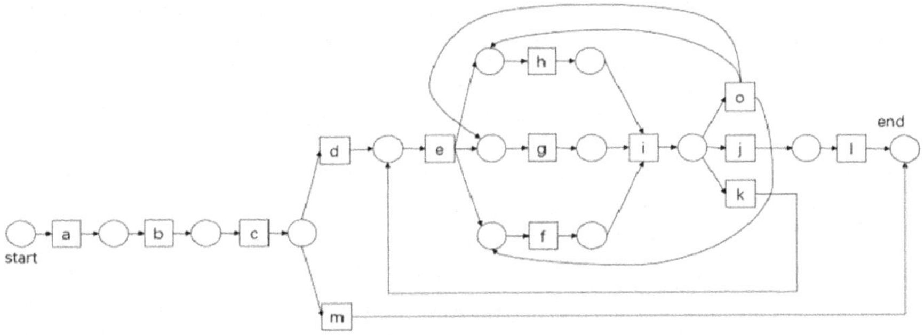


**Fig. 8.** The improved Petri net $N_5'$ of warehousing processing

Enterprises should strictly enforce the details of the standard in the actual implementation of the business process, and regularly take advantage of event log to verify whether there are errors in the actual operation process and fix the loophole in time. At the same time, in the actual operation process, it is recommended that the process that helps to improve efficiency could be modified independently, and the pre-defined business process model should be revised in time, to realize the complementary and joint optimization.

**Table 5.** The Petri net $N_5'$ of warehouse under normal circumstances

| Events | Meanings | Events | Meanings |
|---|---|---|---|
| *a* | Receive purchase order | *g* | Allocation goods |
| *b* | Receive goods | *h* | Dispatch equipment |
| *c* | Goods inspection | *i* | Execute warehouse |
| *d* | Sign | *j* | Print entry |
| *e* | Stack | *k* | Debug major problems |
| *f* | Print and paste labels | *o* | Debug minor problems |
| *m* | Inferior-quality products stop warehousing | *l* | End of warehousing and sorting |

## 5 Conclusion

In this paper, the research framework for warehousing processing mining are put forward. We firstly analyze the general flow of the warehousing and builds the Petri net of the theoretical warehousing processing. Secondly, the event logs recorded in the business system of Enterprise A are pre-processed to build Petri nets of the warehousing processing in the real situation of the enterprise. Then, the event logs in the real situation are replayed on the theoretical Petri net. Finally, the results of consistency check are analyzed, and the latent problems appear. We try to dig out possible reasons and make improvements.

## References


1. Yusuf, B.R., Reddy, P.C.: Mining data streams using option trees. Int. J. Comput. Netw. Inf. Secur. **4**(8), 49–54 (2012)
2. Pei, H.: Research on process mining method based on Petri net and event log. Anhui University of Technology, Hefei (2016). (in Chinese)
3. Pujeri, R.V., Karthik, G.M.: Constraint based periodicity mining in time series databases. Int. J. Comput. Netw. Inf. Secur. **4**(10), 37–46 (2012)
4. Wang, J., Wen, L.: Process Mining - Discovery, Compliance, and Improvement of Business Processes. Tsinghua University Press, Beijing (2014). (in Chinese)
5. Lu, F., Zeng, Q., Duan, H.: A parallel heuristic process mining. J. Softw. **26**(03), 533–549 (2015). (in Chinese)
6. Sun, H., Li, C.: Improved data mining method combining genetic algorithm and association rules. Data Acquis. Process. **34**(05), 863–871 (2019). (in Chinese)
7. Sun, H., Du, Y., Qi, L.: A method for mining process models with indirect dependencies via petri nets. IEEE Access **7**, 81211–81226 (2019)
8. Yu, J., Dong, C., Li, C.: A process mining algorithm based on statistical. J. Beijing Univ. Aeronaut. Astronaut. **44**(05), 895–906 (2018). (in Chinese)
9. Xu, J., Liu, J.: A profile clustering-based event logs repairing approach for process mining. IEEE Access **7**, 17872–17881 (2019)

10. Liu, W., Wang, L., Feng, X.: Soundness analytics of composed logical workflow nets. Int. J. Parallel Prog. **47**, 709–724 (2017)
11. Kalenkova, A.A., van der Aalst, W.M.P., Lomazova, I.A.: Process mining using BPMN: relating event logs and process models. Softw. Syst. Model. **16**(4), 1019–1048 (2017)
12. Hadis, H.: Analysis of reconfigurable processors using Petri net. Comput. Netw. Inf. Secur. **9**, 28–36 (2013)
13. Ahmed, F., Nandi, D., Rahman, M., et al.: Investigating factors that influence rice yields of Bangladesh using data warehousing, machine learning, and visualization. Int. J. Mod. Educ. Comput. Sci. **9**(3), 36–47 (2017)

# Mathematical Methods of Algebraic Biology

Georgy K. Tolokonnikov[1](✉) and Sergey V. Petoukhov[2]

[1] Federal Scientific Agro-Engineering Center VIM, RASciences, Moscow, Russia
[2] IMASh RAS, Moscow, Russia

**Abstract.** For algebraic biology, the main task of which is to predict the properties of organisms from the genome using rigorous algebraic methods, in particular, in the perspective of the properties of intelligence that can be modeled by conventional and strong artificial intelligence (AI), new categorical methods are proposed that are used in categorical systems theory. These methods are based on the theory of categorical splices, with the help of which biomolecules, including RNA and DNA, are modeled as categorical systems. Thus, in addition to the previously used, new algebraic and categorical methods are involved in the analysis of the genome. The substrate of the genetic code in the form of DNA and RNA molecules, influencing the genome, as shown in the work, becomes a direct object of research in algebraic biology.

**Keywords:** Genome · Categorical systems · Categorical splices · Polycategories · Chemical bond · Covalent · Ionic · Hydrogen bond

## 1 Introduction

In algebraic biology, a strictly mathematical analysis of the genome is carried out to predict the consequences of the genetic code for organisms. Among the research methods, matrix calculus and other sections of algebra are mainly used. Recently, the arsenal of algebraic tools has been expanding and in this work, which is a continuation of [1], along this path new methods of the theory of categorical splices are added, a section of category theory, which in fact also refers to algebra. There is no doubt about the importance of studying the properties of the DNA and RNA molecules themselves, the structure of which affects the genetic code, its noise immunity and the transmission of information both by inheritance and during protein synthesis. The emerging category model for molecules and chemical compounds (see [2], as well as Tolokonnikov G.K. Category splices and modeling with their help chemical systems and biomolecules, AIPE2020) makes it possible to start research in this direction from the standpoint of algebraic biology. The next section briefly describes the already used mathematical methods in algebraic biology (tensor product of matrics, modules over U-number algebras, etc.). The third section is devoted to the description of categorical splices, which in the fourth section are used to model biomolecules, including DNA and RNA. Some properties of these models are discussed, and directions of development of categorical modeling that are interesting for algebraic biology are indicated. In the conclusion, the results of the work are briefly summarized.


Z. Hu et al. (Eds.): ICAILE 2021, LNDECT 82, pp. 66–75, 2021.
https://doi.org/10.1007/978-3-030-80475-6_7

## 2 Matrix Genetics and U-Number Algebras

Amino acids include compounds in the molecules of which there is a carboxyl group COOH and an amine group $NH_2$, the simplest of amino acids glycine (Gly) has the formula $NH_2$ - $NH_2$ - COOH. Several hundred amino acids have been synthesized and found in nature. The fundamental question remains open why nature chose 20 known amino acids (Gly, Arg, Pro, Ser, Asp, etc.) from their numerous family for the formation of proteins in organisms. Savings considerations for biological objects justify the reduction of the number of amino acids to the minimum number sufficient to perform the functions of the body. On the other hand, the noise immunity of the transmission of the genetic code requires its significant redundancy. Mathematical regularities are manifested here at the most fundamental level, the connection between the tensor product of matrices and the structure of genetic coding, discovered (see [3, 4] and references there) by the second author, is expressed by a remarkable scheme (C - cytosine, A - adenine, T - thymine, G - guanine) (Fig. 1).

[C A; T G] =

| | |
|---|---|
| C | A |
| T | G |

;

$[C\ A;\ T\ G]^{(2)}$ =

| | | | |
|---|---|---|---|
| CC | CA | AC | AA |
| CT | CG | AT | AG |
| TC | TA | GC | GA |
| TT | TG | GT | GG |

$[C\ A;\ T\ G]^3$ =

| | | | | | | | |
|---|---|---|---|---|---|---|---|
| CCC Pro | CCA Pro | CAC His | CAA Gln | ACC Thr | ACA Thr | AAC Asn | AAA Lys |
| CCT Pro | CCG Pro | CAT His | CAG Gln | ACT Thr | ACG Thr | AAT Asn | AAG Lys |
| CTC Leu | CTA Leu | CGC Arg | CGA Arg | ATC Ile | ATA Met | AGC Ser | AGA Stop |
| CTT Leu | CTG Leu | CGT Arg | CGG Arg | ATT Ile | ATG Met | AGT Ser | AGG Stop |
| TCC Ser | TCA Ser | TAC Tyr | TAA Stop | GCC Ala | GCA Ala | GAC Asp | GAA Glu |
| TCT Ser | TCG Ser | TAT Tyr | TAG Stop | GCT Ala | GCG Ala | GAT Asp | GAG Glu |
| TTC Phe | TTA Leu | TGC Cys | TGA Trp | GTC Val | GTA Val | GGC Gly | GGA Gly |
| TTT Phe | TTG Leu | TGT Cys | TGG Trp | GTT Val | GTG Val | GGT Gly | GGG Gly |

**Fig. 1.** Tensor family of genetic matrices

As you can see, glycine is redundantly encoded by four triplets GGC, GGA, GGT, GGG.

The study of symmetries inherent in genomatrices and, as it turned out, associated with numerous direct manifestations of mathematical regularities in biological objects (phyllotaxis, etc.), and other properties is carried out using hypercomplex numbers, matrices, bisexes and, in the general case, U-numbers. The methods used relate to the theory of modules over the indicated and other algebras (see details in [3, 4]). The system requirements for biological objects in algebraic biology are taken into account on the basis of categorical systems theory [5], generally speaking, taking them beyond the bounds of set-theoretic mathematics and giving appropriate generalizations, in particular, for algebras of U-numbers [1].

Algebraic methods for analyzing the degeneracy of genomatrices show that the set of amino acids is divided into two subsets, consisting, respectively, of 8 and 12 elements, giving a total of 20, and not another number. Thus, the question of the number of required amino acids has moved into the plane of elucidation of the reasons for the existence of two alternative groups of amino acids, 8 and 12, respectively (see [3] pp. 64–65).

The indicated considerations of economy and noise immunity, the division of a set of 20 amino acids into two subsets of 8 and 12 elements, as well as many other characteristics of the genome, are realized due to certain properties of amino acids chosen by nature and the structure of DNA and RNA molecules. The modeling of the molecules themselves, studied in this work, by strictly mathematical objects by categorical splices, as the authors hope, will shed a certain light, including on this problem. This is one of the areas of study of the proposed models of biomolecules, while in addition to the indicated algebraic research methods, methods of the theory of categories are in demand in algebraic biology.

## 3 Categorical Splices

Let's introduce a four-sorted first-order logic language with variables $p, p', \bar{p}, \overline{p'}$ responsible for splices, convolutions, inner splices and inner convolutions, respectively. Convolutions are represented using: specified letters , $p', \bar{p}, \overline{p'}$; functional symbols of co-regions $C_i, C'_i, \bar{C}_i, \overline{C'}_i$; names $\pi_i, \pi'_i, \bar{\pi}_i, \overline{\pi'}_i$; values $ev_\alpha, ev'_\alpha, \bar{ev}_\alpha, \overline{ev'}_\alpha$; predicates $M_b, M_r, M_t, M_l$,; using a graphic representation of formulas, for example

$$x_i \overset{\text{def}}{=} \pi_i(p), b_i \overset{\text{def}}{=} C_i(p),$$
$$x_i = \pi_i(p) \wedge b_i = C_i(p) \wedge x_j = \pi_j(p) \wedge b_j = C_j(p) \wedge x_i = x_j$$

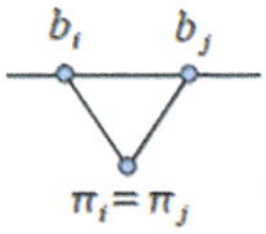


The properties of predicates and other letters of the signature are determined by axioms. Formulas $x_i = \pi_i(p), b_i = C_i(p)$ will be denoted by $\pi_i$, $C_i$.

The convolution S is defined by the ratio, the symbol ev has the opposite and the opposite convolution is determined by the $S = M_b evM_l$. The complete splices can be represented as a conjunction formula $a \wedge b' \wedge \bar{c} \wedge \overline{d'}$, divided into four subformulas in accordance with the sort of variables, located in the $2 \times 2$ quadrants of the table

| | |
|---|---|
| $a$ | $b$ |
| $c$ | $d$ |

Equivalences between co-regions, between names that correspond to the equal sign, are depicted by connecting the right and left letters inequality with dashed lines. The predicates $M_t, M_r, M_l, M_b$ are determined by the following formula transformations. $M_b$ moves to the lower quadrants from the upper quadrants only those pairs $C_k\pi_k$ that have co-area connections, $M_t$ moves to the upper quadrants from the lower quadrants only those pairs $C_k\pi_k$ that have co-area connections, $M_l$ moves vertically and inside a pair from right quadrants to left ones, $M_r$ moves connected pairs from left quadrants to rights.

The function symbol $ev$ translates $\pi_i$, $C_i$ to $ev(\pi_i)$, $ev\left(C_j\right)$ and places the $=$ sign in the appropriate places. If you do not impose restrictions on the choice of splices and convolutions, then we speak of free general splices. Special categorical splices are constructed from a fixed set of convolutions and a fixed set of splices, while the original splices are called generators, and all the rest are obtained by all possible applications of these convolutions to splices.

For clarity, consider the following formula as an example.

$$C_k\wedge\pi_k\wedge C_i\wedge\pi_i\wedge C_m\wedge\pi_m\wedge C_n\wedge\pi_n\wedge C'_j\wedge\pi'_j\wedge C'_w\wedge\pi'_w\wedge C'_v\wedge\pi'_v\wedge$$
$$\wedge\pi_k=\pi_i\wedge\pi_m=\pi_n\wedge\pi'_j=\pi'_v\wedge C_i=C'_j\wedge C_m=C'_v.$$

We will omit the conjunction sign, replace equalities with a line connecting the right and left sides of the equalities. The table for the formula is

| | |
|---|---|
| $C_k\pi_kC_i\pi_iC_m\pi_mC_n\pi_n$ | $C'_j\pi'_jC'_w\pi'_wC'_v\pi'_v$ |
| Ø | Ø |

For convolution, we have a table

| | |
|---|---|
| $C_k\pi_kC_n\pi_n$ | $C'_w\pi'_w$ |
| $C_i\pi_iC_m\pi_m$ | $C'_j\pi'_jC'_v\pi'_v$ |

Applying, $M_l$ we get

| | |
|---|---|
| $C_k\pi_kC_n\pi_n$ $C_w\pi_w$ ¶ ¶ | Ø ¶ |
| $C_i\pi_iC_m\pi_m$ $C_j\pi_jC_v\pi_v$ ¶ | Ø ¶ |

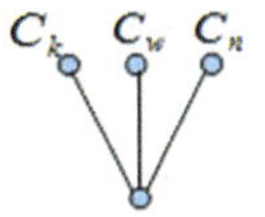


$\pi_k=\pi_w=\pi_n$

A complete restoration of the original splice is possible as follows. We apply a transfer $M_r$ to the final table in relation to $C_w\pi_w\overline{C_j}\overline{\pi}_j\overline{C_v}\overline{\pi}_v$, we will get a table that coincides with the previous final table. Then we apply $M_t$ to it, which gives the original table and, thereby, the original splice.

Let there be k splices with sets of variables of four sorts, functional symbols, and predicates. It is not difficult to construct a unified formal theory of higher categorical k-splices. Let's go to a table with 2k columns, in the first k columns, in order, insert $a_i, c_i, i = 1, \ldots, k$, in the second k columns, insert $b_i, d_i, i = 1, \ldots, k$, as indicated below

| | | | | | |
|---|---|---|---|---|---|
| $a_1$ | ... | $a_k$ | $b_1$ | ... | $b_k$ |
| $c_1$ | ... | $c_k$ | $d_1$ | ... | $d_k$ |

In convolutional polycategories, poly-arrows have inputs and outputs. Their modeling is possible using 2-splices. Let's look at an example of traditional categories. Two arrows, when the beginning of one arrow coincides with the end of the other arrow, and the convolution (corresponding to the composition of the arrows), corresponds to the formula (between the letters the conjunction sign is omitted)

$$C_1^{(1)}\pi_1^{(1)}C_2^{(1)}\pi_2^{(1)}C_3^{(1)\prime}\pi_3^{(1)\prime}C_1^{(2)}\pi_1^{(2)}C_2^{(2)}\pi_2^{(2)}C_3^{(2)\prime}\pi_3^{(2)\prime}\wedge$$
$$\wedge\left(\pi_1^{(1)}=\pi_1^{(2)}\right)\left(\pi_2^{(1)}=\pi_2^{(2)}\right)\left(\pi_3^{(1)\prime}=\pi_3^{(2)}\right)\wedge$$
$$\wedge\left(C_1^{(2)}=C_2^{(1)}\right)\left(C_2^{(1)}=C_3^{(1)\prime}\right)\left(C_1^{(2)}=C_3^{(2)\prime}\right)\left(C_3^{(1)\prime}=C_3^{(2)\prime}\right)$$

The polygraph and convolution are shown in the diagram below (Fig. 2).

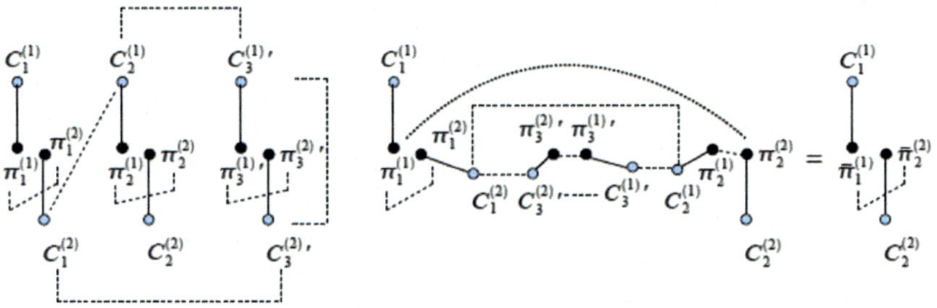

**Fig. 2.** Graphical representation of the splices formula

In order not to write out awkward superscripts, the second pairs $C^{(2)}, \pi^{(2)}$ will be denoted by $D$, $\omega$:

$$C_1\pi_1C_2\pi_2C_3'\pi_3'D_1w_1D_2w_2D_3'\omega_3'\wedge$$
$$\wedge(\pi_1=\omega_1)(\pi_2=\omega_2)\left(\pi_3'=\omega_3'\right)\wedge$$
$$\wedge(D_1=C_2)\left(C_2=C_3'\right)\left(D_1=D_3'\right)\left(C_3'=D_3'\right)$$

We have a table

| | | | |
|---|---|---|---|
| $C_1\pi_1C_2\pi_2$ | $D_1\omega_1D_2\omega_2$ | $C_3'\pi_3'$ | $D_3'\omega_3'$ |
| $\varnothing$ | $\varnothing$ | $\varnothing$ | $\varnothing$ |

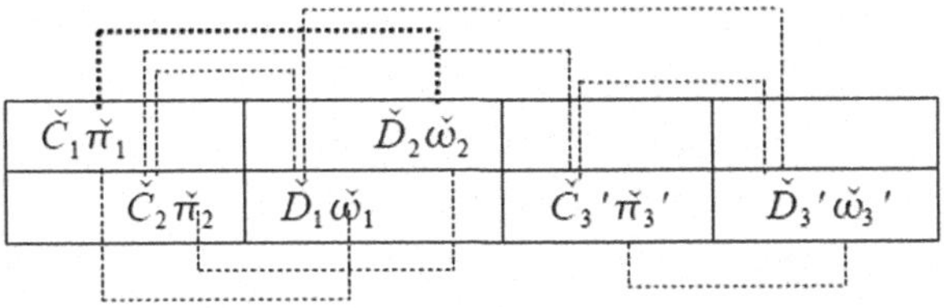


Apply $M_b$ and $ev$ from the convolution $S = M_l ev M_b$, we get $(ev(C, D, \ldots) = \check{C}, \check{D}, \ldots)$.

In this model of the category, simultaneously with the result of the composition, its receipt from the original arrows is recorded. Since convolutional polycategories are constructed in a similar way from splices, the category theory of systems is based on the formalism of higher category splices.

# 4 Categorical Model of Biomolecules

The exact construction of a representation of a molecule as a system in the form of a polycategory arrow requires knowledge of the wave function of the molecule in stationary and other states, therefore, as it is forced to do in quantum chemistry, one will have to be content with approximate methods. For available approximations, you can also build categorical models.

The main approximation in quantum chemistry [6] is the adiabatic approximation, which consists in a separate consideration of the motions of nuclei and electrons. It can be assumed that chemical bonds are determined by the distribution density of the electron charge in the electron cloud, specified by the wave function of the molecule, which ensures the equilibrium position of the nuclei by compensating for their electrical interactions with each other, as justified by the Gelman-Feynman theorem.

An electron cloud in different molecules can have the same substructures, each of these substructures can correspond to one or another type of chemical bond. The main types of chemical bonds in biochemistry are covalent, hydrogen bonds, and hydrocarbon aromatic bonds. After some refinements, the structural chemical formulas become equivalent to the corresponding splices and their convolutions, for which covalent strokes should be juxtaposed with the co-regions of the splices. For example, a carbon atom and its bivalent modification correspond to the splices in Fig. 3. Methane is obtained by simple convolutions (grouped into one convolution).

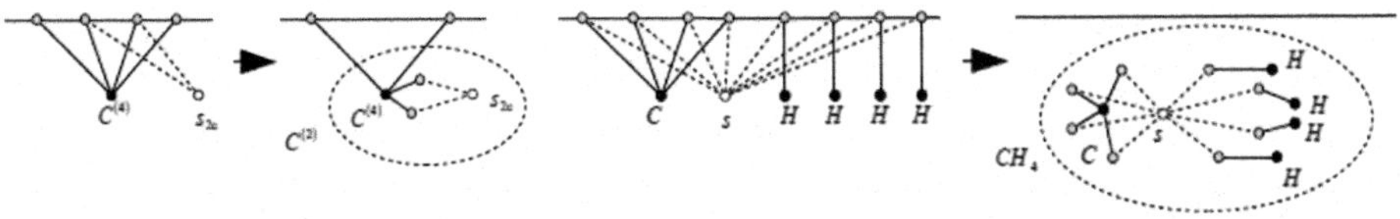


**Fig. 3.** Convolutions for obtaining divalent carbon and methane

Hydrocarbons with double and triple bonds are modeled taking into account different types of convolutions for $\sigma$, $\pi_x$, $\pi_y$ bonds (Fig. 4).

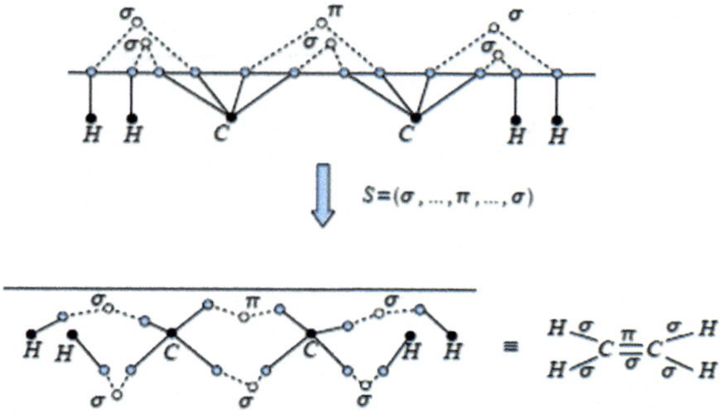


**Fig. 4.** Category splice for ethylene (σ- and π-bonds are marked)

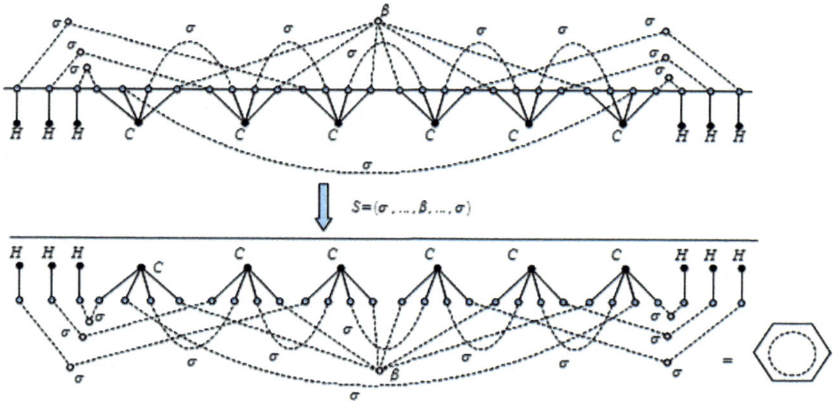


**Fig. 5.** Modeling a benzene molecule by splices

For the aromatic bond, a convolution with six (for benzene) and a large number of H atoms (for other aromatic hydrocarbons) is selected (Fig. 5).

Cyclic hydrocarbons are modeled naturally. The properties of the algebras, PROPs, and splices that appeared in this way, studied by strictly mathematical methods, give the properties of biomolecules.

RNA and DNA biomolecules, which are essential for algebraic biology and its predictions, are also modeled by gluing together. We emphasize the strictly mathematical nature of these models and the operations in them, which equips algebraic biology with a number of categorical algebraic research methods.

Let us consider the modeling of DNA and RNA molecules, the structure of the DNA scheme in a convenient form is shown in Fig. 6 and Fig. 7.

A phosphodiester bond is used in the DNA molecule, see Fig. 7.

Phosphodiester bond is modeled by the following splices (Fig. 8).

The formation of a hydrogen bond is generated by the relatively small size of the hydrogen atom in comparison with other atoms, which allows the hydrogen electron to come close enough to the lone electron pairs of molecules.

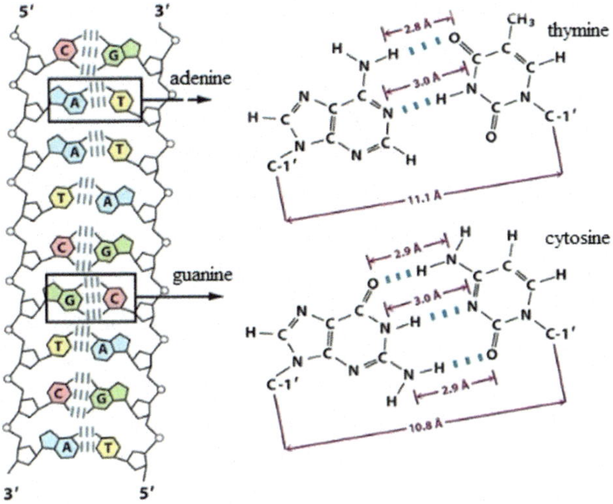


**Fig. 6.** Structural formulas of DNA

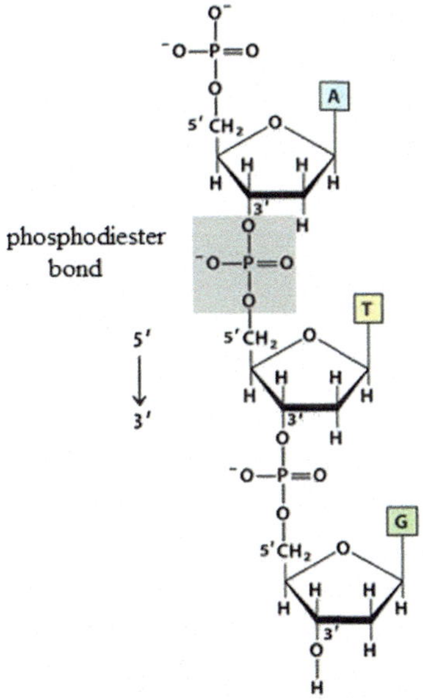


**Fig. 7.** Structural formulas of DNA (phosphodiester bonds)

It is possible to simulate the hydrogen bonds in DNA and RNA responsible for the secondary structure by adding new types of elements (marked with yellow circles) and a convolution between them to the splices co-regions. We also note another option for taking into account the differences in bonds using higher splices, when certain types of bonds are modeled by the corresponding splices included in the higher splices, which

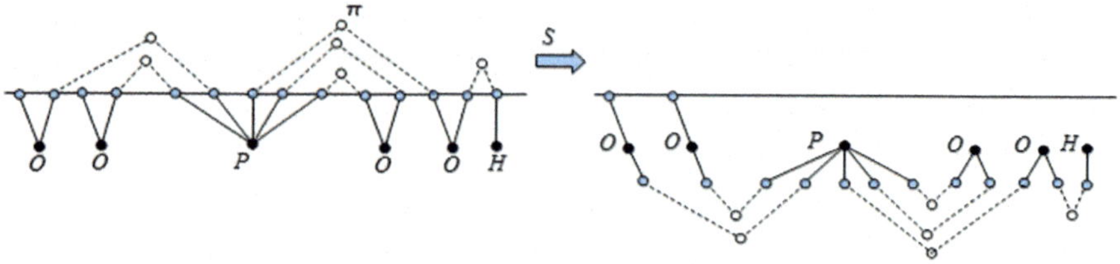


**Fig. 8.** Modeling the phosphodiester bond by splices

completely model the molecule. Let us illustrate the formation of water dimers and their models (Fig. 9).

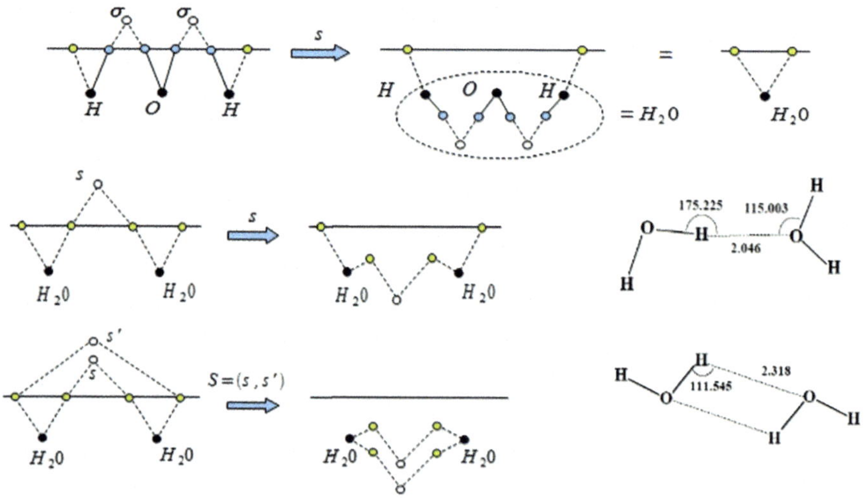


**Fig. 9.** Modeling by splices isomers of water dimers (according to Darinsky BM et al. Sorption and chromatographic processes, V. 13. No. 6, C.902–911 (2013)).

As mentioned, the geometry of molecules, in particular, the angles and positions of atomic nuclei for dimers can also be simulated by adding parameters to the corresponding convolutions. The disassembled types of models are sufficient to represent the entire DNA (or RNA) molecule, including various levels of its structure. The examples given make it possible to construct formulas and their graphic designations for DNA and RNA, due to their cumbersomeness they are not presented here.

## 5 Conclusion

One of the main tasks of algebraic biology is to predict the properties of organisms from the genome using rigorous algebraic methods, in particular, to predict the properties of intelligence that can be modeled by conventional and strong artificial intelligence (AI). Popular methods of artificial neural networks [7–12], as it is gradually becoming clear, are not enough for modeling strong AI, while algebraic biology in the future should provide the necessary models of AI. The report provides an overview of mathematical methods used in algebraic biology, new categorical methods used in categorical systems theory are proposed. These methods are based on the theory of categorical splices. With their help, biomolecules are modeled, in particular, RNA and DNA biomolecules. Thus, RNA and DNA as the carrier of the code, the properties of which, as indicated, affect the genome, becomes an object of study in algebraic biology, and new algebraic and categorical methods are involved in the analysis of the genome. Further works will consider the specified analysis of DNA and RNA.

## References


1. Tolokonnikov, G.K., Petoukhov, S.V.: New mathematical approaches to the problems of algebraic biology. In: Hu, Z., Petoukhov, S., He, M. (eds.) AIMEE 2019. AISC, vol. 1126, pp. 55–64. Springer, Cham (2020). https://doi.org/10.1007/978-3-030-39162-1_6
2. Tolokonnikov, G.K.: Categorical models of artificial neural networks and system information, categorical splices, and the paradigm of categorical systems theory. Biomachsystems **2**(1), 127–174 (2018)
3. Petoukhov, S.V.: Matrix genetics, algebra of genetic code, noise immunity. M., RHD (2008)
4. Petoukhov, S.V.: Hyperbolic Numbers in Modeling Genetic Phenomena, 181p. (2020). https://doi.org/10.20944/prepr201908.0284.v4https://www.preprints.org/manuscript/201908.0284/v4
5. Tolokonnikov, G.K.: Informal categorical theory of systems. Biomachsystems **2**(4), 41–144 (2018)
6. Gribov, L.A., Mushtakova, S.P.: Quantum chemistry, M., 390 p. (1999)
7. Dharmajee Rao D.T.V., Ramana K.V.: Winograd's Inequality: Effectiveness for Effi-cient Training of Deep Neural Networks. IJISA **6**, 49–58 (2018)
8. Karande, A.M., Kalbande, D.R.: Weight assignment algorithms for designing fully connected neural network. IJISA, No **6**, 68–76 (2018)
9. Hu, Z., Tereykovskiy, I.A., Tereykovska, L.O., Pogorelov, V.V.: Determination of structural parameters of multilayer perceptron designed to estimate parameters of technical systems. IJISA, No **10**, 57–62 (2017)
10. Awadalla, M.H.A.: Spiking neural network and bull genetic algorithm for active vibration control. IJISA **10**(2), 17–26 (2018)
11. Abuljadayel, A., Wedyan, F.: An approach for the generation of higher order mutants using genetic algorithms. IJISA **10**(1), 34–35 (2018)
12. Sreedhara, B., Manu, S.: Mandal. Application of Particle Swarm based Neural Network to Predict Scour Depth around the Bridge Pier. IJISA **11**(11), 38–47 (2019)

# Forecast and Analysis of Domestic Iron Ore Shipping Pricing Based on BP Neural Network

Lu Cao(✉) and Meng Wang

School of Logistics, Wuhan Technology and Business University, Wuhan 430065, China

**Abstract.** China is a big iron and steel producer and a big iron ore importer. Over the past few decades, iron ore has mainly imported from major iron ore exporters. Due to the absolute monopoly of iron ore resources by exporting countries, the pricing power of iron ore has always been in the hands of several iron ore monopoly exporters. Iron ore shipping market has also caused a certain impact. This paper will forecast and analyze the shipping rate of iron ore from two aspects: price composition and supply-demand balance of iron ore market. By qualitative analysis, the influencing factors of iron ore shipping freight rate are found. After normalization, as input variables, the output related freight rate is calculated BP the neural network model. By comparing the forecast results, it is found that the method is effective and provides some reference for shipping enterprises to formulate cargo shipping more reasonably.

**Keywords:** Shipping pricing · Supply-demand relationship · BP neural network · Price model

## 1 Introduction

Iron ore is the raw material for smelting iron and steel, which belongs to commodities. It is transported directly into the cabin in the course of transportation without any packing means [1]. Due to the influence of natural conditions, economic scale, development level, industrial structure and other factors, the distribution of iron ore production and demand around the world is not balanced. The world's major iron ore exporters are mainly Australia, Brazil, India and South Africa, the main importing countries China, Japan, the European Union and so on.

In recent years, China's iron and steel industry has developed rapidly, and the demand for iron ore is increasing day by day. Due to the low taste of iron ore and insufficient mining degree, a large number of iron ore can only rely on imports, resulting in a large difference between the import market freight rate and market price of iron ore [2]. In import transportation, the forecast of the market is delayed to a certain extent. The shipping market is also a completely free competition market, and it is also an extremely sensitive market. The uncertainty of freight price fluctuation brings great operating risk to shipping enterprises, cargo owners and other industry investors. The freight rate of iron ore has also become the leading factor affecting the cost of enterprises.


Z. Hu et al. (Eds.): ICAILE 2021, LNDECT 82, pp. 76–84, 2021.
https://doi.org/10.1007/978-3-030-80475-6_8

# 2 Analysis on Influencing Factors of Iron Ore Freight Rate

## 2.1 Impact of Transport Costs

Transportation cost is the main factor affecting freight rate, which accounts for a large proportion of total cost [3]. Every link in transportation will have a certain cost. The sum of the costs of each link constitutes the direct economic cost of iron ore transportation, that is the shipping cost of ships [4]. The main components of shipping cost in ship operation are as follows:

1) *Cost of capital*

The cost of capital refers to the expenses paid for the direct purchase of fixed assets, which refers to the expenses paid by the shipping enterprises of Zhonglian. This also includes ship-related interest and taxes. This is the direct cost of buying a ship and is the most basic cost. Assets can be equivalent to depreciation expenses at the end of the year.

2) *Operating costs*

The operating cost of a ship refers to the expenses incurred by the ship when sailing under suitable climatic conditions, mainly including port charges and fuel charges and so on.

3) *Business management fees*

Business management fee refers to the cost of each department and organization established within the company to complete the business.

## 2.2 Impact of Domestic Shipping Enterprise Competition

Generally speaking, shipping prices are determined by the relationship between supply and demand in the market [5]. But the actual situation is that in order to compete for resources, enterprises deliberately reduce freight rates, resulting in a sharp decline in freight rates on the market, directly lower than the market freight, even below actual costs.

There is also a phenomenon in which enterprises form certain transactions in order to achieve certain results, such as organizing the establishment of shipping chambers of commerce and other organizations to reach an agreed tariff on a certain area or route or a certain cargo transport, which leads to fair competition among enterprises. This leads to a certain stage of shipping pricing may be higher than market rates.

Usually, the proper competition in the shipping industry helps to improve the overall level of the industry, but there will be some malicious competition in the actual competition, which has a certain impact on the price of iron ore shipping. Such as price war, vicious competition and so on, this kind of phenomenon will not only cause both sides to lose, even make the whole shipping market in disorder.

## 2.3 Impact of Routes and Port Conditions

### 2.3.1 Impact of Routes

Different routes have different navigation conditions, and their waterways and ship types are limited. The cost of different ship types is also different [6]. For example, Masteel logistics of Beilun to saddle route, the route is mainly in the Yangtze River Basin. Due to the influence of seasons, different waters have different characteristics. There will be flood season and dry season in the waters of the Yangtze River, so there will be different prices for the same goods on the route from Beilun to saddle at different periods. Different from the Yangtze River basin, the coastal route has no flood season and dry season, but it is easy to send typhoons in July and August, which will also have a certain impact on the sea cargo transportation, which will easily cause the outage and the increase of the price of cargo transportation. So through seasons and waters, the different routes indirectly affect the freight rate.

### 2.3.2 Port Conditions

When a ship docks at a port, the conditions to be met are port loading and unloading equipment, berth conditions, loading and unloading efficiency and so on.

## 2.4 Impact of National Policies

According to statistics, China's iron ore enterprises tax rate than the international iron ore four giants average level is nearly 6. It can be seen that if the country can reduce the tax burden of iron ore import at the policy level, it can greatly improve the competitiveness of domestic iron ore in the international market. Effectively restrain the rapid trend of iron ore dependence in China. Relevant policies introduced in the first half of 2016 require China to use five years to significantly reduce crude steel production capacity. At the end of 2016, the target of resolving 4500 tons of crude steel production capacity was exceeded. The introduction of environmental protection policies has played an important role in the development of iron and steel industry capacity, and played a positive role in promoting the recovery of iron ore prices.

## 2.5 Impact of Supply and Demand

According to NSO data, China produced 844.356 million tons of iron ore in 2019, The cumulative increase was 4.9. As of August 2020, China's ore production was 76.999 million tons, 3.972 million tons more than in July, an increase of 5.4%, The year-on-year increase was 4.8. According to statistics from the General Administration of Customs of China, China's imports of iron ore and concentrate reached 1.06895 billion tons in 2019, the cumulative increase was 0.5. As of August 2020, China's imports of iron ore and concentrate were 100.36 million tons, the year-on-year increase was 5.9%.

China's average iron ore price in 2019 was 663.2 yuan per ton. Since 2013, the average iron ore price and domestic prices have been declining; the average price of imported iron ore rose 159.2 yuan per ton in 2018. Prices for China's iron ore industry for 2013–2019 are shown in Table 1 and Fig. 1 below.

**Table 1.** China iron ore industry prices in 2013–2019

| Year | 2013 | 2014 | 2015 | 2016 | 2017 | 2018 | 2019 |
|---|---|---|---|---|---|---|---|
| Average domestic iron ore price: Yuan/ton | 649.88 | 593.98 | 533.57 | 490.30 | 455.46 | 614.17 | 663.2 |
| Average price of imported iron ore: Yuan/ton | 802.37 | 615.35 | 377.21 | 376.32 | 479.07 | 587 | 708.01 |
| Average iron ore price: Yuan/ton | 704.90 | 602.10 | 469.73 | 433.72 | 466.34 | 600.59 | 685.6 |

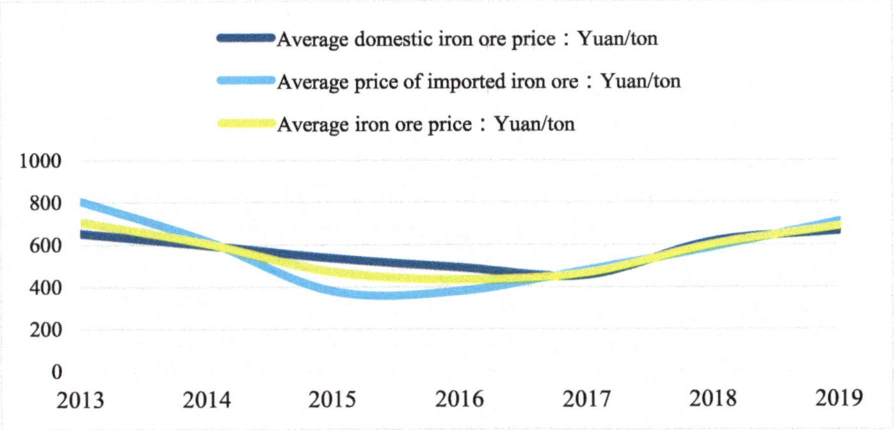


**Fig. 1.** China iron ore industry prices in 2013–2019

In general, the increase or decrease of the demand for ship transportation should be increased or reduced accordingly. But from the use of iron ore and the entire economic market, China's annual demand for iron ore is basically stable growth. However, in the shipping market, when the supply of capacity and volume exceeds the demand, the shipping rate will fall; when the supply is less than the demand, the shipping rate will gradually rise and tend to stabilize, but the supply of capacity can not be reduced in time.

## 2.6 Impact of Uncertainties

First, the influencing factor of iron ore freight rate is the impact of international crude oil price fluctuations. The influencing factors of oil price are higher in the weight of freight change, the profit margin of shipping industry is generally low, and there is no sign of improvement in the short term. In view of this, the fluctuation of oil price will affect the cost of using ships. Driven by the market, the fluctuation of cost will inevitably lead to the fluctuation of freight rate. Second, the factor of iron ore freight rate is affected by bad weather. For example, the impact of typhoons, floods and so on. Due to the bad weather, the ship cannot sail normally, resulting in the reduction of effective market capacity, which can easily lead to a surge in the price of shipping goods [7].

## 3 Forecast of Iron Ore Price Based on BP Neural Network

BP (Back propagation) neural network is a forward propagation neural network with tutors learning [8]. Its learning rules are delta rules that require that the activation function in the network structure be different. Figure 2 is a general three-layer BP network model, in which the $x_i$ represents the input value, the $w_{ij}$ is the connection weight, the middle layer is the activation function, and the output layer is followed [9, 10]. Since there are many influencing factors and weak correlation of iron ore shipping freight rate, the BP neural network model just meets the requirements of iron ore freight rate prediction model, so this paper adopts three-layer BP network to establish iron ore freight rate prediction system, as shown in Fig. 2.

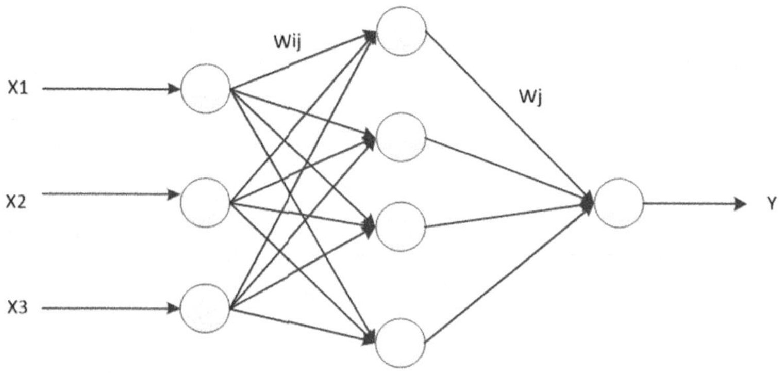


**Fig. 2.** Three layer BP network model

### 3.1 Establishment of Forecast Model of Iron Ore Shipping Freight Rate

According to the previous analysis, we can see that the main influencing factors of iron ore shipping freight rate are transportation cost, enterprise competition, national policy and supply and demand relationship. In the model verification, the predicted variables are domestic iron ore consumption, domestic iron ore average price, domestic iron ore quantity, domestic ore demand, iron ore raw ore production quantity and iron ore import quantity. The data are from 2013 to 2019, as shown in Table 2.

In general, the number of hidden layers is calculated according to formula (1).

$$m = \sqrt{n + l} + a \tag{1}$$

n is the number of input layer nodes, l is the number of output layer nodes, m is the number of hidden layer nodes, a is the constant between 1 and 10. For a general model, usually uses a hidden layer, but for more complex systems, the more hidden layers are created. The higher the accuracy, but the more time it takes, Therefore, increasing the number of input neurons can improve the accuracy of network structure. Based on the impact factors and the predicted values required, it can be determined that in this model, the input node is 6, the output node is 1, according to the above formula, the number of hidden layer nodes is 5, BP neural network architecture diagram is shown in Fig. 3.

**Table 2.** Variational factors for forecasting iron ore shipping prices

| Year | Domestic iron ore consumption/ million tons | Average domestic iron ore price/ yuan | Domestic Iron Ore Volume/ million tons | Domestic ore demand/ million tons | Domestic Iron Ore Volume/ million tons | Iron ore imports/ million tons |
|---|---|---|---|---|---|---|
| 2013 | 22.69 | 649.88 | 14.2 | 22.34 | 14.15 | 8.19 |
| 2014 | 24.42 | 593.98 | 15.1 | 24.46 | 15.14 | 9.32 |
| 2015 | 23.33 | 533.57 | 13.8 | 23.33 | 13.81 | 9.52 |
| 2016 | 23.1 | 490.3 | 12.8 | 23.05 | 12.81 | 10.24 |
| 2017 | 22.9 | 455.46 | 12.3 | 23 | 12.29 | 10.73 |
| 2018 | 18.1 | 614.67 | 7.6 | 18.27 | 7.63 | 10.64 |
| 2019 | 19 | 663.2 | 8.4 | 19.15 | 8.44 | 10.71 |

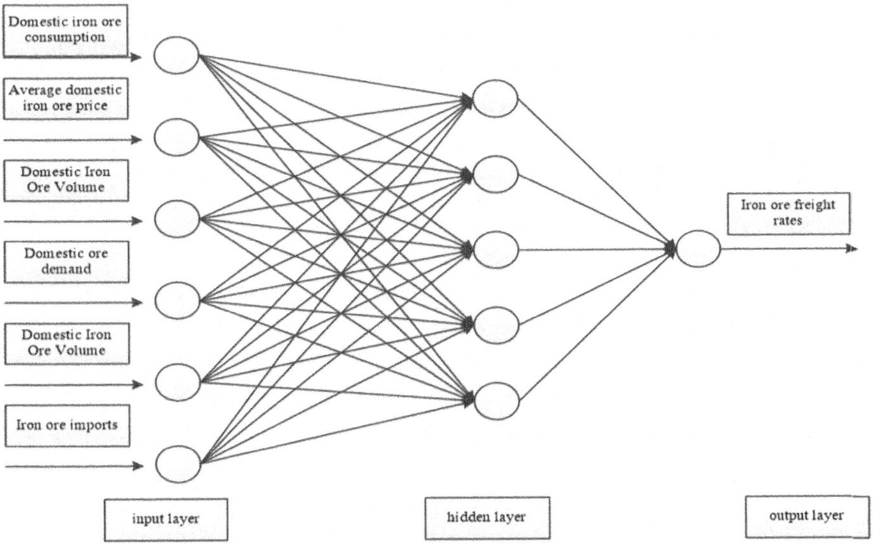


**Fig. 3.** BP neural network architecture

### 3.2 Forecast of Iron Ore Shipping Prices

Because there are many factors affecting the pricing of iron ore shipping, and it is very complex, so there are few alternative prediction methods. The BP neural network model can input more variables and fit the nonlinear system of iron ore shipping pricing well. Therefore, BP neural network is the most suitable model for the prediction of iron ore shipping pricing. Through the research and investigation, it can be found that the operating cost of ship shipping, the competition among shipping enterprises, the influence of route and port and national policy are analyzed theoretically and lack of relevant data.

In this paper, six groups of data were collected from 2013 to 2018, forecast the freight rate of iron ore in 2019. Using these six sets of data as training samples, the optimal 6-5-1 network structure is obtained. That is, the number of neurons in the input layer, the hidden layer and the output layer is 6, 5 and 1, respectively, programming with MATLAB. Finally we set the network training algebra to 5000, the training rate was 0.05, the training target error is 0.0000001. Figure 4 and Fig. 5 show the programming process. By running the MATLAB, it can be concluded that the forecast price of iron ore in 2019 is 27.31 yuan/ton, and the error with the actual value is within the allowable range.

### 3.3 Analysis of Iron Ore Shipping Rates

Iron ore shipping rates are affected by many factors, the most important is the competition within the enterprise and iron ore market supply and demand relationship. The change of these factors is groundless and even very uncertain, which is a hidden danger to the iron ore shipping market, because the influence of these uncertain factors may make the iron ore shipping freight rate change point at any time. Which will increase the iron ore shipping freight rate forecast and formulation difficulty. Therefore, when building a model, we need to consider as much as possible so that the model can adapt to most environments.

Figure 4 and Fig. 5 show that the predicted price of iron ore in 2019 can be obtained by combining BP neural network with MATLAB. The actual value of iron ore freight rate in 2019 is 26.97 yuan/ton, and the predicted value is 0.34 yuan/ton lower than the actual value.

Moreover, through the combination of BP neural network and MATLAB, the value of 2013–2019 is calculated and analyzed as input variable to predict that the shipping freight rate of iron ore in 2020 is 32.31 yuan/ton, which has certain reference value for shipping enterprises.

## 4 Conclusion

The shipping market is a volatile market, and the freight rate has been in a state of constant fluctuation in the shipping market, which indirectly reflects the change of supply and demand. Because the data released by the Shanghai Shipping Exchange relate to the coastal bulk cargo transport market, the freight rate index of China's coastal bulk cargo and coal, these data are aimed at the coastal areas, and in addition to part of the coastal transport, Zhonglian shipping is more inland river transport of the Yangtze River, which is divided into river and sea transport, river and sea direct transport and other modes.

In view of the price index in the Yangtze River field, domestic enterprises do not have a clear reference price, so shipping companies generally can only adjust the shipping index of the company by a small margin according to the coastal freight rate index. However, there is a big gap between Yangtze River freight rate and coastal freight rate due to market delay and uncertainty.

Therefore, this paper hopes to establish a nonlinear system with multiple influencing factors determining the change of a parameter. Through the systematic analysis of iron

**Fig. 4.** MATLAB operations-1

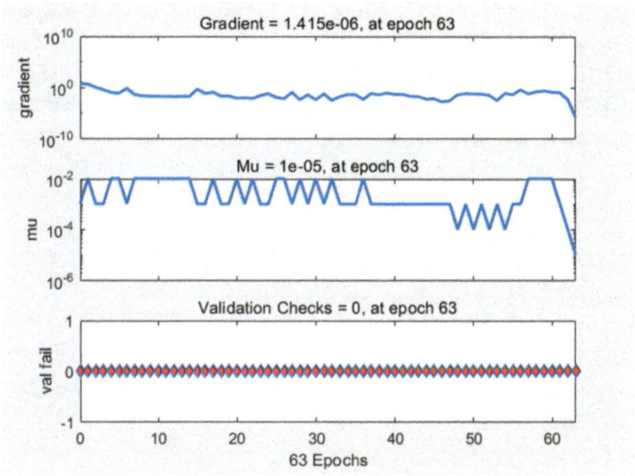


**Fig. 5.** MATLAB operations-2

ore transportation price, the main factors affecting iron ore price under the market economy system are found out, and these influencing factors are taken as input variables. A mathematical model of iron ore transportation price as the direct output object is established. So that enterprises can price according to the actual transportation conditions, thus saving costs and maximizing the interests of enterprises.

**Acknowledgment.** This project is supported by the Project of Science and Technology Innovation Team for Outstanding Young and Middle-aged Students in Hubei Province (T201938).

## References

1. Dong, Y., Lingge, Z., Meifeng, L., et al.: Does shipping market affect international iron ore trade?-an equilibrium analysis. Transp. Res. Part E **1**, 144–149 (2020)
2. Du Sheng, W., Min, C.L., et al.: An intelligent decision-making strategy based on the forecast of abnormal operating mode for iron ore sintering process. J. Process Control **9**(6), 57–66 (2020)
3. Linquan, F.: Dilemma and path of China's striving for international pricing right of commodities. Mod. Comm. Ind. **9**, 57–58 (2014). (in Chinese)
4. Omisore, M.O., Ofoegbu, E.O., Fayemiwo, M.A., Olokun, F.R., Babalola, A.E.: Intelligent mobile application for route finding and transport cost analysis. Int. J. Inf. Technol. Comput. Sci. **9**(8), 73–80 (2016)
5. Li Feng, T., Yijian, G.X.: Who is the free rider in the drop-shipping supply chain? Int. J. Inf. Eng. Electr. Bus. **6**(17), 44–51 (2011)
6. Zhang, J., Zhang, H., Chen, C.: Study on the asymmetric transfer of commodity prices at home and abroad--take soybean, cotton, crude oil and iron ore as example. Price Theory Pract. **2020**(07), 81–84+161 (2020). (in Chinese)
7. Karabutov, N.: Structural identifiability of nonlinear dynamic systems under uncertainty. Int. J. Intell. Syst. Appl. **2**(8), 12–22 (2020)
8. Deqiang, Z.: A new hybrid grey neural network based on grey verhulst model and bp neural network for time series forecasting. Int. J. Inf. Technol. Comput. Sci. **9**(1), 114–120 (2013)
9. Liang, M., Feng, L., Liangliang, C., Ming, H.: Simulation of crop evaportranspiration based on BP neural network model and grey relational analysis. Int. J. Inf. Technol. Comput. Sci. **2**(29), 15–21 (2012)
10. Deqiang, Z.: Optimization modeling for GM(1,1) model based on BP neural network. Int. J. Inf. Technol. Comput. Sci. **2**(27), 24–30 (2012)

# On Trilateral Evolutionary Game and Simulation Analysis of College-Enterprise-Government Collaborative Digital Talent Cultivation in the Context of Digital Economy

Ping Liu(✉), Wei Ma, and Yi Zhang

Business School, Sichuan University, Chengdu 610065, China

**Abstract.** During rapid progress in China's digital economy, there exists a giant gap in digital talents that are deemed as the core competitiveness of innovation driven development and enterprise transformation and upgrading. In this case, college-enterprise-government collaborative digital talent cultivation is an important way to settle such a problem. In order to In order to analyzing a trilateral collaborative cultivation mechanism oriented by colleges, involving enterprises and supported by the government, a trilateral evolutionary game model was established here for collaborative digital talent cultivation oriented by colleges, involving enterprises and supported by the government. Furthermore, strategy selection by colleges, enterprises and the government was also analyzed in the course of collaborative digital talent cultivation. After that, simulation analysis was made to investigate influencing factors on strategy selection. As demonstrated by relevant results, participation willingness of colleges, enterprises and the government produces impacts mutually to diverse extents. Since enterprises are in pursuit of profits, they are more sensitive to earnings and the default penalty. As far as the government is concerned, it needs to boost trilateral cooperation from various perspectives of policies, financial affairs and regulation, etc.

**Keywords:** Digital talent · Collaborative cultivation · Evolutionary game · College-enterprise-government

## 1 Introduction

As proposed in Digital China Construction Progress Report (*2018*) released by the Cyber Safety and Information Office of China in May *2019*, digital economy is in rapid development currently in China; the scale of China's digital economy reached RMB *31.3* trillion in *2018*, taking *34.8%* in national GDP; this signifies that digital transformation of China's economy is going increasingly deep and its demand for professional talents with digital skills is constantly growing. According to relevant research, under the circumstance that digital talents turn into increasingly critical core competitiveness of innovation driven development and enterprise transformation and upgrading in China,


Z. Hu et al. (Eds.): ICAILE 2021, LNDECT 82, pp. 85–95, 2021.
https://doi.org/10.1007/978-3-030-80475-6_9

progress in China's digital economy is also confronted with a huge challenge from digital talent shortage [1].

As an emerging economic industry, the digital economy needs talents who not only possess digital thinking and capabilities, but can master practical operational modes of the industry [2]. Therefore, digital talent cultivation simply relying on traditional training modes in colleges is far from satisfying industrial demands. In line with the existing practice and literature, digital talent cultivation requires collaborative participation of colleges, enterprises and the government. In Japan, for example, a series of government papers are issued, encouraging scientific researchers to participate in innovation and entrepreneurship, or giving relevant support, so as to promote the cultivation of top-notch innovative personnel [3, 4]. Moreover, the EU is well experienced in digital talent cultivation. It cooperates with different interest subjects (e.g., competent administrative department for education, trade associations, social capital sectors and colleges) to facilitate professional digital talent cultivation through cooperation of colleges, enterprises and the government [5]. In China, by summarizing the status quo of innovative talent cultivation in China in a context of digital economy, Wu Huabin et al. Put forward three new talent cultivation patterns from the perspective of three subjects (i.e., colleges, enterprises and the government) in an era of digital economy [6, 7]. But, there are many problems with the cooperative education model in China. Some of these are external environment, and there are also some problems in the process of cooperation, which requires common solutions among government, industry, schools and businesses [8].

Based on the above literature, certain achievements have been made in collaborative talent cultivation in the context of digitalization. However, most relevant studies are still descriptive and based on relevant cases and practice. In several articles, roles played by colleges, enterprises and the government in collaborative cultivation are sorted out and summarized, so are corresponding driving factors. For instance, it is pointed out by some scholars that the government exerts mechanism design and supervision functions in trilateral collaboration; it protects and coordinates the interests of all parties, and also optimizes the environment of collaborative talent cultivation, by dynamically adjusting policies related to taxation, finance and the market, etc.; as for collaborative talent cultivation in colleges and enterprises, it still needs to depend on policy design and supervision of the government, and carries out dynamic adjustment as the case may be [9–11]. However, it is rarely reported how three subjects of collaborative talent cultivation constantly adjust their own positions and also game accordingly; in addition, the influence of the corresponding game process on collaborative talent cultivation is also seldom investigated.

In this study, a trilateral evolution model was constructed for colleges-enterprise-government collaborative digital talent cultivation based on both the evolutionary game theory and the bounded rationality. Moreover, numerical simulation analysis was also made to explore strategy selection factors that affect collaborative digital talent cultivation among colleges, enterprises and the government.

# 2 Construction of the Trilateral Evolutionary Game Mode for Colleges-Enterprise-Government Collaborative Digital Talent Cultivation

## 2.1 Fundamental Assumptions of the Model

According to the above analyses, collaborative cultivation among enterprises, colleges and the government is an optimal path to training digital talents in colleges. Nevertheless, enterprises, colleges and the government have their own functions and benefits in collaborative cultivation. All of them may pursue benefit maximization from their own interests during cultivation [12]. On this basis, the following assumptions are made in this research:

### 2.1.1 Set of Participation Subjects and Cooperation Strategies

Three parties participating in collaborative cultivation of digital talents are the College, the Enterprise and the Government. During gaming of collaborative cultivation, a college may choose to cooperate or not, in which case; an enterprise may also decide to participate in collaborative cultivation or not; and the government may select to provide support, or to provide no support.

### 2.1.2 Cost of Digital Talent Cultivation

Both college and enterprise are practical subjects of collaborative digital talent cultivation; and the total cost incurred by such cultivation is denoted by $A$. Through coordination and guidance as well as service guarantee, the government may achieve its purpose of directing digital industry progress. Moreover, the cost incurred in this process is written into $B$. Once the government selects to support collaborative cultivation of digital talents, it may provide taxation, financial and policy support, etc., enabling the total cost incurred in colleges and enterprises to decline; and the total reduced amount is denoted by $D$. Then, the total investment by colleges and enterprises turn into $A - D$. If the sharing coefficient of the total cost incurred by collaborative cultivation in colleges and enterprises is $\alpha$, the cost the college needs to pay should be $\alpha A$ or $\alpha(A - D)$.

### 2.1.3 Revenues of Digital Talent Cultivation

If the government selects to "Support" digital talent cultivation through cooperation of colleges and enterprises, revenues yielded by the digital talents can be denoted by $R_1$. Moreover, if a ratio of revenues earned in the case where the government selects to "Not Support" to those where the government selects to "Support" is represented by $\beta$, the revenues obtained by the government can be expressed in $\beta R_1$ in a strategy of "Not Support".

As for the original revenues before colleges and enterprises decide to participate in collaborative cultivation of digital talents, they are denoted as $R_2$ and $R_3$, respectively. When both the college and the enterprise is willing to train digital talents collaboratively, that is when the college selects the strategy of "Cooperate" and the enterprise selects the strategy of "Participate", a sum of extra revenues obtained by colleges and enterprises thanks to digital talents can be denoted as $R$, and the sharing coefficient of such extra revenues between the college and the enterprise as $b$; then, the extra revenue incurred by collaborative digital talent cultivation should be $bR$ for the college, and that should be $(1 - b)R$ for the enterprise. Furthermore, $S_1$ represents the revenue yielded in the college that independently trains digital talents since no enterprises are willing to participate in collaborative cultivation; and $S_2$ refers to the revenue yielded in the enterprise that independently trains digital talents since no colleges are willing to participate in collaborative cultivation.

### 2.1.4 Rewards and Punishments

Under the circumstance that the government supports collaborative cultivation of digital talents, it may provide a college that takes an active part in such cultivation with both policy and financial support; and such a reward is denoted as $A_2$. Once a college or an enterprise breaches the cooperation contract after its conclusion, free-riding revenues obtained can be described from the following two aspects. First, for those obtained by an enterprise independently training digital talents since no colleges are willing to participate in such collaborative cultivation, they are denoted as $E_1$; and second, for those obtained by a college that independently trains digital talents since no enterprises are willing to participate in such collaborative cultivation, they are denoted as $E_2$. To prevent such scenarios, any party failing to perform the corresponding contract needs to pay a penalty denoted by $K$.

**Table 1.** A payoff matrix for the game among the college, the enterprise and the government

| | | College | |
|---|---|---|---|
| | | Cooperate ($x$) | Not Cooperate ($1 - x$) |
| Government (Support) ($z$) | Enterprise (Participate) ($y$) | $R_2 + bR + A_2 - \partial(A - D)$<br>$R_3 + (1-b)R - (1-\alpha)(A - D)$<br>$R_1 - B - A_2$ | $R_2 + E_1 - K$<br>$R_3 + S_2 - (1-\alpha)(A - D) + K$<br>$R_1 - B$ |
| | Enterprise (Not Participate) ($1 - y$) | $R_2 + S_2 + A_2 + K - \alpha(A - D)$<br>$R_3 + E_2 - K$<br>$R_1 - B - A_2$ | $R_2$<br>$R_3$<br>$R_1 - B$ |
| Government (Not Support) ($1 - z$) | Enterprise (Participate) ($y$) | $R_2 + \mathrm{bR} - \alpha \mathrm{A}$<br>$\mathrm{R}_3 + (1-b)R - (1-\alpha)A$<br>$\beta R_1$ | $R_2 + E_1 - K$<br>$R_3 + S_2 - (1-\alpha)A + K$<br>$\beta R_1$ |
| | Enterprise (Not Participate) ($1 - y$) | $R_2 + S_1 + K - \alpha A$<br>$R_3 + E_2 - K$<br>$\beta R_1$ | $R_2$<br>$R_3$<br>$\beta R_1$ |

### 2.2 Game Matrix Construction

In the game model, the College, the Enterprise and the Government may select a strategy according to their own actual situations. As assumed, the probability for a college to select collaborative digital talent cultivation is $x$, then the probability produced by the college unwilling to cooperate with such cultivation should be $(1 - x)$; if the probability for an enterprise to participate in collaborative digital talent cultivation is $y$, then its probability of not participating in such cultivation can be denoted as $(1 - y)$; if the probability for the government to support collaborative digital talent cultivation is $z$, then its probability of not supporting such cultivation is $(1 - z)$. Based on the above assumptions, a payoff matrix, as shown in Table 1, is obtained for the college, the enterprise and the government of collaborative digital talent cultivation game.

## 3 Trilateral Evolutionary Game Analysis for Digital Talent Cultivation

### 3.1 Expected Revenue Function Construction

When the college, during the game, selects to "Cooperate", or selects the other strategy of "Not Cooperate", its expected revenues are denoted as $U_{c1}$ and $U_{c2}$, respectively. In this case, the average expected revenue is $\overline{U}_c$.

$$\begin{aligned} U_{c1} &= yz[R_2 + bR + A_2 - \alpha(A - D)] + z(1 - y)[R_2 + S_1 + A_2 + K - \\ &\alpha(A - D)] + y(1 - z)(R_2 + bR - \alpha A) + (1 - y)(1 - z)(R_2 + S_1 + K - \alpha A) \end{aligned} \tag{1}$$

$$U_{c2} = yz(R_2 + E_1 - K) + z(1 - y)R_2 + y(1 - z)(R_2 + E_1 - K) + (1 - y)(1 - z)R_3 \tag{2}$$

$$\overline{U}_c = xU_{c1} + (1 - x)U_{c2} \tag{3}$$

When the enterprise, during the game, selects to "Participate", or selects the other strategy of "Not Participate", its expected revenues are denoted as $U_{e1}$ and $U_{e2}$, respectively. In this case, the average expected revenue is $\overline{U}_e$.

$$\begin{aligned} U_{e1} &= xz[R_3 + (1 - b)R - (1 - \alpha)(A - D)] + z(1 - x)[R_3 + S_2 - (1 - \alpha)(A - D) + K] + \\ &x(1 - z)[R_3 + (1 - b)R - (1 - \alpha)A] + (1 - z)(1 - x)[R_3 + S_2 - (1 - \alpha)A + K] \end{aligned} \tag{4}$$

$$U_{e2} = xz(R_3 + E_2 - K) + z(1 - x)R_3 + x(1 - z)(R_3 + E_2 - K) + (1 - x)(1 - z)R_3 \tag{5}$$

$$\overline{U}_e = yU_{e1} + (1 - y)U_{e2} \tag{6}$$

When the government, during the game, selects to "Support", or selects the other strategy of "Not Support", its expected revenues are denoted as $U_{g1}$ and $U_{g2}$, respectively. In this case, the average expected revenue is $\overline{U}_g$.

$$U_{g1} = xy(R_1 - B - A_2) + y(1-x)(R_1 - B) + x(1-y)(R_1 - B - A_2) + (1-x)(1-y)(R_1 - B) \quad (7)$$

$$U_{g2} = xy\beta R_1 + y(1-x)\beta R_1 + x(1-y)\beta R_1 + (1-x)(1-y)\beta R_1 \quad (8)$$

$$\overline{U}g = \mathrm{x}U_{g1} + (1-x)U_{g2} \quad (9)$$

## 3.2 Evolutionary Stable Strategy Based on the Replicator Dynamics Equation

Through the above analysis, the replicator dynamics equation for the College, the Enterprise and the Government can be written as follows:

$$F(x) = \frac{dx}{dt} = x(U_{c1} - \overline{U}_c) = x(1-x)[y(bR - E_1 - S_1) + z(A_2 + \alpha D) + S_1 + K - \alpha A] \quad (10)$$

$$F(y) = \frac{dy}{dt} = y(U_{e1} - \overline{U}_e) = y(1-y)\{x[(1-b)R - E_2 - S_2)] + z(1-\alpha)D + S_2 + K - (1-\alpha)A\} \quad (11)$$

$$F(z) = \frac{dz}{dt} = z(U_{g1} - \overline{U}_g) = z(1-z)[(1-\beta)R_1 - B - xA_2] \quad (12)$$

Through the method proposed by Friedman, equilibrium points of the above trilateral game system are obtained by local stability analysis based on its Jacobian matrix. According to the above equation expressing the replicator dynamics system, such a Jacobian matrix is acquired below [13, 14].

$$\begin{bmatrix} \begin{matrix}(1-2x)[y(bR - E_1 - S_1)+\\ z(A2 + \alpha D) + S_1 + K - \alpha A]\end{matrix} & x(1-x)(bR - E_1 - S_1) & x(1-x)(A_2 + \alpha D) \\ y(1-y)[(1-b)R - E_2 - S_2] & \begin{matrix}(1-2y)\{x[(1-b)R - E_2 - S_2]+\\ Z(1-\alpha)D + S_2 + K + (1-\alpha)A\}\end{matrix} & y(1-y)(1-\alpha)D \\ -z(1-z)A_2 & 0 & (1-2z)[(1-\beta)R_1 - ]B - xA_1 \end{bmatrix} \quad (13)$$

In the replicator dynamics system, if F(x) = F(y) = F(z) = 0, then 8 local equilibrium points can be obtained and they are respectively denoted as $G_1$(0,0,0), $G_2$(0,0,1), $G_3$(0,1,0), $G_4$(0,1,1), $G_5$(1,0,0), $G_6$(1,0,1), $G_7$(1,1,0), $G_8$(1,1,1). By substituting the above local equilibrium points into the Jacobian matrix, feature values corresponding to it are gained, as shown in Table 2.

## 3.3 Equilibrium Point Analysis

To generalize the research and facilitate analysis on symbols corresponding to various feature values, it is assumed that revenues yielded by college-enterprise-government collaborative digital talent cultivation should be greater than those obtained when they fail to cooperate mutually, that is, $bR - \alpha A > E_1 - K$; $(1-b)R - (1-\alpha)A > E_2 - K$; $R_1 - B - A_2 > \beta R_1$. Subsequently, it is described from the following three scenarios for the convenience of analysis.

**Table 2.** Feature values of the Jacobian matrix

| Equilibrium point | $\lambda_1$ (College) | $\lambda_2$ (Enterprise) | $\lambda_3$ (Government) |
|---|---|---|---|
| $G_1(0,0,0)$ | $S_1+K-\alpha A$ | $S_2+K-(1-\alpha)A$ | $(1-\beta)R_1-B$ |
| $G_2(0,0,1)$ | $A_2+S_1+K-\alpha(A-D)$ | $S_2+K-(1-\alpha)(A-D)$ | $-[(1-\beta)R_1-B]$ |
| $G_3(0,1,0)$ | $bR+K-E_1-\alpha A$ | $-[S_2+K-(1-\alpha)A]$ | $(1-\beta)R_1-B$ |
| $G_4(0,1,1)$ | $bR+A_2+K-E_1-\alpha(A-D)$ | $-[S_2+K-(1-\alpha)(A-D)]$ | $-[(1-\beta)R_1-B]$ |
| $G_5(1,0,0)$ | $-(S_1+K-\alpha A)$ | $(1-b)R+K-E_2-(1-\alpha)A$ | $(1-\beta)R_1-B-A_2$ |
| $G_6(1,0,1)$ | $-[A_2+S_1+K-\alpha(A-D)]$ | $(1-b)R+K-E_2-(1-\alpha)(A-D)$ | $-[(1-\beta)R_1-B-A_2]]$ |
| $G_7(1,1,0)$ | $-(bR+K-E_1-\alpha A)$ | $-[(1-b)R+K-E_2-(1-\alpha)A]$ | $(1-\beta)R_1-B-A_2$ |
| $G_8(1,1,1)$ | $-[bR+A_2+K-E_1-\alpha(A-D)]$ | $-[(1-b)R+K-E_2-(1-\alpha)(A-D)]$ | $-[(1-\beta)R_1-B-A_2]$ |

**Scenario 1:**
$A_2+S_1+K<\alpha(A-D)$; and $S_2+K<(1-\alpha)(A-D)$. As can be seen, three feature values at points $G_2(0,0,1)$ and $G_8(1,1,1)$ are all negative in this system. For this reason, there exist two asymptotically stable points, $G_2(0,0,1)$ and $G_8(1,1,1)$, in this scenario; and the corresponding revolutionary strategies turn out to be (Not Cooperate; Not Participate; Support) and (Cooperate; Participate; Support).

**Scenario 2:**
$S_1+K>\alpha A$; and $S_2+K>(1-\alpha)A$. According to this table, three feature values at the point $G_8(1,1,1)$ are all negative in this system. Therefore, there is only one asymptotically stable point $G_8(1,1,1)$ in this scenario; and the corresponding revolutionary strategies should be (Cooperate; Participate; Support).

**Scenario 3:**
$A_2+S_1+K>\alpha(A-D)$, $S_1+K<\alpha A$; and $S_2+K>(1-\alpha)(A-D)$, $S_2+K<(1-\alpha)A$. Based on this table, three feature values at the point $G_8(1,1,1)$ are all negative in this system. Therefore, there is only one asymptotically stable point $G_8(1,1,1)$ in this scenario; and the corresponding revolutionary strategies should be (Cooperate; Participate; Support) (Table 3).

## 4 Numerical Simulation of College-Enterprise-Government Collaborative Cultivation of Innovative Talents

To verify the above analysis and analyze the actual influences of correlation variables on evolutionary game, numerical simulation is carried out here by virtue of Matlab. Firstly, relevant parameters were set as follows based on assumptions in Sect. 3.3 above, that

**Table 3.** Local stability results

| Equilibrium | Scenario 1 | | Scenario 2 | | Scenario 3 | |
|---|---|---|---|---|---|---|
| | $\lambda_1, \lambda_2, \lambda_3$ | Condition | $\lambda_1, \lambda_2, \lambda_3$ | Condition | $\lambda_1, \lambda_2, \lambda_3$ | Condition |
| $G_1(0, 0, 0)$ | $-, -, +$ | Unstable | $+, +, +$ | Unstable | $-, -, +$ | Unstable |
| $G_2(0, 0, 1)$ | $-, -, -$ | ESS | $+, +, -$ | Unstable | $+, +, -$ | Unstable |
| $G_3(0, 1, 0)$ | $+, +, +$ | Unstable | $+, -, +$ | Unstable | $+, +, +$ | Unstable |
| $G_4(0, 1, 1)$ | $+, +, -$ | Unstable | $+, -, -$ | Unstable | $+, -, -$ | Unstable |
| $G_5(1, 0, 0)$ | $+, +, +$ | Unstable | $-, +, +$ | Unstable | $+, +, +$ | Unstable |
| $G_6(1, 0, 1)$ | $+, +, -$ | Unstable | $-, +, -$ | Unstable | $-, +, -$ | Unstable |
| $G_7(1, 1, 0)$ | $-, -, +$ | Unstable | $-, -, +$ | Unstable | $-, -, +$ | Unstable |
| $G_8(1, 1, 1)$ | $-, -, -$ | ESS | $-, -, -$ | ESS | $-, -, -$ | ESS |

is, $R = 15, D = 2, K = 6, A = 10, S_1 = 6, S_2 = 5, E_1 = 4, E_2 = 3, R_1 = 15, B = 4, A_2 = 3$. In addition, the initial willingness of colleges, enterprises and the government was designed to be $x = y = z = 0.5$, which was used for relevant numerical simulation in line with the above assumptions.

### 4.1 Influence of Variations in Collaborative Willingness of Enterprises, Colleges and the Government on the Trilateral Evolutionary Game Tendency

According to Fig. 1, collaborative cultivation willingness of colleges, enterprises and the government is increased or reduced provided that other parameters remain the same; respectively, their willingness is set as follows: $x = y = z = 0.1$, $x = y = z = 0.5$ and $x = y = z = 0.9$. As can be observed from this figure, strategy selection of colleges, enterprises and the government all converges at the equilibrium point $G_8(1, 1, 1)$ under the circumstance that collaborative cultivation willingness of them is proved to be different. Such an outcome is consistent with the analytical result of equilibrium points above. On the premise of assuming the revenue yielded by college-enterprise-government collaborative digital talent cultivation to be greater than that obtained when no cooperation is reached, colleges, enterprises and the government are all willing to support collaborative cultivation of digital talents. However, in case of $x = y = z = 0.1$, the initial participation willingness of enterprises may decline; however, as participation willingness of colleges and the government keeps rising, the participation willingness of enterprises may also gradually begin to increase and eventually converges at the equilibrium point $G_8(1, 1, 1)$.

### 4.2 Influence of Free-Riding Revenues on the Trilateral Evolutionary Game Tendency

Free-riding revenue refers to spillover revenue obtained by a single party that cultivates talents independently without the help of other parties. As shown in Fig. 2, under the circumstance that other parameters remain unchanged, increases in $E_1$, $E_2$ lead to a gradual

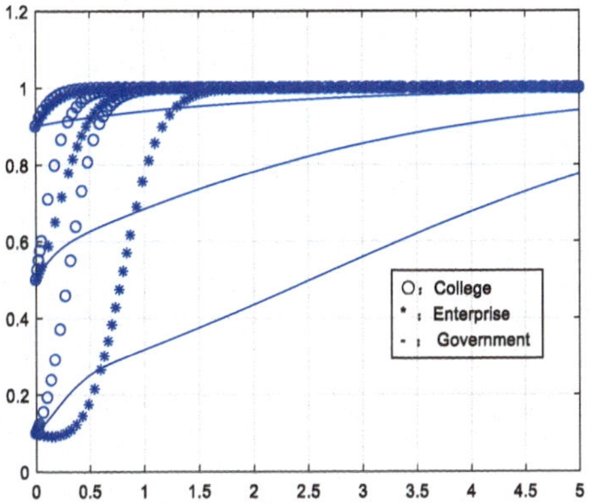


**Fig. 1.** Participation willingness variations of colleges, enterprises and the government

decline of their participation willingness. This may exert a more obvious influence on enterprises. Especially in case of $E_1 = E_2 = 9$, enterprises' participation willingness becomes weak before $t = 0.5$, and does not rise gently until the willingness of colleges begins to gradually go up.

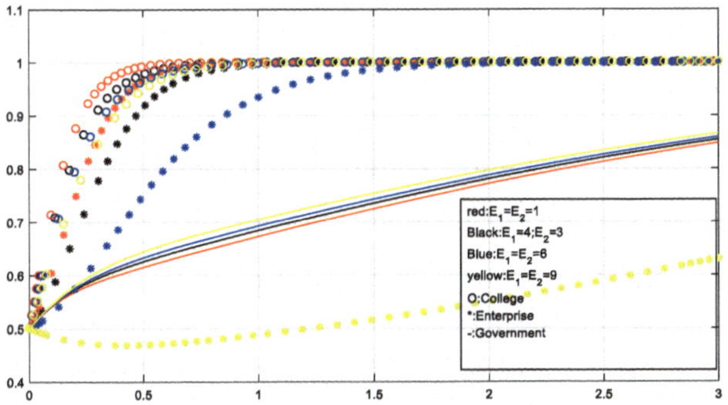


**Fig. 2.** Influence of free-riding revenues on the trilateral evolutionary game tendency

### 4.3 Influence of Penalty Variations on the Trilateral Evolutionary Game Tendency

Figure 3 presents how variations in default penalty K of colleges or enterprises affect the trilateral evolutionary game tendency provided that other parameters remain the same. As can be observed, penalty variations have a great influence on colleges and enterprises, but a minor influence on the government. If $K = 2$, the initial participation willingness of enterprises is inclined to drop. Considering that penalty variation insignificantly affects the government, the enterprises' willingness begins to become stronger once the willingness of colleges directed by the government has been raised to a certain level. With increases in the penalty, convergence rates of colleges and enterprises gradually climb; and, the influence on enterprises becomes apparently greater than that on colleges.

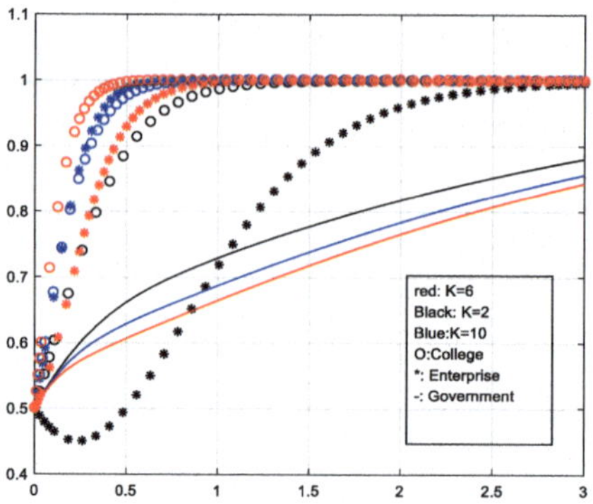


**Fig. 3.** Influence of penalty variations on the trilateral evolutionary game tendency

## 5 Conclusions

Cultivating digital talents to expand and improve the digital economy is an important way to boost economic growth, transformation and upgrading at present in China. Here, the evolutionary game theory is selected to analyze behaviors of colleges, enterprises and the government in facilitating digital talent cultivation in China. Moreover, Matlab based numerical simulation is conducted to analyze strategy behavior and influence factors of college-enterprise-government collaborative digital talent cultivation. Finally, the following conclusions can be drawn.

First, participation willingness of colleges, enterprises and the government may affect one another during the collaborative cultivation of digital talents; however, specific influence degrees are different. The reason why participation willingness of colleges has a great influence on that of enterprises is that the college is the subject of digital talent cultivation and more inclined to be affected by government intervention and restricted by government funds and policies. In terms of enterprises, variations in their participation willingness apply minor influences on another two parties, because they have different interest appeals.

Then, participation willingness of colleges and enterprises may decline along with increases in free-riding revenues, but go up as the default penalty is raised. In addition, the influence on enterprises is also more obvious than that on colleges. In both cases, the government is subjected to minor influences. Therefore, both the college and the enterprise need to clarify their own rights and responsibilities on one hand; and on the other hand, penalty on their breaching behavior should be properly elevated, ensuring that neither party violates the cooperation contract due to their pursuit of extra free-riding revenues. Moreover, the government should reinforce its own regulatory functions in this.

At last, college-enterprise-government collaborative digital talent cultivation needs to always abide by a principle of being college oriented, encouraging enterprise participation and acquiring government support. Thanks to incentives by relevant measures taken by the government, participation willingness of colleges, as the subject of collaborative cultivation, can be enhanced, and enterprises are also encouraged to participate in such cultivation. Eventually, the purpose of training digital talents and accelerating China's digital economy transformation can be achieved.

## References


1. Yubo, C., Yefeng, M.: Digital talents-the core driving force for digital transformation of China's economy. Tsinghua Bus. Rev. **1**, 30–40 (2018)
2. Haikun, D., Jianmin, L.: On digital talent cultivation from the perspective of EU experience. China High. Educ. **22**, 61–62 (2018)
3. Feng, G., Hengliang, Y., Gao Dongmei, Y., Yuelong.: A comparative study on entrepreneurship p education between Chinese and Japanese college students. J. Innov. Enterpr. Educ. **2**(06), 89–93 (2011)
4. Wang, L., Zhang, Y., Wu, Y.: Exploration on Japanese experience in government-industry-university-institute collaboration. World Sci-tech R & D **2006**(04), 91–95+90 (2006)
5. Beiqun, L., Yuzhu, H.: Collaborative talent cultivation mode reform in colleges with industrial characteristics: transformation and paths. Jiangsu High. Educ. **4**, 22–25 (2018)
6. Wu Huabin, X., Qingrui, C.Z.: Research on cultivation mode and countermeasures of innovative talents under background of digital economy. Sci. Technol. Manage. Res. **39**(8), 116–121 (2019)
7. Yabin, X.: A new education mode of engineering specialty graduate students—Co-founding by college and enterprises, uniting of production, learning and researching. Int. J. Educ. Manage. Eng. **2**(12), 590–593 (2012)
8. Lina, Y.: Research on the "Cooperative Education" model cultivating in higher vocational education. Int. J. Educ. Manage. Eng. **2**(1), 521–524 (2012)
9. Drongelen, K.V., Bilderbeek, I.C.: R&D performance measurement: more than choosing a set of metrics. R&D Manage. **29**(1), 25 (1999)
10. Guixiang, C.: On the formation of innovative and entrepreneurial talents cultivation mechanism with linkages among university, government and enterprise. Univ. Educ. Sci. **1**(1), 42–47 (2015)
11. Liqin, Z., Dandan, W., Lehui, H.: The research on the program of educational technology ability building for vocational education teachers. Int. J. Educ. Manage. Eng. **2**(11), 46–51 (2012)
12. Wang, X.: Study on game behaviors and evolutionary model of HRD's multi-subjects. Dongbei University of Finance and Economic (Liaoning) (2013)
13. Friedman, D.: Evolutionary game in economics. Econometrica **59**(3), 637–666 (1991)
14. Beltadze, G.N., Giorgobiani, J.A.: The stability of equilibrium situation in lexicographic strategic games. Int. J. Mod. Educ. Comput. Sci. **8**(12), 38–45 (2016)

# Design and Development of Mobile Learning Platform Based on WeChat Mini Programs

Kai Zeng[1], Yuyan Zhang[1](✉), Yong Liu[1,2], and Fuan Wen[1,2]

[1] Network Education School, Beijing University of Posts and Telecommunications, Beijing 100088, China

[2] Beijing Key Laboratory of Network System and Network Culture, Beijing University of Posts and Telecommunications, Beijing 100876, China

**Abstract.** Traditional online learning relies heavily on computer equipment, which is inconvenient for moving and not suitable for fragmented and informal learning. To solve this problem, a mobile learning platform based on WeChat Mini Programs is proposed. Firstly, according to features of mobile learning, introduce the design background of the mobile learning platform. Then, the system architecture is designed after analyzing the functions and interaction requirements of the mobile learning platform, and there are five functional modules and an overall design process proposed. Finally, based on the functional modules and design process, use WeChat Mini Programs as the front end and the Cloud Base of WeChat as the back end to develop the mobile learning platform. Practice has proved that the mobile learning platform based on WeChat Mini Programs can realize mobile learning and exercising without installation on any devices, and solve the constraints of learners' time and space.

**Keywords:** Mobile learning · Platform design · WeChat Mini Program

## 1 Introduction

With the rapid development of Internet technology and the popularization of intelligent terminals such as smart phones and tablet computers, the learner's learning philosophy has changed and updated. Therefore, online education has entered a stage of rapid development [1]. The 44th "Statistical Report on China's Internet Development Status", published by the China Internet Network Information Center, shows that by June 2019, the number of online education users in China has reached 232 million. Among them, the number of mobile online education users has an increase of 5.3 million from the end of 2018 and reached 199 million [2]. Mobile learning will become the mainstream of online education with its personalized learning scenarios and the functional advantages of mobile devices [3].

WeChat has become an important tool for mobile learning with the advancement of the mobile trend of online education. The 2019 WeChat data report shows that in 2019, WeChat had more than 1.1 billion monthly active users [4]. WeChat has strong advantages in terms of traffic acquisition, community operation, user scale and stickiness,


Z. Hu et al. (Eds.): ICAILE 2021, LNDECT 82, pp. 96–106, 2021.
https://doi.org/10.1007/978-3-030-80475-6_10

and more and more education products are beginning to use the WeChat ecosystem to explore new service models [5]. Mobile learning based on WeChat has also received more and more researches.

However, traditional online mobile learning platforms are mostly large and bloated, with cumbersome use processes, and are not suitable for fragmented and informal learning. With the emergence of WeChat Mini Programs in 2017, new solutions for mobile learning have been provided [6–9]. Different from using a website, an app or a WeChat Official Account as a mobile learning platform in the past, WeChat Mini Programs have many advantages: Firstly, relying on the WeChat platform, WeChat Mini Programs also have a considerable number of users. The stability is guaranteed without restricting user's equipment. Secondly, WeChat Mini Programs don't need to be installed. You can find them by directly searching for the applet name or just scanning the QR code. In addition, for developers, the front-end visualization of the WeChat Mini Program allows developers to design more conveniently, and the Cloud Base environment of WeChat provides convenience for back-end database deployment. These advantages make the WeChat Mini Program an ideal mobile learning platform suitable for mobile learning [10–12].

In this paper, based on existing learning resources, we have developed a mobile learning platform by using an interactive design method to assist in-class teaching and self-test after class. We use the WeChat Mini Program as the front-end interactive interface of the mobile learning platform, and use the WeChat Cloud Base database as the server-side database to interact with the WeChat Mini Program for learning resources and user information. In this way, we abandon traditional front-end application development, multi-platform compatibility testing, and back-end server construction and maintenance. Therefore, it is a positive attempt to develop educational functions under the WeChat Official Accounts Platform.

# 2 Platform Design

## 2.1 Module Design

The mobile learning platform designs functional modules from the perspective of learners, aiming to develop a WeChat Mini Program that is convenient for the experience of users. The main module design of the platform is shown in Fig. 1.

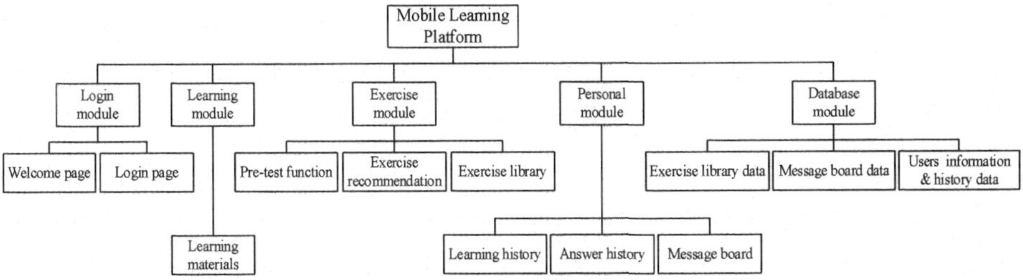


**Fig. 1.** The main modules structure

#### 2.1.1 Login Module

The login module includes a welcome page and a login page. The first time you enter the Mini Program, you will directly enter the welcome page, and follow the instructions to perform the authorized login operation of the WeChat account. Users can learn after logging in, and use the history record function and exercise recommendation function.

#### 2.1.2 Learning Module

The learning module includes some learning materials related to this learning theme, and users can view the learning materials anytime and anywhere. It is convenient for users to assist in-class teaching and to supplement the knowledge after class.

#### 2.1.3 Exercise Module

The exercise module is composed of three parts: exercise recommendation, pre-test function and exercise library. Based on the Zone of proximal development theory (ZPD), the exercise recommendation part is to recommend exercises that are suitable for the learner, so that the learner will not be confused because the topic is too difficult or tired because the topic is too easy. Therefore, in order to obtain the learner's initial ability, the pre-test function performs a pre-test to determine the learner's current level when he uses the exercise recommendation function for the first time, so as to recommend appropriate exercises. The function of the exercise library is that after the learner has completed all the exercises recommended for him, he can check all the exercises at any time in order to check for omissions.

#### 2.1.4 Personal Module

The personal module includes various personalized information of learners. The learning history function can save the learning progress for the learner so that the learner can continue learning next time. The answer history function saves all answers for the learner and marks out the wrong answers, in order that the learners can check the missing questions at any time. The message board function provides a platform for learners to communicate with each other, so that learners can ask questions about difficulties encountered in learning.

#### 2.1.5 Database Module

The database module is the back-end module of the entire mobile learning platform. It stores all the data of the mobile learning platform, including all exercise data, message board data, and all users' basic information and history of learning and answering.

### 2.2 User Interface Design

In order to implement the functions of the modules mentioned above, meet the needs of users and improve user experience, we formulate the following interactive interface design principles:

### 2.2.1 Concise and Focused

The design interface of the WeChat Mini Program should be simple and clear. Every page must have a clear key focus so that the user can quickly understand the content of the page whenever a new page is opened. Meanwhile, minimize the disruption to users from unnecessary elements that are not relevant to user decisions and operations on the page.

### 2.2.2 User-Friendly and Provide Clear Processes

In order to protecting users from getting lost when browsing through pages, navigation must inform the users of matters including the current location, possible destinations, and how to get back. When the user performs an operation, avoid interrupting users with unnecessary elements that are not related to the user's goals to create a smooth user journey. At the same time, clear feedback should be provided with regard to the results of an operation, and when an exception occurs, users must be given the necessary prompts and offered a solution.

### 2.2.3 Consistency, Stability and Scalability

WeChat Mini Programs should always pay attention to the consistency and continuity between different interfaces, and use consistent controls and interactive modes as much as possible on different pages. The designed code should be standardized and easy to understand, and the interface should be coordinated with cloud development, so that learning topics can be easily changed.

Based on the above principles, the overall design process of WeChat Mini Programs is shown in Fig. 2.

# 3 Platform Implementation

## 3.1 Overall Architecture

The development principle of WeChat Mini Programs lies between the web and native apps. The framework of Mini Program is based on the MINA framework, as shown in Fig. 3.

The development process of our WeChat Mini Program includes two parts: front-end interface and back-end database.

In terms of the front-end interface, it is mainly developed in the WeChat developer tool called Weixin DevTools. It helps to design different pages and jump logic to realize the user interaction process. The MINA framework used by the WeChat Mini Program is similar to the traditional front-end development framework. Each page consists of a ".wxss" style sheet file, a ".wxml" page structure file, a ".js" script file, and a ".json" configuration file. These four files are used to render the page layout and define the logic code. Among them, the first two belong to the rendering layer, responsible for rendering the interface of the Mini Program, and the latter two belong to the logic layer, responsible for running JS scripts. When the Mini Program is running, the Mini Program will convert WXML into the corresponding JS object in the rendering layer. When the data changes in the logic layer, the data is transferred from the logic layer to the rendering layer through the "setData" method provided by the system, and then comparing the difference before and after, and apply the difference to the original Dom tree to render the correct UI interface [13].

In terms of back-end database development, our mobile learning platform builds cloud functions and a cloud database on the basis of WeChat Cloud Base environment, and uses Tencent Cloud Object Storage (COS) to store unstructured data, such as pictures, documents, etc. WeChat Cloud Base provides developers with complete native cloud support, does not need to build a server, weakens the concept of back-end and operation and maintenance, and uses the API provided by the platform to develop core functions. Among them, cloud functions are codes that run in the cloud, with natural authentication using WeChat private protocol, and their usage is similar to traditional server deployment.

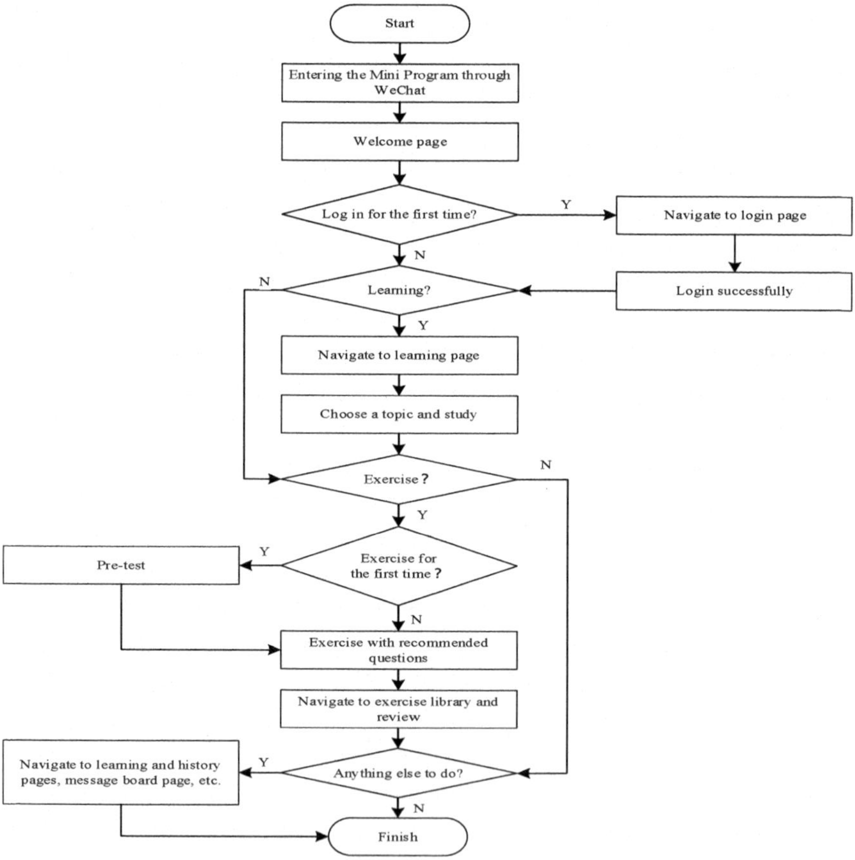


**Fig. 2.** Interface design flowchart

The cloud database is a kind of JSON database, which can be both directly operated by the front-end script of the Mini Program, and read and written in the cloud function.

Tencent COS does not have directory structure or data format restrictions, and can accommodate massive data and supports HTTP/HTTPS access. It has a wide range of APIs, and can be used in the front-end of the Mini Program and the background web pages it provided directly [14–16].

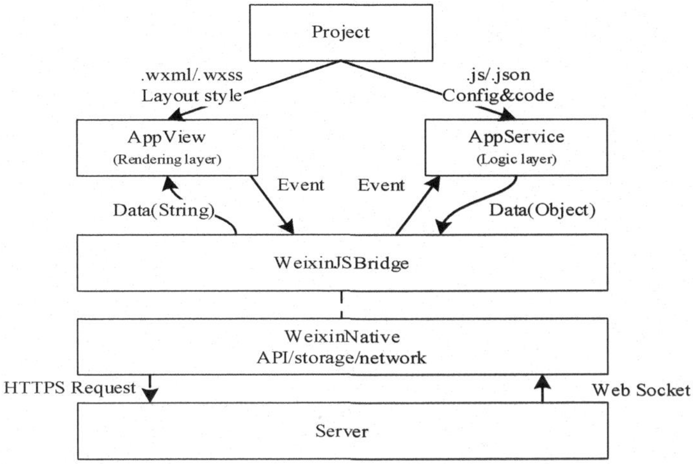


**Fig. 3.** WeChat Mini Program structure

## 3.2 Module Implementation

### 3.2.1 Login Module

The login module contains the welcome page and the login page, as shown in Fig. 4(a, b). After entering the mini program, the user can see the welcome page, which includes the recommended content which is in the carousel. The user can directly tap the relevant content to learn, or tap the "Start Learning" button to enter the main page of the system. If it is the first time for the user to login, he needs to follow the instructions to enter the login page, follow the instructions to perform authorization and other operations. After successfully logging in, you can return to the main page for the next step. The main pages of the system are shown in Fig. 4(c), including the “Course” page, “Exercise” page and “My” page, which correspond to the learning module, exercise module and personal module.

### 3.2.2 Learning Module

The learning module includes some learning materials related to the course and after-school expansion materials, which are divided according to chapters. The user can tap to enter the relevant learning page for reading and learning. After the user has studied for a certain period of time, the current learning state will be recorded in the database for subsequent review.

(a).Welcome page (b). Login page (c). Main page

**Fig. 4.** Login module

### 3.2.3 Exercise Module

The exercise module includes the pre-test page, recommendation page and exercise library page, as shown in Fig. 5. On the pre-test page, the user answers the questions provided by our platform on the page to help the platform determine the user's current ability level. After the user finished the pretest, the exercise recommendation page can be used to answer the recommended exercises.

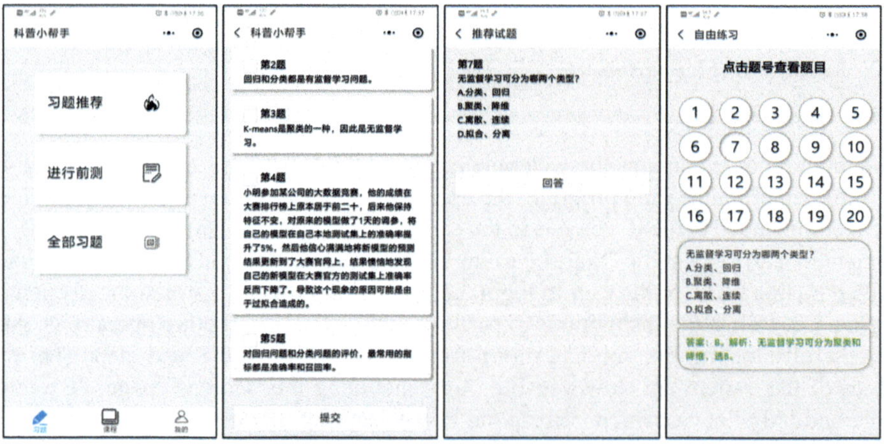


(a). Main page of exercise module (b). Pre-test page (c). Recommendation page (d). Exercise library page

**Fig. 5.** Exercise module

In the exercise recommendation page, the platform will recommend exercises for the learners in turn according to their current ability level. After the user answers, the correct answer keys are displayed for the user, and the user's answer history is recorded through cloud functions and cloud databases. If all the recommended exercises are completed, the exercise library page will provide users with the whole questions and answer keys of all exercises. In this page, users can tap the corresponding question number to view and study the specified exercises.

### 3.2.4 Personal Module

The personal module includes learning history, answer history and message board pages, as shown in Fig. 6. The learning history page records the latest 5 learning contents read by the user. The user can be navigated to the corresponding page by tapping the history. The answer history page records all the questions answered by the user on the exercise recommendation page. Among them, the wrong answers have been marked in red. Long tap the answer history will jump to the corresponding question analysis on the exercise library page, which is convenient for users to review. On the message board page, users can swipe up to view more previous messages, or click the I want to leave a message button to leave a message. At the same time, the page includes a pull-down refresh function, which is convenient for users to view messages in real time.

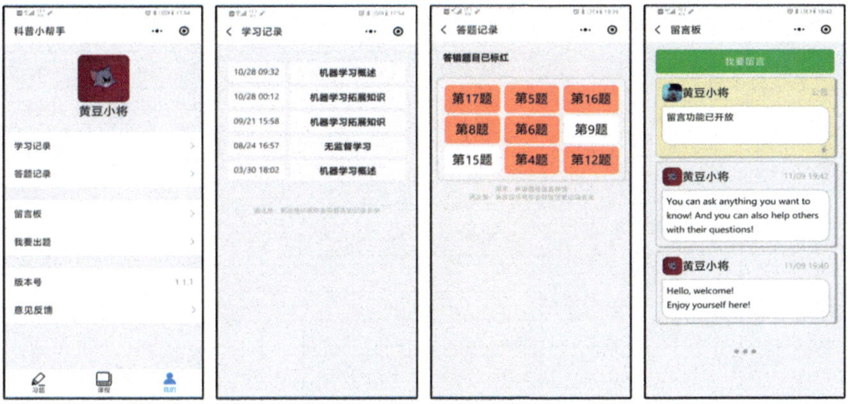


(a). Main page of 'my' module (b). Learning history page (c). Answer history page (d). Message board page

**Fig. 6.** Personal module

### 3.2.5 Database Module

The database module includes all the data of the mobile learning platform. Since we use a file-based JSON database as the database of this platform, the comparison between it and a relational database is shown in Table 1.

In the management part of the exercise library, a collection named "test" is designed to store all exercise data. In the interactive communication management part, the "comments" collection is designed to store the message board information of the entire platform. In the user information management part, two collections named "userlearn" and "userdata" are designed to store the user's learning and answer history, and the "pretest" set is designed to manage the user's pre-test level information.

**Table 1.** The comparison between a relational database and a JSON database

| Relational | File-based |
|---|---|
| Database | Database |
| Table | Collection |
| Row | Record/Doc |
| Column | Field |

In addition, for the entire platform, we designed the "sources" collection to manage the resource links used by the entire platform, in order to link to Tencent COS for unstructured data scheduling. In terms of database management, a variety of cloud functions for creating, retrieving, updating and deleting are designed as the data manipulation language, which is convenient for front-end scripts to perform convenient database operations. The relationship between the collections of the database is shown in Fig. 7.

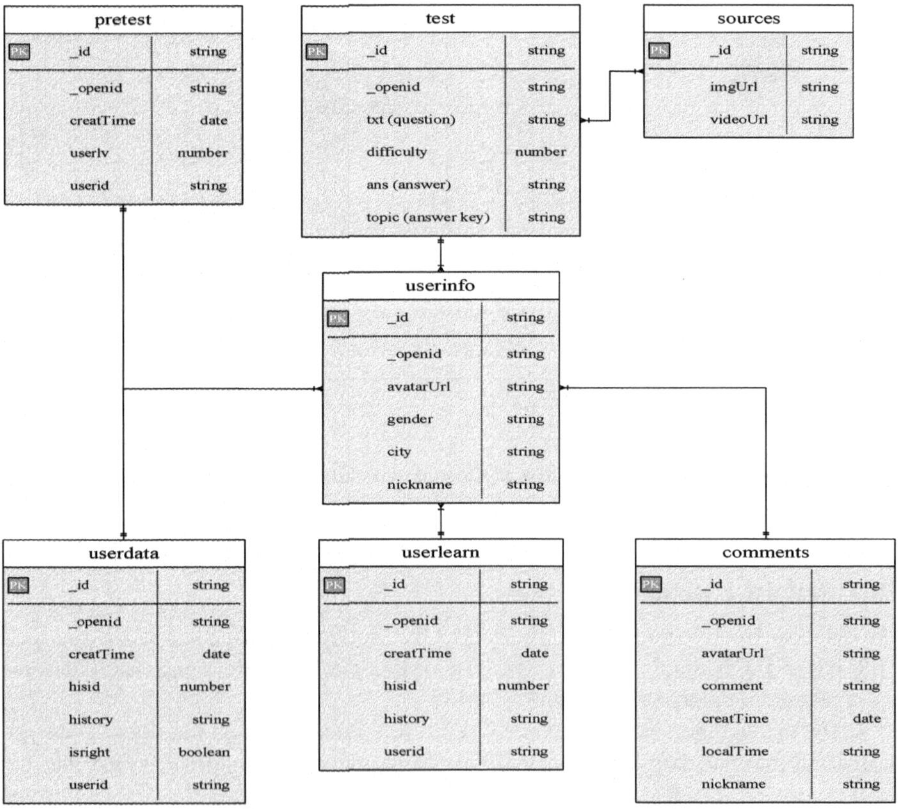


**Fig. 7.** Database collection relationship diagram

## 4 Conclusion

In this paper, we designed a mobile learning platform based on the WeChat Mini Program, and developed a WeChat Mini Program with exercise recommendation and learning interactive functions. With the popularization of portable mobile terminal designs such as smart phones, mobile learning based on mobile learning platforms will surely become an important learning method in the future. Since it is able to be used in any operating system and devices, the WeChat Mini Program can easily realize mobile learning and exercising without installation which overcomes the time and space constraints faced by learners in traditional online learning, and enables learners to change their concept from passive learning to active learning. It not only provides practical guidance for the development of mobile learning, but also provides a good reference for the development of the educational function of the WeChat Official Accounts Platform.

## References


1. Jie, L., Hao, S., Dongxu, G., et al.: Design and implementation of online education platform based on Spring MVC and MyBatis framework. J. Shenyang Normal Univ. (Natural Science Edition) **37**(03), 268–273 (2019)
2. China Internet Network Information Center: The 44th China Statistical Report on Internet Development[EB/OL] (2019). http: //www.cac.gov.cn/2019-08/30/c_1124938750.htm. (in Chinese)
3. Adejo, O.W., Ewuzie, I., Usoro, A., Connolly, T.: E-learning to m-learning: framework for data protection and security in cloud infrastructure. Int. J. Inf. Technol. Comput. Sci. **10**(4), 1–9 (2018). https://doi.org/10.5815/ijitcs.2018.04.01
4. Weixin, P.: WeChat Data Report[EB/OL] (2019). https: //mp.weixin.qq.com/s/vmho iRzpBs7-JK_x2a7gZw. 2020–01–09. (in Chinese)
5. China Internet Network Information Center: The 43th China Statistical Report on Internet Development[EB/OL] (2019). http: //www.cac.gov.cn/2019-02/28/c_1124175677.htm. (in Chinese)
6. Li, H., Liu, Z., Xu, S., et al.: How to find it better?: cross-learning for WeChat mini programs. In: The 28th ACM International Conference, pp. 2753–2761 (2019)
7. Wan, F., Bo, X., Chang, N.: Discussion on the application of WeChat mini program in the lifelong education of basic computer knowledge in the age of mobile learning. In: Sun, G., Gan, J., Liu, S., Lang, F., Zeguang, L. (eds.) e-Learning, e-Education, and Online Training: 5th EAI International Conference, eLEOT 2019, Kunming, China, August 18–19, 2019, Proceedings, pp. 48–55. Springer, Cham (2019). https://doi.org/10.1007/978-3-030-35095-6_5
8. Ni, Z.: Research on design and development of mobile learning platform based on WeChat mini program. Comput. Knowl. Technol. **15**(04), 94–96 (2019). (in Chinese)
9. Wu, X., Zhang, Y., Zhang, J., et al.: Design and implementation of college online learning system based on WeChat mini program. In: 2018 9th International Conference on Information Technology in Medicine and Education, pp. 447–451 (2018)
10. Shuo, T., Xiaohua, Y., Kemin, J., et al.: WeChat: An applicable and flexible social app software for mobile teaching: WeChat official account of biochemistry teaching. Biochem. Mol. Biol. Educ. **46**(5), 555–560 (2018)
11. Gu, X., Yang, L., Cao, S.: Design and implementation of knowledge sharing system based on WeChat small program. In: Electronic and Automation Control Conference, pp. 2242–2246 (2018)

12. Nwokoye, C.N., Umeh, I., Mbeledogu, N.: GeoNaija: enhancing the teaching and learning of geography through mobile applications. Int. J. Educ. Manage. Eng. **9**(6), 11–24 (2019)
13. Cheng, A., Ren, G., Hong, T., et al.: An exploratory analysis of travel-related WeChat mini program usage: affordance theory perspective. Information and Communication Technologies in Tourism 2019, pp. 333–343 (2019)
14. Lixiang, Z., Fawang, G., Jianguo, B.: Score query system based on WeChat mini programs and cloud development. Inner Mongolia Sci. Technol. Econ. (13), 59–60 (2019). (in Chinese)
15. Lixuan, L.: Design and implementation of digital asset protection system based on multi-cloud terminals. Hebei University of Science and Technology, Shijiazhuang (2019). (in Chinese)
16. Yan, W., Fang, W., Yanying, Z., et al.: Push management platform based on wechat small program and cloud development. Int. J. Educ. Manage. Eng. **10**(1), 20–26 (2020)

# A New Fog Enabled Sensor Cloud Platform for Smart Logistics Park

XueJiang Wei[1,2(✉)] and Meng Wang[2]

[1] State Key Laboratory of Software Engineering, Wuhan University, Wuhan 430072, China
[2] School of Logistics, Wuhan Technology and Business University, Wuhan 430065, China

**Abstract.** With the rapid development of smart logistics, Internet of Things (IoT) has been widely used in information perception in smart logistics park. Traditional IoT-based information platform for smart logistics park has some issues need to be concerned, such as the storage and computing of massive sensor data, the real-time access to sensor nodes, the sharing of sensor resources and so on, so sensor cloud has been proposed to provide on-demand sensing service just like the cloud computing, but it also has some problems need to be deal with such as the effective support of real-time application, the high cost of communication bandwidth, and the high redundancy of sensing data. To resolve these problems, in this work, we propose a new fog enabled sensor cloud platform for smart logistics based on our previous research, design the architecture, functional modules and service modes of the platform in combination with the scene of smart logistics park.

**Keywords:** Sensor cloud · IoT · Fog computing · On-demand service · Smart logistics park

## 1 Introduction

Logistics parks are key nodes in modern logistics system, and are important information hubs too, which provide a various information services, such as demand information, transport capacity information, goods information, logistics facilities information, logistics network information, and so on [1]. With the development and application of IoT, Cloud Computing, Big Data and Artificial Intelligence, the traditional logistics park is gradually transformed into the smart logistics park [2].

In the context of smart logistics park, IoT is widely used in information perception. However, inherited the characteristics of IoT, there are still some issues need to be concerned. First, due to the limited storage and computing capacity of wireless sensor network (WSN) nodes, it is difficult to store and analyze the massive sensor data in the WSN domain [3]. Second, in order to reduce the energy consumption and extend the WSN life cycle, some WSN nodes supplied by batteries need to be applied sleep mechanism which influence the real-time access of these nodes [4]. Last, for WSN is closely bound to specific application scenarios and applications such as warehouse management, transportation management, most WSN systems have little or no interoperability and sensor nodes only can be used by specific application within the management domain


Z. Hu et al. (Eds.): ICAILE 2021, LNDECT 82, pp. 107–117, 2021.
https://doi.org/10.1007/978-3-030-80475-6_11

and cannot be shared by other applications outside the management domain, which lead to the redundant deployment of sensor nodes [5].

Sensor cloud, as a combination of IoT and cloud computing, inherited the characteristics of cloud computing [6], can resolve above mentioned problems encountered by the IoT-based application in smart logistics park effectively [7]. Nevertheless, it brings some new problems. First, a large amount of sensing data is directly transmitted to the data center (DC) before it is processed, which results in the waste of network transmission bandwidth and storage resources [8]. Second, local applications accessing the DC located in the cloud may result in long transmission delays, which influence the support of real-time application, such as smart warehouse management, smart vehicle scheduling, smart warehouse environment monitoring, and so on [9].

Based on our previous research [8], fog computing [10] can be used to solve above mentioned problems in traditional sensor cloud platform. So in this work, we propose a new fog enabled sensor cloud platform for smart logistics park, design the architecture, functional modules and service modes of the platform in combination with the scene of smart logistics park.

## 2 Overview of the Proposed Platform

Fog computing is a kind of edge computing [11]. Compared with cloud computing, it is closer to terminals and can effectively support distributed intelligence and real-time applications [12]. Our idea is applying fog nodes to preprocess sensing data to achieve data fusion, so as to reduce the consumption of network bandwidth, and deploying some simple application or some functions of these applications in the fog nodes to reduce the network transmission delay.

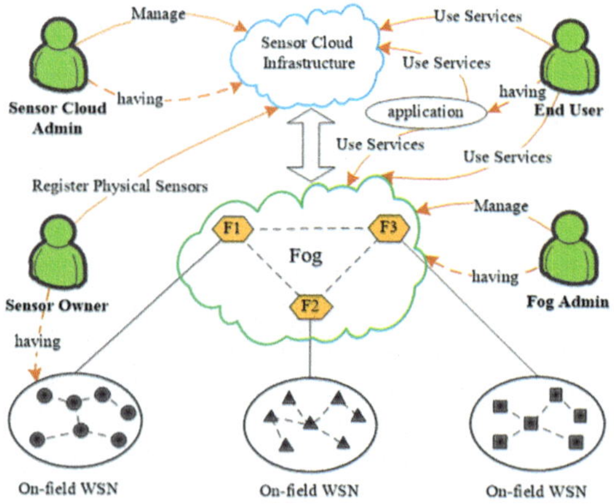


**Fig. 1.** Fog enabled sensor cloud platform framework

As shown in Fig. 1, the fog enabled sensor cloud platform includes three layers: the physical sensor layer, the fog layer and the cloud layer. The physical sensor layer is responsible for collecting the information of objects or the context in smart logistics park, and consists of several on-field WSNs which are composed of heterogeneous wireless sensors, such as temperature sensor, humidity sensor, gas sensor, flame sensor and so on. The fog layer is just like a data relay station that connects the cloud layer to the front-end physical sensor layer. It includes several fog nodes which can be common devices with limited capability such as access points, routers, switches, base stations and terminal devices or special fog devices such as Cloudlet and IOx. The cloud layer where the sensor cloud infrastructure (SCI) locates in provides powerful storage and computing power to the applications of smart logistics park, such as inventory management, logistics tools and vehicles management, security management, environment monitoring and so on.

There are four roles in the platform: the sensor owner (SO), the sensor cloud administrator (SCA), the fog administrator (FA) and the end user (EU). In the context of smart logistics park, a sensor owner can be the builder of smart logistics park or third-party sensor provider, deploys and manages various types of physical sensors in smart logistics park according to the requirements of end users. For users' convenience, sensor owners need to register the physical sensors' metadata (sensor type, sensor function, location, and so on) in the SCI. The sensor cloud administrator who can be the administrator of smart logistics park, is responsible for the construction and management of the SCI, including related hardware and software infrastructure, such as virtual servers and virtual sensors. The fog administrator who can be the administrator of smart logistics park or the sensor owner, is responsible for the construction and management of the fog layer, and provide local service to local terminals. End users enjoy the on-demand services provided by the platform without caring about the construction and management of the platform, and pay corresponding fees according to the usage amount, thus they can focus more on the development of their own business.

## 3 Architecture of the Proposed Platform

According to the framework of the fog enabled sensor cloud platform, we design the architecture of the platform based on the idea of modularity. As shown in Fig. 2, the key components of the fog enabled sensor cloud platform include portal server, provisioning server, management server, virtual server, monitoring server and fog node. Here are the main functions of each component.

### 3.1 Portal Server

The portal server is the interface through which the users log into the platform. In our work, we use Browser/Server mode, so the users can access the platform directly through the Web browser. The functions of the platform are divided into common functions and special functions. The common functions are authorized to all users, including login, logout, personal information management, access statistics and so on. The special functions vary with the role of the user shown by Table 1, so the dashboard shows different menus to different role. To sensor owners, they are authorized to register or delete their physical sensors, monitor the status and usage of the physical sensors, check the rental

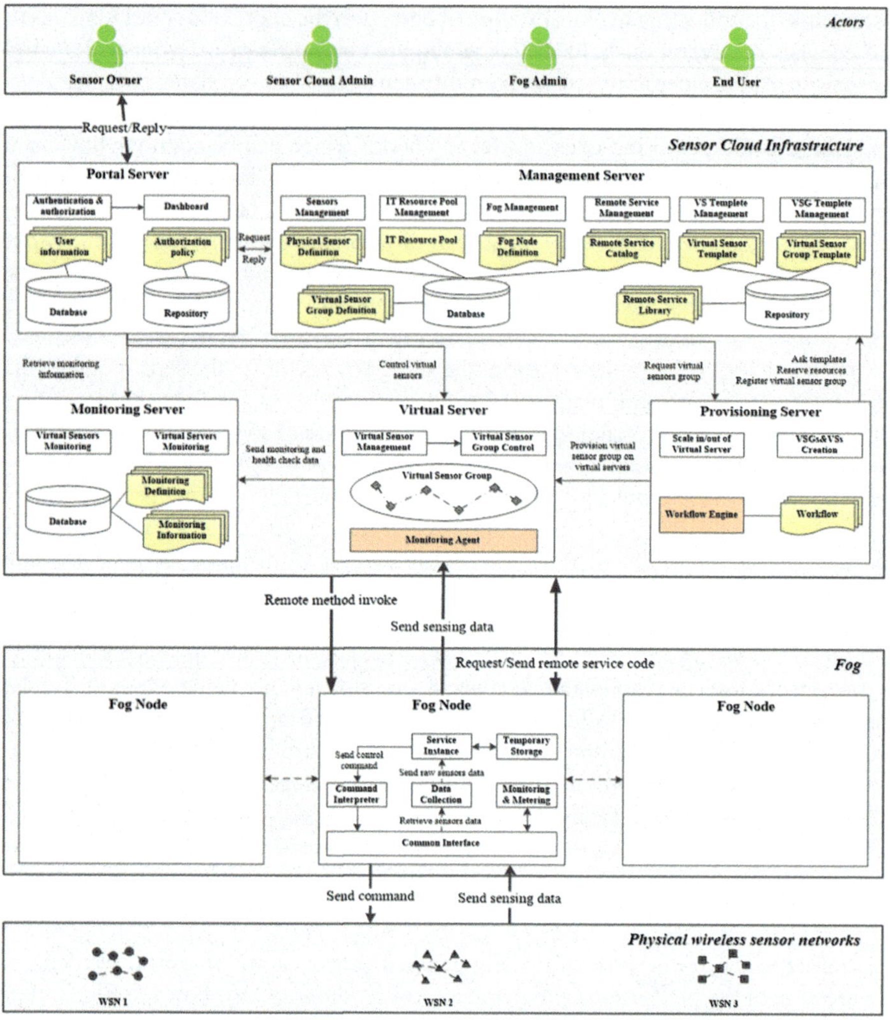


**Fig. 2.** Fog enabled sensor cloud platform architecture

fee, and modify the registration information when necessary. To fog administrators, they are authorized to manage their fog nodes and service instances running on fog nodes, including registering, monitoring or deleting fog nodes, and creating, modifying or deleting service instances. To end users, they are authorized to request service from the SCI or fog nodes. They can create, control, monitor or delete virtual sensors (VSs) and virtual sensors groups (VSGs) according to their own requirements, and also can create new VSG templates to meet their special requirements, but cannot modify or delete the existing templates created by others. To local end users, they can invoke service instance from fog nodes to get service. To sensor cloud administrators, as super users, they own all the functions provided by the platform and are able to authorize them to other users.

The most important function of sensor cloud administrators is the management of IT resources. They are able to monitor the usage of the IT resource pool and register or remove servers based on the real-time usage status of the IT resource pool.

**Table 1.** Function allocation matrix

| Object | Operation | SO | SCA | FA | EU |
|---|---|---|---|---|---|
| VSG | Create, control, monitor, delete | | √ | | √ |
| VSG template | Create | | √ | | √ |
| | Modify, delete | | √ | | |
| VS | Create, control, monitor, delete | | √ | | √ |
| VS template | Create, modify, delete | | √ | | |
| Physical sensor | Register, monitor, modify, delete | √ | √ | | |
| Service instance | Create, modify, delete | | √ | √ | |
| | Invoke | | √ | | √ |
| Fog node | Register, monitor, modify, delete | | √ | √ | |
| IT resource | Register, monitor, modify, delete | | √ | | |
| User | Authorize | | √ | | |

## 3.2 Management Server

The management server is composed of management application and the relative database and repository. The database store the data of the physical sensor definition, IT resource pool, fog node definition, remote service catalog and VSG definition, and the repository store VS templates and VSG templates. The management server receives the request from the portal server created by different roles, and provides corresponding management operation of physical sensors, IT resources, fog nodes, remote services, VSGs, VS templates and VSG templates. While receiving the request of providing templates of VS and VSG from the provisioning server or the request of providing remote service codes from fog nodes, the management server responds automatically according to predefined process.

## 3.3 Provisioning Server

The provisioning server mainly provides sensing-as-a-service to end users. End users send request of sensing service to the provisioning server through the portal server, and the provisioning sever asks the corresponding VSG(s) template to the management server, and create VSG(s) automatically according to the predefined workflow driven by the workflow engine. At the same time, the provisioning server calculates hardware resources needed for the VSG(s) to run, and then allocates required resources to it/them from the IT resource pool.

### 3.4 Virtual Server

The virtual server provides the base running environment for the VSs and VSGs, which are created according to the IT resources requirement calculate by the provisioning server. Similar to cloud computing, the virtual server is built on the IT infrastructure of sensor cloud, which provides independent operation space to each end user, and each server is completely independent, so as to ensure the security and stability of the sensor cloud.

The virtual server is composed of a VS manager, a monitoring agent, and one or more VSGs. Each VSG consists of one or more VSs, and has a controller which manages the VSs of the VSG in their life-cycle. Like the physical sensor, the VS can be activate or sleep by the end user, and he/she also can set the data acquisition frequency and check the state of the VS through the controller. While one VS is created, there is a map be created too between the VS and a service instance in the fog node. Each VS only provides service to a specific user, but the service instance can be shared by multiple users, so thus realizing the sharing of physical sensor resources while ensuring the operational independence of end users.

### 3.5 Monitoring Server

The monitoring server provides monitoring function of VSs and virtual servers to end users and sensor cloud administrators. The monitoring server is composed of the monitoring program and the corresponding database. The program receives the information of VSs and virtual servers, and stores them into the database. The information is collected by monitoring agents that running on virtual servers, automatically sent to the monitoring server periodically. If the status of a virtual sensor or virtual server violates the predefined rules, the program automatically sends an alarm to the end user or sensor cloud administrator.

### 3.6 Fog Node

The fog node is a bridge connecting the cloud and physical sensors, responsible for sensing data relay. The main functions of fog node include data acquisition, data preprocessing, data temporary storage and simple sensing service. The fog node receives sensing data collected by physical sensors through common interface, and the raw data is sent to service instance for processing. Due to the limited computing capacity of the fog node, the service instance only can carry out simple data processing which will be introduced in next part. To reduce the consumption of network bandwidth between the cloud and the fog, the service instance only uploads the result rather than raw sensing data to the corresponding VS that running on the virtual server. If the connection between the fog node and the cloud is interrupted, the fog node can also use the temporary storage module to temporarily store the processing result. Meanwhile, the service instance can also respond to the sensing requests of local end users, and provides simple sensing service, such as the temperature and humidity in the warehouse, the status of logistics facilities, the location of the goods, and so on.

In the context of smart logistics park, the front-end WSN not only includes a large number of sensors, but also includes some controllers such as temperature regulator, fan controller, motor controller and so on, so it is need that the fog node can also transmit the control instructions of the cloud applications or local end users to front-end controllers. Considering the diversity of users and applications, the control instructions are translated into standard control instructions before being sent to the specified device.

## 4 Service Modes of the Proposed Platform

In this section, we focus on the service modes of the fog enabled sensor cloud, which is based on the virtualization of physical sensors.

### 4.1 Virtualization Scheme

Virtualization of physical sensors is one of the key issues of sensor cloud, and it is the basis for physical sensors' sharing too. In traditional sensor cloud, physical sensors are directly virtualized into VSs, and VSs are composed of VSGs to provide sensing services to end users. Due to the addition of the fog layer, it is not an ideal solution to completely rely on fog nodes or cloud platform to complete the virtualization of physical sensors. Based on the requirements of smart logistics park, we propose a multi-level physical sensor virtualization scheme.

The key issue of the scheme is how to realize the load balance between the cloud and the fog. Considering the limitation of the computing and storage capacity of the fog node, we only assign virtualization tasks with small workload to the fog node, while assign those with large workload to the SCI. As shown in Fig. 3, in the fog node, physical sensors are virtualized as service instances. The service instance can be taken as a special kind of VS, and only preprocesses the raw sensing data and uploads the results to the SCI. Meanwhile, to reduce the consumption of physical sensor energy, network bandwidth, the service instance is temporarily connected to physical sensors and the SCI when necessary. In the SCI, physical sensors are virtualized as VSs. The VS is temporarily created based on the predefined VS template according to the user's request. Multiple VSs can form a VSG, and the same VS can be shared by different VSG.

### 4.2 Service Modes

Based on the multi-layer virtualization scheme, the proposed platform apply eight modes to virtualize physical sensors and provide corresponding services, as shown in Table 2. Set $R$ and $V$ represent physical sensor and virtual sensor set, $r_i$ and $v_i$ respectively represent physical and virtual sensor, $B \in R$ includes one or more physical sensors, $h$, $f$, $p$ and $c$ respectively represent aggregation function, qualification function, prediction function and analysis function.

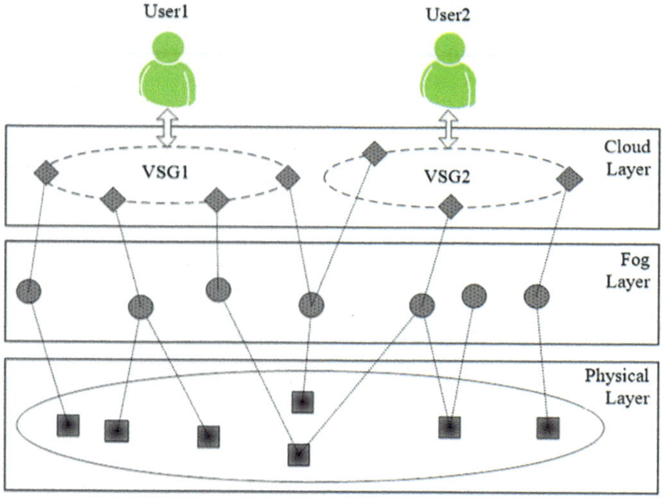


**Fig. 3.** Multi-level physical sensors virtualization scheme

**Table 2.** Relations between virtual sensor and physical sensor

| Virtualization level | Relation | Type |
|---|---|---|
| Fog level | $v_i \rightarrow r_i$ | Singular |
| | $v_i \rightarrow \{r_i \mid r_i \in B\}$ | Selector |
| | $v_i \rightarrow B$ | Accumulator |
| | $v_i \rightarrow h(B)$ | Aggregator |
| | $v_i \rightarrow \{r_i \mid \forall r_i \in B \rightarrow f(r_i)\}$ | Qualifier |
| | $v_i \rightarrow \{r_i \mid \forall r_i \in B \rightarrow f(r_i, B)\}$ | Context qualifier |
| Cloud level | $v_i \rightarrow p(r^t{}_i : t \in T)$ | Predicted |
| | $v_i \rightarrow c(B)$ | Computed |

### 4.2.1 Singular

In the singular mode, the virtual sensor $v_i$ is virtualized from only one physical sensor $r_i$. This mode is the simplest form of virtualization and requires the least amount of computing resources, so it can be implemented in the fog node. In the context of smart logistics park, this mode can be widely used. For example, in the logistics facility positioning scenario, the sensors can be virtualized according to the one-to-one correspondence.

### 4.2.2 Selector

In the selector mode, the virtual sensor $v_i$ is virtualized from a group of physical sensors $B$, but it does not mean that any group of physical sensors $B$ is suitable for this virtualization approach. There are two prerequisites have to be met. These physical sensors are homogeneous, and any one physical sensor $r_i$ can represent the bundle $B$. This mode only needs to run a simple selection algorithm and does not require high computing resources, so it can also be implemented in the fog node. For example, in the warehouse temperature monitoring scenario, any sensor's value of a region can represent the temperature of this region, so these sensors can be virtualized as a virtual sensor.

### 4.2.3 Accumulator

In the accumulator mode, the virtual sensor $v_i$ is virtualized from a group of homogeneous or heterogeneous sensors $B$. Unlike the previous mode, the virtual sensor $v_i$ is an accumulation of the bundle $B$, so it requires more computing resources, but it's still within the capacity of the fog node. For example, in the logistics personnel monitoring, the deployed body sensor network can be virtualized as a single virtual sensor.

### 4.2.4 Aggregator

In the aggregator mode, there is an function $h$ applied to aggregate the raw sensing data collected by all physical sensors in the bundle $B$, and the result of the function $h$ is virtualized as a single virtual sensor $v_i$. Compared with afore-mentioned modes, although this mode requires more computing resources, it can obtain higher data fusion efficiency, so it is also suitable for deploying in the fog node, thus reducing the consumption of network transmission bandwidth. For example, in the traffic flow statistics scenario of smart logistics park, count sensors deployed in a certain area of the park can be aggregated to obtain the total traffic flow or average traffic flow.

### 4.2.5 Qualifier

In some application scenarios of smart logistics park, it is unnecessary to monitor the object status all the time, but only need to know its state when some unexpected situations occur. It means that the virtual sensor $v_i$ only comes active when sensed value coming from the physical sensor $r_i$ qualifies a qualifier function $f$. For example, in warehouse fire monitoring scenario, we only need to activate the virtual sensor when there is a fire. This mode can effectively filter the raw sensing data, but the requirement of computing resource is determined by the complexity of function $f$, so only those application scenarios with small amount of calculation are suitable for implementation in the fog node.

### 4.2.6 Context Qualifier

This mode is similar to the previous, but the qualifier function $f$ observes one or more other sensor(s) $\{r_j | \forall\ r_j \in B \text{ and } r_j \neq r_i\}$ present inside bundle $B$. For example, in the application of warehouse anti-theft, the infrared sensor will not alarm once it detects the activities of personnel, but needs to judge whether the personnel is authorized or whether unauthorized goods are taken out of the warehouse, so the assistance of other sensors is needed. In this mode, some simple virtualization operations can also be implemented in the fog node.

#### 4.2.7 Predictor

The communication link from the front-end physical sensors to the cloud is not 100% reliable. When the communication link is interrupted or the front-end physical sensor fails, the current state of the monitored object can be predicted according to the previously accumulated data in order to ensure the normal operation of the application of smart logistics park. For these situations, it is necessary to create a new kind of virtual sensor, named predictor. In this mode, the value of the virtual sensor $v_i$ is not the actual value of the monitored object $r_i$, but the outcome calculated by the prediction algorithm. Therefore, it consumes a large amount of storage and computing resources, so it can only be deployed in the cloud rather than in the fog node.

#### 4.2.8 Computer

This mode is similar to the aggregator mode in the fog node, but the former is much more computationally intensive than the latter. For example, in the application of access control of smart logistics park based on image analysis or video analysis, the virtual sensor $v_i$ for identity authentication needs strong computing and storage capacity as support, so it can only be deployed in the cloud platform.

## 5 Conclusion

In this work, we propose a new fog enabled sensor cloud platform for smart logistics park based on our previous research, design the architecture, functional modules and service modes of the platform in combination with the scene of smart logistics park. Compared with the traditional smart logistics park information platform, the propose platform can provide good data fusion performance under the condition of realizing the sharing of physical sensor resources in the context of smart logistics park.

Our work is a preliminary consideration of the application of sensing cloud and fog computing in smart logistics park. Although we have designed the architecture, functional modules and service mode of the platform in detail according to the requirements of smart logistics park, there is still a big gap between the practical application. The feasibility and effectiveness of the platform need to be further verified through experiments, and the data fusion algorithm also need to be further designed.

**Acknowledgment.** This project is supported by the Natural Science Foundation of Hubei Province (2019CFC930), the Scientific Research Project of Hubei Education Department (B2019278), the Special Funding Assignment of Wuhan Technology and Business University (S018003), and the Distinguished Young and Middle-aged Team Program for Scientific and Technological Innovation in Higher Education of Hubei (T201938).

## References

1. Yaqiong, L., Shangjia, X., Tianyi, Z.: Data-driven design and optimization for smart logistics parks: towards the sustainable development of the steel industry. Sustainability **12**(17), 7034 (2020)

2. Yuanyuan, L., Shuxian, L.: Research on key technologies of smart logistics dynamic positioning based on NB-IOT. J. Phys. Conf. Ser. **1634**(1), 86–91 (2020)
3. Rambabu, C., Prasad, V.V.K.D.V., Prasad, K.S.: Multipath cluster-based hybrid MAC protocol for wireless sensor networks. Int. J. Wirel. Microw. Technol. **10**(1), 1–16 (2020)
4. Huang, Q., Yang, Y., Wang, L.: Secure data access control with ciphertext update and computation outsourcing in fog computing for internet of things. IEEE Access **5**, 12941–12950 (2017). https://doi.org/10.1109/ACCESS.2017.2727054
5. Sanka, S., Srinivasan, P.: Fuzzy logic based energy aware routing protocol for internet of things. Int. J. Intell. Syst. Appl. **10**(10), 11–19 (2018)
6. Kumar, V., Laghari, A.A., Karim, S.: Comparison of fog computing & cloud computing. Int. J. Math. Sci. Comput. **5**(1), 31–41 (2019)
7. Alamri, A., Ansari, W.S., Hassan, M.M., Shamim Hossain, M., Abdulhameed Alelaiwi, M., Hossain, A.: A survey on sensor-cloud: architecture, applications, and approaches. Int. J. Distrib. Sensor Netw. **9**(2), 917923 (2013). https://doi.org/10.1155/2013/917923
8. Wei, X., Wu, L.: A new proposed sensor cloud architecture based on fog computing for Internet of Things. IEEE **2019**, 615–620 (2019)
9. Geetha, S., Deepalakshmi, P.: Rapid retrieval of secured data from the sensor cloud using a relative record index and energy management of sensors. Int. J. Intell. Enterp. **6**(3), 1–12 (2019)
10. Chaudhry, S.A., Yahya, K., Al-Turjman, F.: A secure and reliable device access control scheme for IoT based sensor cloud systems. IEEE Access **99**, 1–10 (2020)
11. Rahman, M.T., Salan, M.S.A., Shuva, T.F.: Efficient sensor-cloud communication using data classification and compression. Int. J. Inf. Technol. Comput. Sci. **9**(6), 9–17 (2017)
12. Ooi, B.Y., Beh, W.L., Lee, W.-K., Shirmohammadi, S.: Using the cloud to improve sensor availability and reliability in remote monitoring. IEEE Trans. Instrum. Measur. **68**(5), 1522–1532 (2019). https://doi.org/10.1109/TIM.2018.2882218

# The Location of Company's Logistics Hub Based on Principal Component Analysis

Zhuoni Zhang(✉) and Zhengguo Wang

School of Logistics Engineering, Wuhan University of Technology, Wuhan 430063, China
zhangzhuoni@whut.edu.cn

**Abstract.** In recent years, China has promulgated a large number of policies proposed national logistics hub. The national logistics hub has core infrastructure, large scale of land occupation, wide radiation range and high degree of resource concentration. The national logistics hub node network with clear positioning and structure will become an important platform and key support for China's modern logistics service network and system. According to Z company's main business market, this paper puts forward the demand of hub layout construction, and carries out strategic layout planning of enterprise logistics hub. Relying on the city with national logistics hub, combining with its specific function positioning, the paper uses principal component analysis method for quantitative analysis, and makes qualitative analysis based on growth pole theory and gradient theory, select the functional logistics hub with higher degree of agglomeration, better service function, higher operation efficiency, then expand the radiation area, so that the hub can play the role of key node and backbone hub in company logistics network.

**Keywords:** Principal component analysis · Location model · Logistics infrastructure

## 1 Instruction

The demonstration and driving role of national logistics hub on industrial innovation and social development emphasizes that the advantages of wide radiation range, low cost and high operation efficiency [1]. National logistics hub should be fully exerted to promote the development of regional agriculture, manufacture, commerce and other industries [2]. Logistics hub has natural hub economic conditions. Through the centralized layout and efficient coordination of transportation and logistics facilities, it attracts the collection of resource elements, helps the spatial organization of the upstream and downstream of the supply chain and industrial clusters, objectively speeds up the integration and efficient integration of logistics, business flow, information flow and capital flow, and promotes the development of new economic growth paradigm and new business forms [3]. Therefore, company can rely on the layout of national logistics hubs, the construction of international logistics channels and the construction of modern logistics service system, effectively attract the agglomeration of relevant industrial elements, find a new path of high-quality economic development, accumulate and develop hub


Z. Hu et al. (Eds.): ICAILE 2021, LNDECT 82, pp. 118–127, 2021.
https://doi.org/10.1007/978-3-030-80475-6_12

economy, promote the collaborative linkage between logistics industry and related businesses, and achieve the goals of financing, optimizing structure and self transformation and upgrading.

## 2 Layout of National Logistics Hub and Existing Main Logistics Network Layout of Z Company

The national logistics hub is the core infrastructure of the logistics system. Like the company's logistics hub, it is a logistics activity organization center and facility group which can realize multiple functions such as the storage, transfer and distribution of goods, but it has the characteristics of wider radiation range, better service function and stronger resource concentration. The National Development and Reform Commission and the Ministry of Transport jointly issued the "Notice on Doing a Good Job in the Construction of a National Logistics Hub in 2019" in September 2019 to continue to implement the decision and deployment of the affiliated Party Central Committee and the State Council on strengthening the construction of logistics and other infrastructure networks [4–6].

In order to strengthen the system layout at the macro level, we take into account the implementation of major national strategies, regional economic development, optimization and upgrading of industrial structure according to the regional economic aggregate, industrial spatial layout, infrastructure connectivity, population distribution, and basic pattern of domestic transportation channels. A number of cities with certain basic conditions are selected as the carrying cities of national logistics hubs. 212 national logistics hubs are planned and constructed, including 41 loading cities of land port type, 30 port type, 23 airport type, 47 production service type, 55 trade service type and 16 land border port type. Relying on railways, highways, airports, transportation corridors, stations, trade gathering areas, large factories, mines and manufacturing bases, etc., it connects trunk and branch line transportation in inland areas, mainly to ensure regional production and living, optimize enterprise layout, enhance regional economic competitiveness, and provide logistics organization and regional distribution services that connect domestic and international.

This paper states the location of company's logistics hub based on the Z company. Z company is a leading comprehensive logistics service provider in China, with a extensive and comprehensive service network. It has seven regional subsidiaries, covering seven regional plates in China, as well as a number of professional companies, covering cold chain, chemical industry, technology electronics, medical and other business sectors.

The company's shipping agency business mainly provides customers with multi-link logistics services related to waterway transportation, such as packing, warehousing, collection and distribution and delivery. In terms of air transportation agency, Z company mainly provides customs declaration and inspection, packaging, booking and operation and other professional air freight forwarding services for customers. With sufficient transport capacity resources, Z company can provide customers with visualization and standardized logistics services. In terms of railway agency, it can provide customers with various modes of railway transportation and information services, including domestic or international railway transportation.

# 3 Positioning Scheme of Company's Logistics Hub Adapting to the Layout of National Logistics Hub

Based on the national logistics hub, we can initially form a phased planning which is suitable for the national strategy. After selecting the key nodes in the region, combining with the functional positioning of the national logistics hub, the paper gives the function and mode positioning to give full play to the existing advantages.

According to the current business market division of Z company in seven geographical plates of East China, North China, South China, central China, Northwest China, Southwest China and Northeast China, the evaluation system is established based on various types of cities with national logistics hub. According to the distribution density of national logistics hub cities in different regions, the main market areas are further determined. Combined with the spatial layout requirements of the city or regional master plan, the overlap and intersection of various transportation modes, logistics resources, regional logistics characteristics and the development of urban or regional industrial layout, etc., then, under the premise of considering the existing regional construction conditions of the enterprise, the key logistics nodes of the enterprise are selected from the candidate cities as the logistics hub of Z company [7, 8].

## 3.1 Quantitative Analysis

The selection method of enterprise logistics hub which adapts to the national logistics hub with specific functions and undertakes the main functions such as distribution and transshipment can be used to select the key nodes of company suitable for other types of national logistics hubs, so as to carry out the overall logistics network planning. According to the distribution of six types of national logistics hub cities, and establish an evaluation system for quantitative scoring. First of all, select the relevant indicators that affect the regional logistics development [9]. Then, use the principal component analysis to extract the main influencing factors in the indicators. Finally, according to the score of the impact factors and combined with other influencing factors, the comprehensive evaluation function is constructed to obtain the comprehensive score of the logistics development of each city, and the suitability degree of the logistics hub layout of the candidate cities is screened out (Fig. 1).

Through the principal component analysis method whose flow chart is shown in the figure above, the extracted principal component factors can be named according to the unique meaning reflected by professional knowledge and indicators, so as to obtain reasonable explanatory variables. The linear transformation model of each principal factor is as follows:

$$F_i = U_i T_x \tag{1}$$

$$Z = 2CR_i F_i \tag{2}$$

X is relevant variables. U is the standardization of the eigenvalues of the covariance matrix corresponding to the feature vector. In the comprehensive evaluation value, according to the number of eigenvalues greater than 1 set main factor. CR is the weight

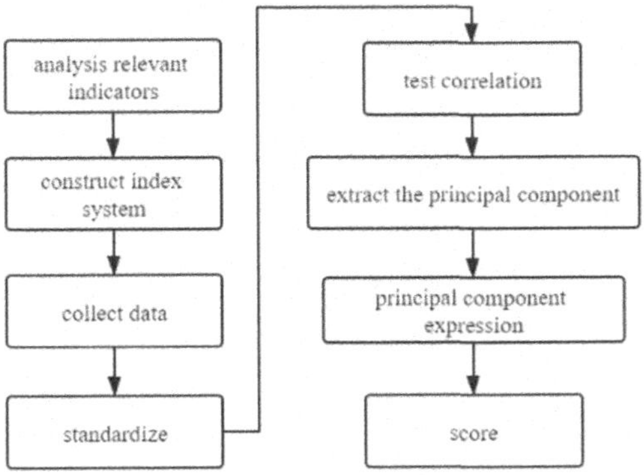


**Fig. 1.** The flow chart of quantitative analysis by principal component analysis

of each index, namely determine according to the variance contribution rates of the principal component. The value of Z is calculated for the final score.

### 3.2 Qualitative Analysis

#### 3.2.1 Growth Pole Theory

Taking cities or regions with superior economic conditions as the point, we can also regard them as growth poles. We should give full consideration to the differences of logistics development conditions and development potential in different regions, and give priority to the development of regions with superior logistics development conditions. The higher the level of economic development, this will be more conducive to the logistics enterprise cluster, forming a higher level of logistics node facilities. Taking the traffic and information context between "point" and "point" as the axis, the point axis development planning is carried out to drive the logistics development of surrounding areas [10]. When the point axis planning is further extended, it can gradually form a network organization, and then obtain good resource space allocation benefits.

#### 3.2.2 Gradient Theory

Gradient propulsion theory is now used in the national or large-scale regional economic development. According to the level of regional economic and technological development, the macro mode of gradual development is staged from high to low. High gradient areas with conditions give priority to the introduction of advanced technology for development, and then gradually move to a lower level gradient. With the acceleration of the speed of the process, the gap between gradients is gradually reduced, and finally the relative balance of regional logistics development can be achieved. Therefore, after the quantitative analysis and selection of the national logistics hub bearing cities with superior economic conditions and high logistics level, combined with the specific functions to be undertaken and the main business of Z company's seven major plates, the main market areas are further determined by the density of the carrying cities in various regions. Taking the national logistics hub carrying city in the selected region as an alternative, combining with the relevant logistics situation, such as external policies, economic conditions and other factors and the existing construction conditions of enterprises in the region, this paper makes a qualitative analysis on the development mode of logistics node facilities, then select the key logistics nodes from the candidate cities as the logistics hub of Z company.

## 4 Case Study on the Layout of Logistics Hubs in Port Type Cities

### 4.1 Establishment of Index System

**Table 1.** Index system of different levels

| Primary level index | Secondary level index | Third level index |
|---|---|---|
| Comprehensive evaluation index system of port hub competitiveness | Port operation capacity | Cargo throughput |
| | | Container throughput |
| | Economic development capacity of port city | GDP of port cities |
| | | Consumer price index |
| | | Retail sales of consumer goods |
| | Port development potential | Growth rate of cargo throughput |
| | | Container throughput growth rate |
| | Economic development potential of port city | Proportion of tertiary industry in GDP of port cities |
| | | Growth rate of total retail sales of consumer goods |
| | | Growth rate of fixed assets investment in the whole society |

Cities with port type logistics hub should rely on coastal and inland ports, connect domestic and international routes and port collection and distribution network, realize the organic connection of land and water transportation, and provide logistics services and other value-added services such as cargo distribution, international transit, bonded supervision, etc. for the port hinterland and its radiation areas [11]. In order to evaluate the candidate cities to build port type national logistics hub more scientifically, completely and effectively, this scheme constructs the competitiveness evaluation index system and model of port type national logistics hub on the basis of reading a lot of relevant literature, understanding and mastering the main relevant theories and technical methods [12]. The basic principle of port type national logistics hub carrying city evaluation is that it cannot only reflect the market effect, but also reflect the potential long-term stability and forward development ability of the port. Three levels of index system are shown in Table 1.

When the principal component analysis method is used, the city is taken as the alternative point, and the representative influencing factors of site selection evaluation.

### 4.2 Scoring by Principal Component Analysis

The data used in this case mainly come from "global port development report (2020 Beijing)" and "China Port Development Report (2019–2020)", Beijing Tianjin, Harbin, Shanghai, Nanjing, Hangzhou, Ningbo, Xiamen, Qingdao, Changsha and other cities' national economic and social development statistical bulletin and its Statistical Yearbook, traffic Yearbook and official website, etc. SPSS software was used for principal component analysis and the output results were obtained [13].

**Table 2.** The KMO and Bartlett test

| The KMO and Bartlett Test | |
|---|---|
| KMO sampling suitability quantity | 0.614 |
| Bartlett sphericity test | 125.757 |
| Freedom | 45 |
| Significance | 0.000 |

**Table 3.** Explanation of total variance

| Component | Total | Variance percentage of initial eigenvalue | Cumulative/% | Total | Percentage sum of squares variance | Cumulative/% |
|---|---|---|---|---|---|---|
| 1 | 3.768 | 37.678 | 37.678 | 3.768 | 37.678 | 37.678 |
| 2 | 1.673 | 16.727 | 54.406 | 1.673 | 16.727 | 54.406 |
| 3 | 1.093 | 10.932 | 65.337 | 1.093 | 10.932 | 65.337 |
| 4 | 0.985 | 9.846 | 75.183 | | | |
| 5 | 0.889 | 8.890 | 84.073 | | | |
| 6 | 0.686 | 6.858 | 90.931 | | | |
| 7 | 0.372 | 3.720 | 94.652 | | | |
| 8 | 0.282 | 2.820 | 97.472 | | | |
| 9 | 0.164 | 1.640 | 99.112 | | | |
| 10 | 0.089 | 0.888 | 100.000 | | | |

Table 2 is the result of the KMO and Bartlett test. KMO = 0.614 > 0.6. The value of chi-squared approximation is 125.757. $p = 0.000 < 0.5$. The data is suitable for factor analysis. The common factor variance table shows the degree of commonality. Except the growth rate of total retail sales of social consumer goods, the value of each common degree is larger than 0.5, indicating that the interpretation degree of extracting principal components in the original quantity is relatively high.

The principle of principal component number extraction is the principal components whose eigenvalues are greater than 1. Table 3 is about the explanation of total variance. According to it, three principal components are extracted. The variance contribution rates of the three principal components were 37.678%, 16.727% and 10.931% respectively, and their eigenvalues were 3.768,1.673 and 1.093, respectively.

**Table 4.** Component matrix

| | Component 1 | Component 2 | Component 3 |
|---|---|---|---|
| Port cargo throughput | 0.755 | 0.223 | – 0.178 |
| Container throughput | 0.852 | 0.043 | – 0.183 |
| Consumer price index | 0.026 | 0.419 | – 0.614 |
| GDP of port cities | 0.842 | 0.028 | 0.230 |
| Total retail sales of consumer goods | 0.887 | – 0.087 | 0.145 |
| The proportion of tertiary industry in GDP | 0.831 | – 0.236 | – 0.036 |
| Growth rate of fixed assets investment in the whole society | 0.159 | – 0.434 | 0.453 |
| Growth rate of port cargo throughput | – 0.116 | 0.836 | 0.243 |
| Container throughput growth rate | 0.107 | 0.649 | 0.530 |
| Growth rate of total retail sales of consumer goods | – 0.484 | – 0.270 | 0.176 |

From Table 4, we can get each index's influence on each component. By dividing the data in the principal component matrix by the square root of the eigenvalue corresponding to the principal component, the coefficient corresponding to each index in the three principal components can be obtained. After multiplying the load matrix with the normalized data, three principal component expressions can be obtained as follows:

$$Y1 = 0.39 * ZX1 + 0.44 * ZX2 + 0.01 * ZX3 + 0.43 * ZX4 + 0.46 * ZX5+ \\ 0.43 * ZX6 + 0.08 * ZX7 - 0.06 * ZX8 + 0.06 * ZX9 - 0.25 * ZX10 \tag{3}$$

$$Y2 = 0.17 * ZX1 + 0.03 * ZX2 + 032 * ZX3 + 0.02 * ZX4 - 0.07 * ZX5- \\ 0.18 * ZX6 + 0.34 * ZX7 + 0.65 * ZX8 + 0.50 * ZX9 - 0.21 * ZX10 \tag{4}$$

$$Y3 = -0.17 * ZX1 - 0.18 * ZX2 - 0.59 * ZX3 + 0.22 * ZX4 + 0.14 * ZX5- \\ 0.03 * ZX6 + 0.43 * ZX7 + 0.23 * ZX8 + 0.51 * ZX9 + 0.17 * ZX10 \tag{5}$$

$$Z = 0.37678 * Y1 + 0.16727 * Y2 + 0.10931 * Y3 \tag{6}$$

The comprehensive evaluation score y value can be obtained from Table 5.

Although the selected indicators are not comprehensive, they generally cover the economic level and logistics level of port cities. According to the score, the logistics development environment of port type national logistics hub cities can be preliminarily compared, which lays a foundation for enterprises to choose specific cities. In addition to the port type national logistics hub, there are other types of national logistics hub. But it should be clear that the evaluation indicators of different types of hubs should be slightly different, we should consider function positioning and market.

**Table 5.** Quantitative scoring results of each city

| Port city | Score | Port city | Score |
|---|---|---|---|
| Shanghai | 2.303 | Qinhuangdao | – 0.085 |
| Guangzhou | 1.295 | Yantai | – 0.086 |
| Shenzhen | 0.904 | Yingkou | – 0.196 |
| Tianjin | 0.900 | Tangshan | – 0.288 |
| Qingdao | 0.690 | Fuzhou | – 0.299 |
| Suzhou | 0.585 | Xiamen | – 0.318 |
| Chongqing | 0.542 | Lianyungang | – 0.361 |
| Nantong | 0.503 | Qinzhou-Beihai-Fangcheng Port | – 0.435 |
| Nanjing | 0.392 | Zhanjiang | – 0.436 |
| Wuhan | 0.229 | Jiujiang | – 0.497 |
| Dalian | 0.225 | Wuhu | – 0.500 |
| Vingbo Zhoushan | 0.117 | Yangpu | – 0.615 |
| Rizhao | 0.013 | Yueyang | – 0.620 |

## 4.3 Qualitative Analysis Based on Business Market and Functional Positioning

After sorting out and summarizing the candidate cities in each region, it is concluded that the port cities in East China and South China are relatively dense. The distribution of them is shown in the figure below. So qualitative analysis is made on the development mode of logistics node facilities according to the port related logistics situation of Z company from the two plate regions (Fig. 2).

East China occupies a leading position in the country and is the highest comprehensive technical level in China economic region. The advantages of economic location and traffic location are the basis of the development of port logistics industry, and the powerful industrial economy is the driving force for the rapid development of port logistics. The important policies issued at the national level provide fundamental guarantee for the development of port logistics. After analyzing the economic strength, port freight volume and other related factors of port type national logistics hub, combined with the business market development situation and functional demand of Z company in East

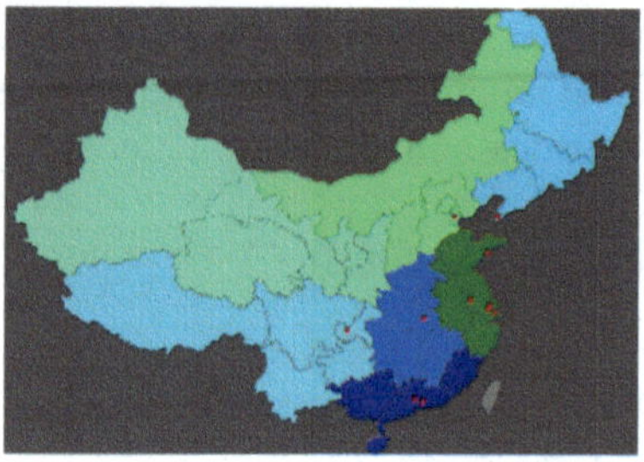

**Fig. 2.** Distribution of port cities with a high score in different regions

China, Shanghai and Suzhou are determined as the key hubs, as the logistics facilities group and logistics activity organization center to realize the functions of cargo distribution and transfer. They radiate many candidate cities.

With Z company's South China regional business strategy, the center of gravity will shift to the south. Based on the Pearl River Delta economic circle, the service of Z company South China Co., Ltd. radiates the whole South China region. Guangzhou, as one of the seven regional distribution centers established by Z company in China, has Guangzhou port, the largest comprehensive hub port in South China. Considering the natural conditions, policies and economic advantages, Guangzhou can be established as a logistics hub in South China.

In order to adapt to the layout of national logistics hubs, Z company focuses on improving the hub functions of the established logistics facilities. If necessary, it can integrate, relocate or build new hub facilities in combination with regional economic development and relevant policies. The selection method of enterprise logistics hub, which adapts to the national logistics hub, can be used to select the key nodes of enterprises suitable for other types of national logistics hub, so as to carry out the overall logistics network planning.

## 5 Conclusion

The national logistics hub is the core infrastructure of the logistics system. Logistics enterprises can rely on the layout of national logistics hubs, drive business development by the construction of enterprise hubs, and effectively attract the rapid agglomeration, financing, structural optimization and transformation and upgrading of relevant industrial elements. In order to adapt to the layout of national logistics hub, Z company mainly focuses on improving the hub function of the established logistics facilities. It also can integrate, migrate or build new hub facilities according to the needs of regional economic development. Based on the analysis and summary of relevant literature and logistics hub location examples, this project adopts the method of combining qualitative and quantitative analysis, and studies the enterprise logistics hub according to the characteristics of Z company. Finally, the feasibility of the whole set of site selection and layout scheme is verified by the example of enterprise hub layout suitable for port type national logistics hub.

## References


1. Ting, P.-H.: An efficient and guaranteed cold-chain logistics for temperature-sensitive foods: applications of RFID and sensor networks. Int. J. Inf. Eng. Electron. Bus. **5**(6), 1–5 (2013)
2. Weihua, G., Tingting, Z., Yuwei, Z.: On RFID application in the information system of rail logistics center. Int. J. Educ. Manage. Eng. **3**(2), 52–58 (2013)
3. Qin, C., Shuyi, T.: Manufacturer's pricing strategy for supply chain with service level-dependent demand. Int. J. Eng. Manuf. **4**(4), 1–13 (2014)
4. Chongbin, X.: Focusing on the planning and construction of national logistics hub. Logist. Technol. Appl. **25**(07), 98–99 (2020). (in Chinese)
5. Ming, W.: Discussion on high quality construction and development of national logistics hub. Continent. Bridge Vision **09**, 57–60 (2019). (in Chinese)
6. Weihua, G., Wenpei, Y., Zheng, L.: Evaluation of the impact of logistics industry on regional economic vitality from the perspective of national logistics hub construction. Logist. Technol. **39**(04), 16–22 (2020). (in Chinese)
7. Lambert, D.M., Cooper, M.C.: Issues in supply chain management. Indust. Mark. Manage. **29**(1), 65–83 (2000). https://doi.org/10.1016/S0019-8501(99)00113-3
8. Nan, L., Yuangao, C., Yumin, L.: Logistics hub, the optimal location of the comprehensive decision-making method. J. Southeast Univ. **S1**, 71–75 (2007). (in Chinese)
9. Waller, M.A., Fawcett, S.E.: Data science, predictive analytics, and big data: a revolution that will transform supply chain design and management. J. Bus. Logist. **34**(2), 77–84 (2013). https://doi.org/10.1111/jbl.12010
10. Shi Xiaolan, X., Zongzhen, Z.X.: Hub facility point positioning of logistics distribution model of a preliminary study. J. Logist. Technol. **12**, 1–4 (2006). (in Chinese)
11. Hongmei, C., Duwei, L., Jing, Y., Zhen, Z.: Port type national logistics hubs: the concept, characteristics and its status in the global supply chain. J. Supply Chain Manage. **1**(9), 62–73 (2020). (in Chinese)
12. Robak, S., Franczyk, B., Robak, M.: Research problems associated with big data utilization in logistics and supply chain design and management. Ann. Comput. Sci. Inf. Syst. **3**(1), 245–249 (2014)
13. Wenlin, Z.: Principal component analysis in SPSS operation application. J. Mark. Res. **12**, 31–34 (2005). (in Chinese)

# Advances in Technological and Educational Approaches

# Deep Learning for Melanoma Detection with Testing Time Data Augmentation

Volodymyr Doms(✉), Yuri Gordienko, Yuriy Kochura, Oleksandr Rokovyi, Oleg Alienin, and Sergii Stirenko

National Technical University of Ukraine "Igor Sikorsky Kyiv Polytechnic Institute", Kyiv, Ukraine

**Abstract.** Recently, the numerous deep learning (DL) approaches were successfully applied for analysis of the melanoma that is one of the most dangerous skin cancer types. This work is devoted to investigation of various DL neural networks (DNNs) applied to melanoma images to classify the skin state like in malignant or benign condition. The obtained results demonstrate the possibility to improve melanoma classification results due to application of testing time augmentation (TTA) for various DNNs with different accuracy and model size. The common tendency to increase of AUC mean values was observed for all investigated DNNs (EfficientNetB0, EfficientNetB6, DenseNet121, Mobile NetV2, InceptionV3, ResNet101, ResNet101V2, Inception ResNetV2, VGG16, and Xception). But it is unavoidably followed by decrease of inference times due to TTA-related time overheads. It can be significant especially for the fastest DNNs deployed on specialized devices based on tensor processing units (TPU) with the limited computational resources for Edge Computing layer.

**Keywords:** Deep learning · Melanoma · Convolutional neural network · Data augmentation · Testing time data augmentation · TPU · Edge Computing

## 1 Introduction

During last years the rapid improvement of information technologies in the domain of artificial intelligence allowed their application in health care, especially for the analysis of medical data like tabulated data [1, 2], 1D time sequences (heart rate, pressure, etc.) [3, 4], 2D images (for example, chest X-ray images [5–7], melanoma images [8–11], etc.), and 3D images (magnetic resonance imaging [12], computer tomography [13], etc.). Recently, the numerous deep learning (DL) approaches were successfully applied for analysis of the melanoma that is one of the most dangerous skin cancer types that is globally accounting up to 40% of cancer cases [8–11]. This research is devoted to investigation of various DL neural networks (DNNs) applied to melanoma images to classify the skin state like in malignant or benign condition.

This paper has the following structure: Sect. 2. *Background and Related Work* contains a brief review of some similar approaches to characterize melanoma by DNNs,


Z. Hu et al. (Eds.): ICAILE 2021, LNDECT 82, pp. 131–140, 2021.
https://doi.org/10.1007/978-3-030-80475-6_13

Sect. 3. *Methodology* describes the dataset, DNNs, and metrics used, Sect. 4. *Experimental* gives the results obtained for different DNNs, Sect. 5. *Discussion* summarizes all results with analysis of advantages and disadvantages of the methods used, and Sect. 6. *Conclusions and Future Work* outlines the results obtained and the propose some possible directions for the future researches.

## 2 Background and Related Work

Among numerous DNNs that have been investigated and applied for melanoma classification, identification, and segmentation [8–11], the special interest is related with deep transfer learning (DTL) attempts on the basis of some classic DNNs pre-trained on the open datasets. For example, the DTL for VGG-type DNNs was successfully applied for various pre-training paths [14, 15]. The similar DTL was applied for the classic DNNs like AlexNet, ResNet101, and GoogleNet last three layers were dropped out and replaced with new layers to be more appropriate with the task of lesion classification [16–18]. The performance of the proposed method has been evaluated using three different datasets with classification accuracy $> 98\%$ [19]. The additional important aspects are shortage of data and imbalanced datasets that can be alleviated by means of data augmentation (DA) [20, 21].

The basic idea of image DA is that the image transformations (like rotation, flipping, contrast enhancement, and others) can be applied to increase the number of available image versions without changing and losing the characteristic features of data. DA can be carried out with and without loss of information [22, 23]. Recently several attempts were applied, for example, AlexNet, ResNet50, VGG16, VGG19, and Xception models demonstrated accuracy 81% (79%), 92% (86%), 88% (84%), 89% (85%), and 90% (66%) when were trained using DA (and without DA) [24].

These results are supported by data hyperspectral data (HSD) acquisition by a hyperspectral imagery (HSI), which can measure color information in detail. GoogLeNet was pre-trained by ImageNet and then TDL was used to analyze the HSD by 5-fold cross-validation. The respective sensitivity and specificity were improved up to 79.9% and 82.4% with DA (from 77.4% and 75.6% without DA) for HSI wavelengths combination of 530, 560, and 590 nm, and up to 76.7% and 82.2% with DA (from 73.1% and 80.6% without DA) for HSI wavelengths combination of 500, 620, and 740 nm [25]. In addition to these results on effect of DA for training data subset, this research was targeted on investigation of additional effect of DA that could be performed before prediction stage also, namely, effect of testing time augmentation (TTA) on tensor processing units (TPU) for various DNNs with different accuracy and model size.

## 3 Methodology

Here the dataset, some standard models, metrics, and general workflow are described. They were used to reach the main purpose: to investigate the effect of TTA on performance of various DNNs applied to melanoma images obtained by standard cameras to classify malignant or benign skin condition.

**Dataset.** In this research the ISIC 2020 Challenge Dataset was used that was provided for the related SIIM-ISIC Melanoma Classification Challenge.

The typical image and its DA transformations are given in Fig. 1. As it was mentioned above, DA is applied for covering all situations of posed camera. It is especially important for testing without warranty that data acquisition at field conditions are the same like during preparation of training dataset. That is why additional effect of TTA was of interest.

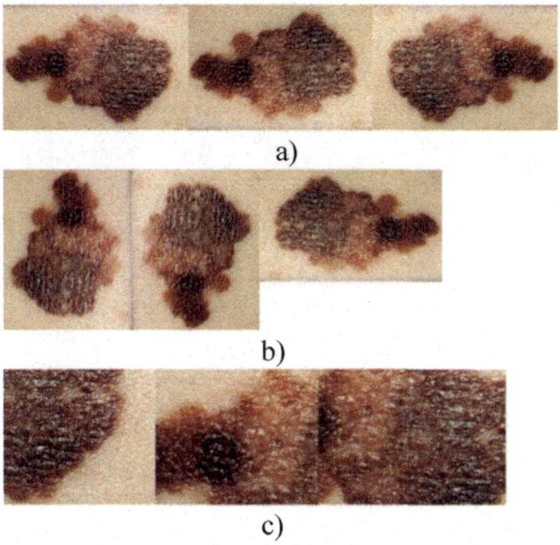


**Fig. 1.** Data augmentation operations obtained by: (a) horizontal and vertical flips, (b) rotations, (c) cropping and cutting.

**Models.** In this paper, the following standard DNNs (with the different model sizes given in Fig. 2) were used for DTL:

- Xception (presented at CVPR 2017) [27],
- VGG16 (presented at ICLR 2015) [14],
- ResNet101/ResNet101V2 (presented at CVPR 2015) [17],
- InceptionV3 (presented at CVPR 2016) [28],
- InceptionResNetV2 (presented at AAAI 2017) [29],
- MobileNetV2 (presented at CVPR 2018) [30],
- DenseNet121 (presented at CVPR 2017) [31],
- EfficientNet (presented at ICML 2019) [32].

**Metrics.** The following standard metrics were used: model size, inference time with/without TTA, accuracy, loss, area under curve (AUC) for receiver operating characteristic (ROC), and their statistical parameters like mean and standard deviation.

**Workflow.** The experimental workflow included training, validation and testing phases for all above mentioned DNNs. The number of epochs was equal to 12 only because the highest accuracy was not the aim of this research. Then after each training iteration the best models with the minimal validation loss were saved as checkpoints for

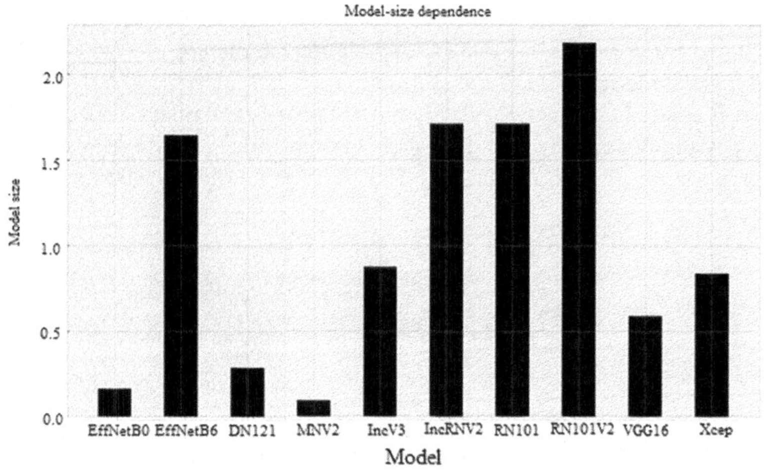


**Fig. 2.** The sizes of different model architectures

the further testing phase. The 5-fold cross validation was used to estimate the statistical parameters like mean and standard deviation of some metrics presented below in Figs. 3, 4, 5 and 6. The training with DA, validation, and testing with and without TTA were performed iteratively for all DNN models. It should be noted that all runs were performed on Google Cloud TPUv3–8 (tensor processing unit) infrastructure (with 8 TPUv3 cores and the total TPU memory 128 GiB) in the free of charge regime (https://www.kaggle.com/docs/tpu), where the availability of resources cannot be warranted.

## 4 Experimental Analysis

As a result of training/validation/testing without TTA, the AUC mean values were calculated that were different for various DNNs with regard to the estimated standard deviations (Fig. 3). For example, the smallest MobileNetV2 demonstrated the lowest AUC, and the biggest InceptionResNetV2 – the highest AUC.

In addition to DA on training stage, usage of testing with TTA led to the higher AUC values for all above mentioned DNNs (in comparison to testing without TTA), but with the different level of improvement (Fig. 4). The highest increase was observed for MobileNetV2 (from 0.797 to 0.841), and the lowest for VGG16 (negligible from 0.877 to 0.878).

The testing times without TTA shown the quite different values in the limits of the measured standard deviations that is explained by the various complexity of these DNN architectures (Fig. 5).

The testing times with TTA shown the different values in the limits of the measured standard deviations also (Fig. 6), but these small differences (except for the biggest DNNs like EfficientNetB6 and InceptionResNetV2) are explained by the nearly the same large overhead time on TTA for all DNNs.

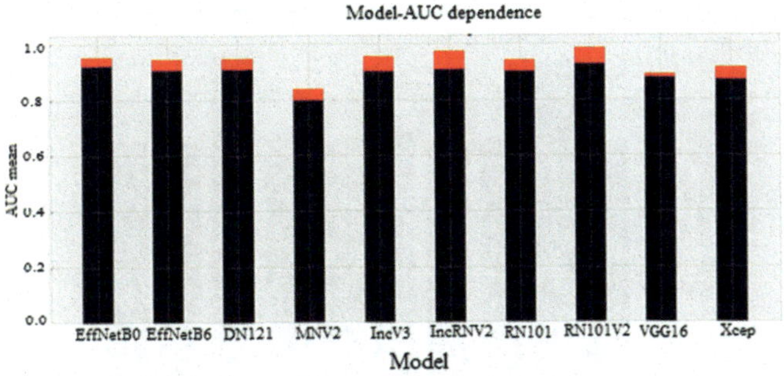


**Fig. 3.** AUC values without TTA (mean values – black, standard deviations – red) for the different model architectures

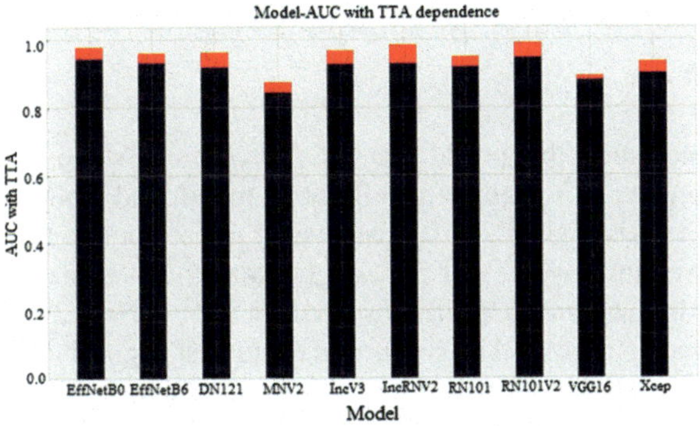


**Fig. 4.** AUC values with TTA (mean values – black, standard deviations – red) for the different model architectures

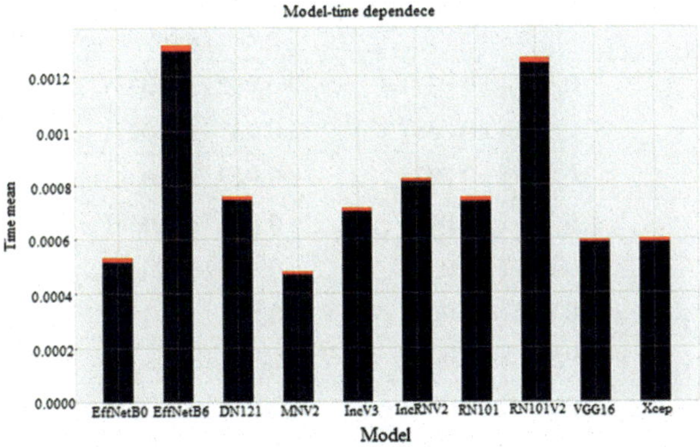


**Fig. 5.** The testing times without TTA (mean values – black, standard deviations – red) for the different model architectures

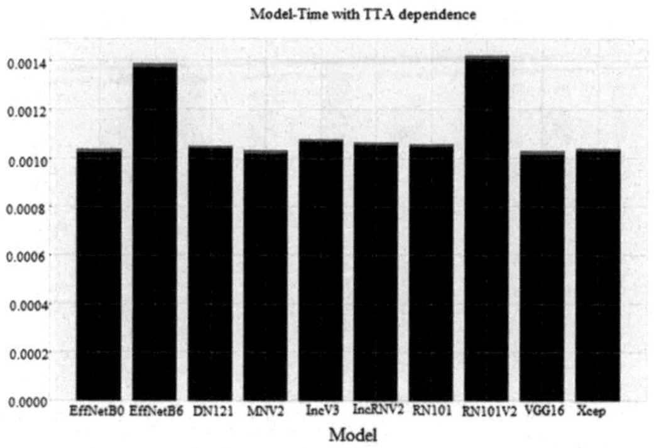


**Fig. 6.** The testing times with TTA (mean values – black, standard deviations – red) for the different model architectures

## 5 Discussion

The results obtained show that in addition to DA on training stage, usage of testing with TTA led to the higher AUC values for all above mentioned DNNs (in comparison to testing without TTA), but with the different level of improvement. But it should be noted that not all observed improvements can be higher than the measured standard deviation. For example, the highest increase was observed for MobileNetV2 from 0.797 (with the standard deviation 0.040) to 0.841 (0.030), and the lowest negligible increase for VGG16 from 0.877 (0.011) to 0.878 (0.012).

**Table 1.** AUC values for testing with/without TTA

| DNN | AUC without TTA, mean (stdev) | AUC with TTA, mean (stdev) | AUC increase |
|---|---|---|---|
| EfficientNetB0 | 0.920 (±0.031) | 0.942 (±0.033) | 0.022 |
| EfficientNetB6 | 0.905 (±0.038) | 0.931 (±0.027) | 0.026 |
| DenseNet121 | 0.910 (±0.037) | 0,917 (±0.044) | 0.007 |
| **MobileNetV2** | **0.797 (±0.040)** | **0,841 (±0.030)** | **0.044** |
| InceptionV3 | 0.902 (±0.030) | 0,926 (±0.040) | 0.024 |
| ResNet101 | 0.907 (±0.066) | 0,928 (±0.054) | 0.021 |
| ResNet101V2 | 0.901 (±0.040) | 0,917 (±0.030) | 0.016 |
| InceptionResNetV2 | 0.926 (±0.056) | 0,944 (±0.043) | 0.018 |
| VGG16 | 0.877 (±0.011) | 0,878 (±0.012) | 0.001 |
| Xception | 0.869 (±0.043) | 0,898 (±0.035) | 0.029 |

It should be noted the general tendency to increase of AUC after application of TTA, but the careful analysis demonstrated that all these increases are in the limits of the standard deviations (Table 1) except for the smallest MobileNetV2 DNN (emphasized by the bold font in Table 1). Despite the common tendency to increase of AUC mean values, these conclusions cannot be considered as statistically strong and reliable with regard to DNNs other than MobileNetV2 and additional research for other DNNs will be needed. Nevertheless, MobileNetV2 is the smallest one and such DNNs are of great importance for deployment of DNN models on TPU-equipped devices with the limited computational resources [33, 34] for Edge Computing layer [35, 36]. That is why the additional measures to increase their potential performance are of great importance. But the additional aspects of the TTA-related time overheads should be taken into account for practical applications as far as they can be significant: > 2 times more from 0.00042 to 0.00102 s per image for the fastest MobileNetV2 (Fig. 5, 6) and compromise between accuracy and speed should be determined for the specific Edge Computing device and application demands [36].

## 6 Conclusions and Future Work

These results demonstrate the possibility to improve melanoma classification results due to application of TTA procedures. But it is unavoidably followed by decrease of inference times due to TTA-related time overheads. It can be significant especially for the fastest DNNs deployed on TPU-equipped devices with the limited computational resources for Edge Computing layer. The common tendency to increase of AUC mean values was observed for all above mentioned DNNs: EfficientNetB0, EfficientNetB6, DenseNet121, MobileNetV2, InceptionV3, ResNet101, ResNet101V2, InceptionResNetV2, VGG16, and Xception. But this conclusion can be considered as statistically strong and reliable for MobileNetV2 only, because the AUC increase was higher than standard deviation (determined by 5-fold cross-validation) for MobileNetV2 only. And the additional research for other DNNs and on other open datasets with image of various sizes and quality will be needed in the nearest future. That is why the future work will be targeted on these aspects of the problem stated with additional analysis of optimal DNN configurations for the better performance with regard to accuracy and inference time like in similar other works [37–40]. It could be potentially useful for their application in health care, especially for the analysis of medical data like melanoma on Edge Computing layer, namely by ordinary people with mobile gadgets equipped by camera and GPU/TPUs for real-time distant computer-aided detection (CADe) and Computer-Aided Diagnosis (CADx) of some sign of such dangerous diseases like melanoma and others.

**Acknowledgment.** The work was partially supported by the National Research Foundation of Ukraine by the grant 2020.01/0490 "Artificial Intelligence Platform for Distant Computer-Aided Detection (CADe) and Computer-Aided Diagnosis (CADx) of Human Diseases".

## References

1. Chen, Y.W., Jain, L.C.: Deep Learning in Healthcare, vol. 171. Springer, Heidelberg (2020)

2. Esteva, A., et al.: A guide to deep learning in healthcare. Nat. Med. **25**(1), 24–29 (2019). https://doi.org/10.1038/s41591-018-0316-z
3. Faust, O., Hagiwara, Y., Hong, T.J., Lih, O.S., Acharya, U.R.: Deep learning for healthcare applications based on physiological signals: a review. Comput. Methods Programs Biomed. **161**, 1–13 (2018)
4. Coutts, L.V., Plans, D., Brown, A.W., Collomosse, J.: Deep learning with wearable based heart rate variability for prediction of mental and general health. J. Biomed. Inform. **112**, 4–14 (2020)
5. Rajpurkar, P., et al.: Chexnet: radiologist-level pneumonia detection on chest x-rays with deep learning, 2–7 (2017). arXiv preprint: arXiv:1711.05225
6. Gordienko, Y., et al.: Deep learning with lung segmentation and bone shadow exclusion techniques for chest x-ray analysis of lung cancer. In: Zhengbing, H., Petoukhov, S., Dychka, I., He, M. (eds.) Advances in Computer Science for Engineering and Education, pp. 638–647. Springer International Publishing, Cham (2019). https://doi.org/10.1007/978-3-319-91008-6_63
7. Gang, P., et al.: Dimensionality reduction in deep learning for chest X-ray analysis of lung cancer. In: 2018 Tenth International Conference on Advanced Computational Intelligence (ICACI), pp. 878–883 (2018)
8. Sultana, N.N., Puhan, N.B.: Recent deep learning methods for melanoma detection: a review. In: Ghosh, D., Giri, D., Mohapatra, R.N., Savas, E., Sakurai, K., Singh, L.P. (eds.) Mathematics and Computing, pp. 118–132. Springer Singapore, Singapore (2018). https://doi.org/10.1007/978-981-13-0023-3_12
9. Cui, X., et al.: Assessing the effectiveness of artificial intelligence methods for melanoma: a retrospective review. J. Am. Acad. Dermatol. **81**(5), 1176–1180 (2019). https://doi.org/10.1016/j.jaad.2019.06.042
10. Naeem, A., Farooq, M.S., Khelifi, A., Abid, A.: Malignant melanoma classification using deep learning: datasets, performance measurements, challenges and opportunities. IEEE Access **8**, 110575–110597 (2020)
11. Adegun, A., Viriri, S.: Deep learning techniques for skin lesion analysis and melanoma cancer detection: a survey of state-of-the-art. Artif. Intell. Rev. **54**(2), 811–841 (2020). https://doi.org/10.1007/s10462-020-09865-y
12. Knoll, F., et al.: Deep-learning methods for parallel magnetic resonance imaging reconstruction: a survey of the current approaches, trends, and issues. IEEE Signal Process. Mag. **37**(1), 128–140 (2020)
13. Wang, J., Zhu, H., Wang, S.-H., Zhang, Y.-D.: A review of deep learning on medical image analysis. Mob. Netw. App. **26**(1), 351–380 (2020). https://doi.org/10.1007/s11036-020-016 72-7
14. Simonyan, K., Zisserman, A.: Very deep convolutional networks for large-scale image recognition, 1–14 (2014). arXiv preprint arXiv:1409.1556
15. Menegola, A., Fornaciali, M., Pires, R., Bittencourt, F. V., Avila, S., Valle, E.: Knowledge transfer for melanoma screening with deep learning. In: IEEE 14th International Symposium on Biomedical Imaging (ISBI 2017), pp. 297–300 (2017)
16. Hinton, G.E., Krizhevsky, A., Sutskever, I.: Imagenet classification with deep convolutional neural networks. Adv. Neural. Inf. Process. Syst. **25**, 1106–1114 (2012)
17. He, K., Zhang, X., Ren, S., Sun, J.: Deep residual learning for image recognition. In: Proceedings of the IEEE Conference on Computer Vision and Pattern Recognition, pp. 770–778 (2016)
18. Szegedy, C., et al.: Going deeper with convolutions. In: Proceedings of the IEEE Conference on Computer Vision and Pattern Recognition, pp. 1–9 (2015)

19. Hosny, K.M., Kassem, M.A., Foaud, M.M.: Skin melanoma classification using ROI and data augmentation with deep convolutional neural networks. Multimed. Tools App. **79**(33–34), 24029–24055 (2020). https://doi.org/10.1007/s11042-020-09067-2
20. Hoang, V.D., Hoang, V.T., Jo, K.H.: Hybrid deep learning and data augmentation for disease candidate extraction. In: Frontiers of Computer Vision, Vol. 1212, pp. 274–286 (2020)
21. Li, Y., Shen, L.: Skin lesion analysis towards melanoma detection using deep learning network. Sensors **18**(2), 556 (2018)
22. Stirenko, S., et al.: Chest X-ray analysis of tuberculosis by deep learning with segmentation and augmentation. In: IEEE 38th International Conference on Electronics and Nanotechnology (ELNANO), pp. 422–428 (2018)
23. Gang, P., et al.: Effect of data augmentation and lung mask segmentation for automated chest radiograph interpretation of some lung diseases. In: Gedeon, T., Wong, K.W., Lee, M. (eds.) Neural Information Processing: 26th International Conference, ICONIP 2019, Sydney, NSW, Australia, December 12–15, 2019, Proceedings, Part IV, pp. 333–340. Springer International Publishing, Cham (2019). https://doi.org/10.1007/978-3-030-36808-1_36
24. Kassani, S.H., Kassani, P.H.: A comparative study of deep learning architectures on melanoma detection. Tissue Cell **58**, 76–83 (2019)
25. Kato, K., et al.: Performance improvement of automated melanoma diagnosis system by data augmentation. Adv. Biomed. Eng. **9**, 62–70 (2020)
26. Rotemberg, V., et al.: A patient-centric dataset of images and metadata for identifying melanomas using clinical context, 1–12 (2020). arXiv preprint: arXiv: 2008.07360
27. Chollet, F.: Xception: deep learning with depthwise separable convolutions. In: Proceedings of the IEEE Conference on Computer Vision and Pattern Recognition, pp. 1251–1258 (2017)
28. Szegedy, C., Vanhoucke, V., Ioffe, S., Shlens, J., Wojna, Z.: Rethinking the inception architecture for computer vision. In: Proceedings of the IEEE Conference on Computer Vision and Pattern Recognition, pp. 2818–2826 (2016)
29. Szegedy, C., Ioffe, S., Vanhoucke, V., Alemi, A.: Inception-v4, inception-RESNET and the impact of residual connections on learning, 1–12 (2016). arXiv preprint arXiv:1602.07261
30. Sandler, M., Howard, A., Zhu, M., Zhmoginov, A., Chen, L.C.: Mobilenetv2: inverted residuals and linear bottlenecks. In: Proceedings of the IEEE Conference on Computer Vision and Pattern Recognition, pp. 4510–4520 (2018)
31. Huang, G., Liu, Z., Van Der Maaten, L., Weinberger, K.Q.: Densely connected convolutional networks. In: Proceedings of the IEEE Conference on Computer Vision and Pattern Recognition, pp. 4700–4708 (2017)
32. Tan, M., Le, Q.: EfficientNet: rethinking model scaling for convolutional neural networks. Int. Conf. Mach. Learn. **2019**, 6105–6114 (2019)
33. Cao, K., Liu, Y., Meng, G., Sun, Q.: An overview on edge computing research. IEEE Access **8**, 85714–85728 (2020)
34. Gordienko, Y., et al.: Augmented coaching ecosystem for non-obtrusive adaptive personalized elderly care on the basis of cloud-fog-dew computing paradigm. In: 2017 40th International Convention on Information and Communication Technology, Electronics and Microelectronics (MIPRO), pp. 359-364 (2017)
35. Kochura, Y., et al.: Batch size influence on performance of graphic and tensor processing units during training and inference phases. In: Zhengbing, H., Petoukhov, S., Dychka, I., He, M. (eds.) Advances in Computer Science for Engineering and Education II, pp. 658–668. Springer International Publishing, Cham (2020). https://doi.org/10.1007/978-3-030-16621-2_61
36. Gordienko, Y., et al.: Scaling analysis of specialized tensor processing architectures for deep learning models. In: Pedrycz, W., Chen, S.-M. (eds.) Deep Learning: Concepts and Architectures, pp. 65–99. Springer International Publishing, Cham (2020). https://doi.org/10.1007/978-3-030-31756-0_3

37. Hanon, A., Alasadi, B., Alsafy, M.: Early detection and classification of melanoma skin cancer. I. J. Inf. Technol. Comput. Sci. **7**(12), 67–74 (2015)
38. Emuoyibofarhe, J.O., Ajisafe, D.: Early skin cancer detection using deep convolutional neural networks on mobile smartphone. Int. J. Inf. Eng. Electron. Bus. **12**(2), 21–27 (2020). https://doi.org/10.5815/ijieeb.2020.02.04
39. Neela, A.G.: Implementation of support vector machine for identification of skin cancer. Int. J. Eng. Manuf. **9**(6), 42–52 (2019). https://doi.org/10.5815/ijem.2019.06.04
40. Sharma, A., Kulshrestha, S., Daniel, S.B.: Machine learning approaches for cancer detection. Int. J. Eng. Manuf. **8**(2), 45–55 (2018). https://doi.org/10.5815/ijem.2018.02.05

# Research and Development of Interactive Virtual Disassembly Teaching System Considering Difficulty Gradient

Qi Wang[1], YanFang Yang[1](✉), Qiang Bai[2], XiaoShu Zhang[1], YaHui Xing[1], and ZhiPing Liu[1]

[1] School of Logistics Engineering, Wuhan University of Technology, Wuhan, China
yangyanfang@whut.edu.cn
[2] China Hangfa Hunan Mechanical Power Research Institute, Zhuzhou, China

**Abstract.** In order to improve the immersion and reality of virtual disassembly process and stimulate the enthusiasm and creativity of learners, this paper designs a virtual disassembly teaching system that considers the gradient of difficulty. Firstly, the difficulty of disassembly is improved through rich interactive operations. Based on the virtual disassembly of mouse interaction, one-handed disassembly and two-handed collaborative disassembly of the HTC Vive handle interactive disassembly are realized, which improves the sense of reality of disassembly experience. Secondly, the virtual disassembly guided by the free mode is realized by importing component constraint relations into the system. Finally, this article uses typical deceleration devices as examples to build a virtual disassembly learning system, and uses Unity 3D as a development platform to complete the development of a virtual teaching system.

**Keywords:** Difficulty gradient · One-hand disassembly · Two-hand collaborative disassembly

## 1 Instruction

As a prerequisite step, disassembly of parts is an indispensable operation in the process of maintenance and recycling, and it is particularly important for the training and teaching of mechanical equipment disassembly operations [1]. The virtual disassembly teaching system uses virtual reality technology to create real disassembly scenes, guide learners to learn disassembly knowledge and experience the disassembly process [2]. It has the characteristics of high safety, low cost, and freedom from site restrictions, which can effectively save economic costs. At present, virtual disassembly technology has been widely used in aerospace, military machinery and equipment, medical treatment, education and other fields, and has attracted the attention of researchers at home and abroad.

For the research of virtual disassembly teaching system, the University of Pennsylvania designed the AVIS-MS virtual maintenance simulation system, focusing on the


Z. Hu et al. (Eds.): ICAILE 2021, LNDECT 82, pp. 141–151, 2021.
https://doi.org/10.1007/978-3-030-80475-6_14

visualization of the system, and showing the maintenance disassembly and assembly technical procedures to the operator through visualization technology [3].

The German Fraunhofer Institute of Industrial Engineering realized human-computer interaction by establishing an operator model, and the system won the Best System Award at the Munich Computer Exhibition [4]. Yang proposed a new idea of combining disassembly sequence planning with tactile disassembly to build a comprehensive virtual disassembly platform. Greatly improves the authenticity of the disassembly [5]. Wang proposed a new evaluation method based on the visibility of parts, disassembly Angle, number of tools, direction change and stability of subassembly for the evaluation of disassembly operation in virtual environment [6]. Hu used genetic algorithm to obtain the disassembly sequence of virtual disassembly and applied it to the disassembly system of multifunctional machine tools [7]. Zheng realized the virtual disassembling and equipping of power equipment, which has practical significance for the staff training of thermal power plants [8]. Zhou proposed a virtual engine disassembly training and evaluation system based on virtual reality simulation and network technology, which achieved the teaching purpose of classroom experiment. Effectively cultivate the practical ability of students [9]. Yang Pu designed a portal crane virtual disassembly system, considering the disassembly process and disassembly sequence, which can be used for the training of disassembly personnel and the design of recovery process [10]. Chen Hao used virtools to design a virtual disassembly system, introduced the disassembly process to guide disassembly, and designed the corresponding collision detection technology [11]. Li Jia used the PHANTOM Omni force feedback device to introduce physical properties during the disassembly process and designed the force feedback disassembly system [12]. In the construction of virtual disassembly teaching system, the difficulty gradient is not considered, and the stimulation of learning mode, degree of freedom and innovative thinking of learners needs to be further strengthened [13].

In order to solve the problems existing in the current virtual disassembly teaching system, this paper uses Unity 3D as the development platform to realize the virtual disassembly teaching system considering the difficulty gradient, which improves the difficulty from the interactive mode [14]. Two interactive modes of mouse interaction and HTC Vive are set up. On the basis of mouse interaction, handle and helmet interaction are added to realize single-handle disassembly and two-handed cooperative disassembly operation for complex devices, which promotes the realism of the interactive experience. At the same time, the component constraint relationship is imported into the system to realize a free disassembly mode, giving the operator more freedom experience.

## 2 System Framework Design

This paper designs the overall framework of the virtual teaching system, which has the following characteristics: 1) It has a variety of interaction methods, using mouse and HTC VIVE for interaction, which can realize multi-style human-computer interaction and reflect the objective world more truly; 2) It has a multi-guided disassembly mode. Based on the interaction between the mouse and HTC Vive, it provides a virtual disassembly sequence guide and free disassembly mode.

Figure 1 shows the framework of the virtual disassembly teaching system. The main scenes of the virtual disassembly system are mainly divided into three modules:

1) The first is the basic demonstration module, which displays the disassembly theory and parts information. It has four parts: design criteria, standard parts, working principles, and disassembly sequence;
2) The second is the mouse disassembly module. The mouse is used for disassembly interaction to complete the two disassembly modes of disassembly sequence guidance and component constraint relationship guidance;
3) The third is the VR interactive disassembly module. Use HTC Vive to disassemble the handle interactively, which can realize three disassembly modes: single-handed, dual-handle coordination, and free disassembly.

For the same interaction mode, there are two disassembly modes, which are divided according to different guidance modes, namely disassembly sequence guidance and component constraint relationship guidance. The former has low degree of freedom, while the latter has high degree of freedom and can derive various difficulty gradients, which improves the richness of the system.

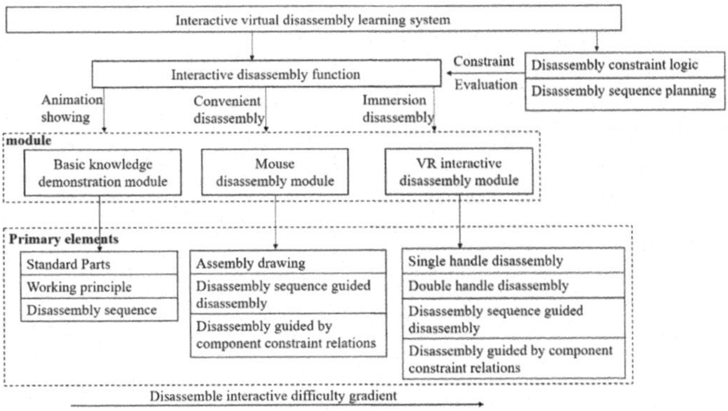


**Fig. 1.** Virtual disassembly of the teaching system framework

## 3 The Key Technology

Figure 2 shows the technical roadmap of the virtual disassembly teaching system. Firstly, the model is established and optimized, and some similar nodes are grouped to simplify the disassembly process. Secondly, collision bodies are added to parts to realize collision detection. Subsequently, the interactive module was built in the Unity 3D engine, including disassembly auxiliary functions, animation demonstration functions, mouse interaction functions and HTC Vive interaction functions, as the main body of the disassembly system. After the disassembly interaction is realized, the fixed optimized disassembly sequence is obtained through the optimization algorithm, and the disassembly optimization sequence is imported into the virtual disassembly teaching system to guide the HTC Vive single handle disassembly and dual handle disassembly. Import component constraints into the teaching system to achieve free disassembly mode.

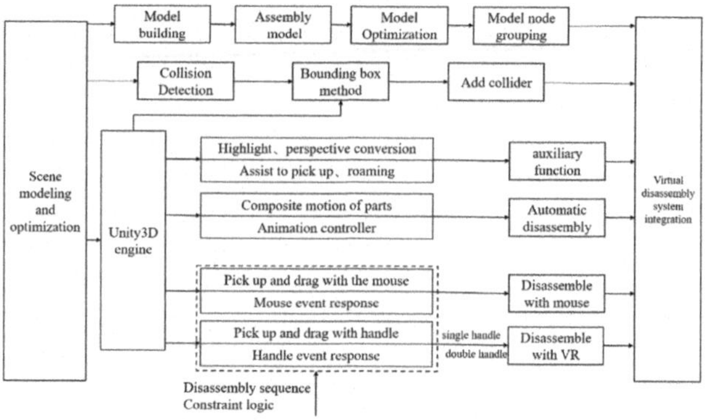


**Fig. 2.** Technical route of virtual disassembly teaching system

## 3.1 Single Handle Disassembly

In the virtual disassembly based on HTC Vive, users mainly use the handle to complete the steps of picking, moving, and storing parts, and realize the single handle disassembly operation.

Create a project in Unity3d, import VRTK, and complete the preparations for VR development. Add an information storage script to the parts to feed back the disassembly information and initial position information of the parts. In order to increase the reliability and realism of disassembly and avoid the collision between parts caused by the shaking of the handle during disassembly, recoding is adopted instead of the built-in picking function. The flow chart of parts picking is shown in Fig. 3.

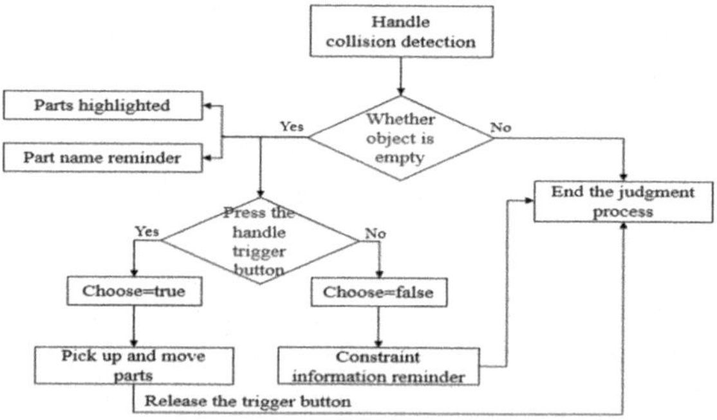


**Fig. 3.** Parts picking process

The system obtains feedback information from the parts through the handle collision detection, and detects whether the handle is idle. Use the variable "Object" to record the state of the handle currently picking up parts. Only when the value of "Object" is "Empty", the handle can pick up the current part. After the handle picks up a part, change the Object value to lock the part. The handle cannot pick up any other part until the handle puts down the part. Record the initial position of the part to prepare for subsequent part movements. Releasing the trigger button will stop the movement.

The handle picks up the part to obtain information. After judging the feasibility of handle disassembly, the initial movement direction of the part is obtained, and the component moves along the initial direction. Most parts initially move along the coordinate axis, and the initial direction of some parts may not follow the direction of the coordinate axis. Special initial directions need to be defined by the other vector. In order to facilitate uniform coding, an interpolation method is introduced, moving along the vector direction, and using the coefficient k to determine the moving distance. Unity 3D works by using the Update function to call frame by frame. The system is set to call this function once every 0.02 s to achieve real-time synchronization between the handle and the part. As shown in Eqs. (1) and (2).

$$k = ((trigger - device)/(move.x)) \tag{1}$$

$$position = intial + move * k \tag{2}$$

*trigger* is the real-time coordinates of the handle.
*devices* is the initial coordinate of the handle.
*move* is the component direction vector.
*intial* is the initial position of the component.
*position* is the real-time position of components.

In order to prevent the disassembled part from blocking the line of sight, when the part moves in the corresponding direction for a certain distance, it is determined to have been disassembled and stored on the storage platform.

### 3.2 Collaborative Disassembly with Both Hands

Two-handed cooperative disassembly refers to a single person using a VR device to operate two handles for cooperative disassembly. It can be divided into two disassembly modes.

1) For some parts that are too large and difficult to disassemble, the part can be disassembled only when two handles collide with the part at the same time.
2) Simulate the parallel disassembly. When the same user is disassembling, in the same disassembly cycle, two handles can respectively disassemble two parts to simulate the parallel disassembly process.

The logic process of double handle disassembly is shown in Fig. 4 Logic processing and disassembly function modules: This module realizes the basic functions of judging the feasibility of parts disassembly and picking up and moving parts. The judgment of the feasibility of part disassembly is divided into two processes. It is judged whether the part is constrained by other parts, and whether the part needs to be disassembled by two handles.

Part information module: This module is used to store part information, including the disassembly status and initial movement direction of the part, and judge whether the part has been disassembled.

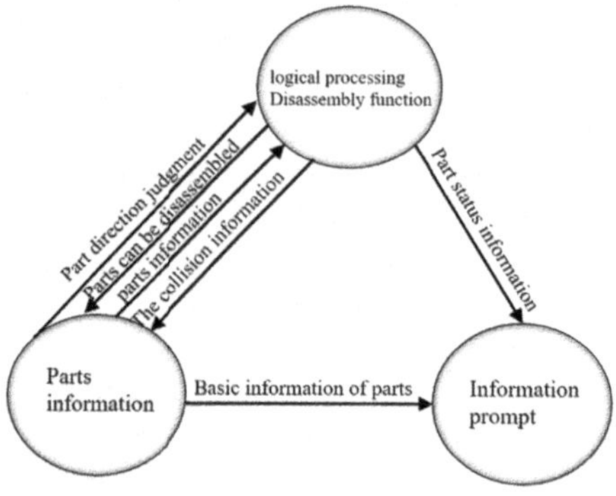


**Fig. 4.** Two-hand collaborative disassembly logic flow

Information prompt module: This module displays the basic information and status information of the part on the helmet display, including the name of the picked part, the status of the part disassembly, and the constraint information of the part.

Main steps are listed:

1) When the handle collides with the part, the logic processing module will send a signal to the identified part, and the part will feed back the part name and part number information to the logic processing module. At the same time, the information is passed to the information prompt module.
2) The logic processing module judges the feasibility of the disassembly of the part. The feasibility information will be passed to the other two modules. The information prompt module processes the feasibility information. If the part cannot be disassembled, it will display the reason why the disassembly cannot be carried out, including the state of the part, the constraints, whether it needs to be picked up together with both hands and other information.
3) In the part information module, the initial movement direction of the part is returned when the part can be disassembled, and the subsequent disassembly function is carried out.

Using the Unity 3D engine, logic processing and disassembly function modules are added to the two handles, so that both handles can independently process part information and realize the disassembly function. In order to avoid the logic conflict when the handle picks up the same part, when any handle picks up the part, the part information module will only feedback the basic information of the part's disassembly to the handle, so that the other handle cannot meet the preconditions of disassembly and cannot pick up and disassemble the same part. At the same time, the part information module feedback information to the information prompt module, displayed in the helmet display. For some parts that need to be disassembled together with double handles, when judging the feasibility of disassembling parts, constraints for picking up together with double handles are added.

### 3.3 Free Disassembly Mode

The component constraint relation is imported into the system in the form of logical constraint to guide the user's disassembly process.

1) In the Unity 3D engine, the component constraints of the deceleration device are described in C# language, a component set is established, and then the parent transition set of each component is established, and stored in the logic judgment script to facilitate subsequent component constraint logic Judge, define the disassembly progress *i*, initially $i = 1$, and the total number of disassembly steps is n;
2) Create part information storage script for storing and feedback part information, including part name and sequence number, for subsequent information transfer and prompt for picking up information. Define part serial number *j*;
3) When the handle and the component collide, the component information storage script satisfies the trigger condition, and the script transfers the collision component information to the logic judgment script. At the same time, highlight the parts and display the name information of the parts on the display device;
4) According to the part constraint logic stored in (1), judge whether the current part parent change set is empty. When the parent change set is empty, the part can be disassembled. If it is not empty, prompt on the display device.
5) Judge whether i is equal to n + 1, if not, return step (3), exit the disassembly process when it is equal.

By importing the constraint relationship between components, the free disassembly mode can be realized. In order to realize the part disassembly prompt, the parent object information of each part should be stored in advance. When the picked up part cannot be disassembled, the parent object information of the part should be displayed in the helmet display through GUI function.

The disassembly feasibility condition of the parts is that the pre-transition must have taken place. Although the constraint relation between the parts is linear and convenient for logical description and coding, the number of logical constraint relation between the parts is large and the description of the relation is too complex. The parent-child relationship is adopted to reflect the priority of parts and the coding process is also simplified in this way.

## 4 System Implementation

In order to verify the feasibility and effectiveness of the method proposed in this paper, a typical reduction gear was used as a case for verification. As shown in Fig. 5, the system interaction interface is designed with Unity 3D as the development platform. Three disassembly modes including automatic disassembly, mouse disassembly and HTC Vive interactive disassembly are integrated to develop an interactive virtual disassembly. At the same time, in order to closely combine the virtual disassembly system with the engineering practice, disassembly sequence planning is used to generate disassembly sequence that conforms to the actual disassembly process, so as to guide the disassembly process. Finally, the system completes the virtual disassembly system with reducers and gearboxes as the disassembly objects. The difficulty is gradually increased in terms of the complexity of the interaction mode and the freedom of disassembly, which is convenient for users to choose learning methods of different difficulties and learn disassembly process from different disassembly methods.

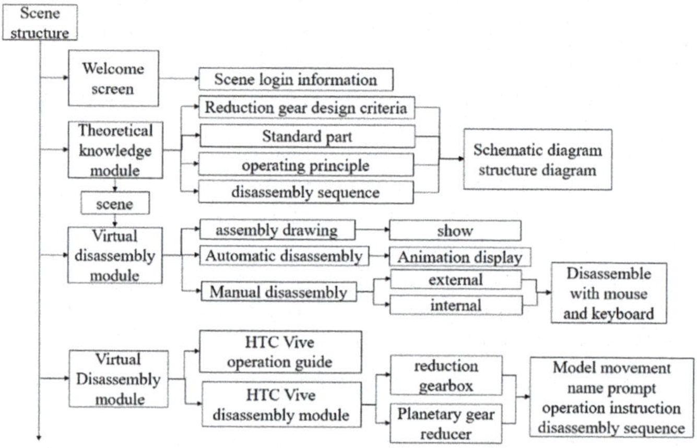


**Fig. 5.** Virtual disassembly of teaching system scene structure

## 4.1 Automatic Disassembly Mode

Through Unity's Animator Controller component, combined with the mouse and the event response mode of button, trigger the composite motion animation that has been made in advance.

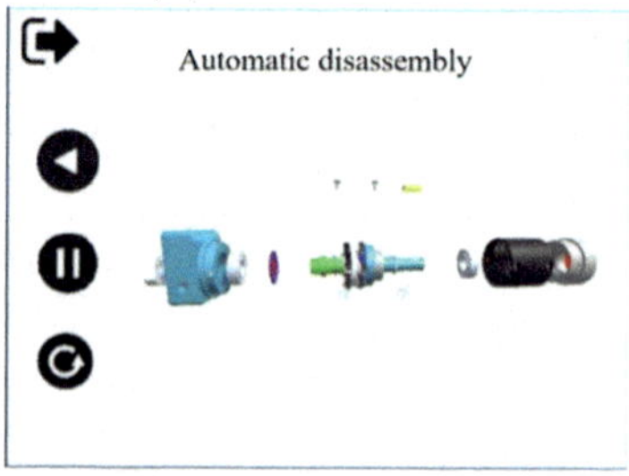


**Fig. 6.** Automatic disassembly process of reducer

Therefore, click the button to achieve some functions such as play, pause and replay to achieve Automatic disassembly with interactive functions. Figure 6 is the automatic disassembly process of the reducer.

### 4.2 Disassembly with Mouse Mode

Control disassembly tools, disassembly sequence, path planning, etc. through disassembly Petri nets. Use the mouse to complete the disassembly of all parts one by one and then complete the disassembly of the entire mechanical structure. The process of disassembling the reducer through the mouse is shown in Fig. 7.

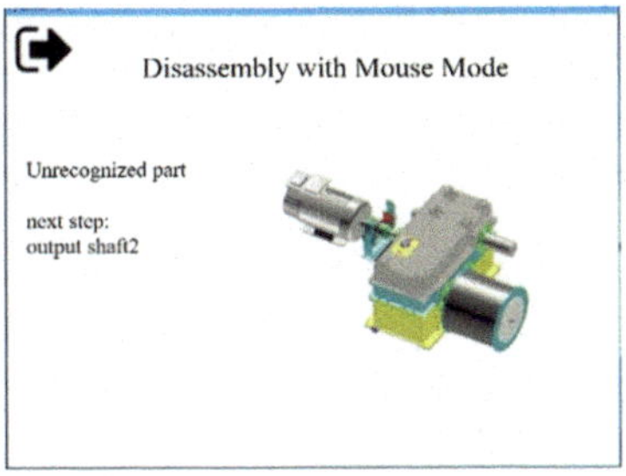


**Fig. 7.** Reducer disassembly with mouse

### 4.3 HTC Vive Disassembly Method

When the moving handle collides with the part to be removed, the trigger button will be pressed and the part will follow the handle to move the part to the appropriate position. The screen will show whether the components are picked, and the constraint relationship of the components will be displayed. Figure 8 shows the disassembly of the gearbox in the two-hand disassembly mode. In the picture, the left handle has picked up the oil collecting hood of the input shaft. The right handle collides with the middle box, the middle box is constrained and cannot be picked up, and the collection of parent objects of the middle box is displayed. The right handle needs to disassemble other parts. When the left and right hands are completely disassembled, the disassembly step is ended.

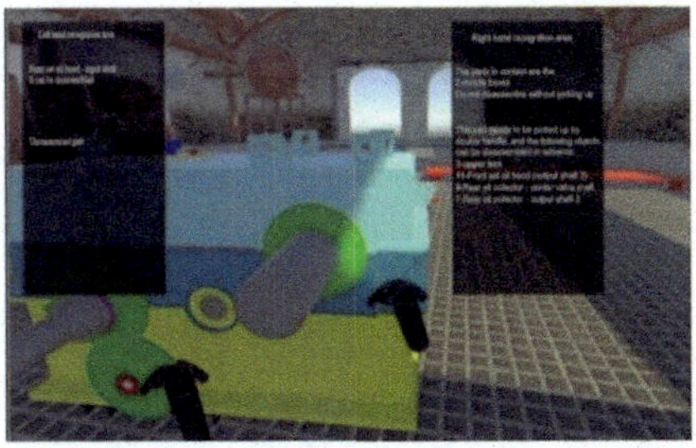

**Fig. 8.** Reduction gearbox disassembly with both hands

## 5 Conclusion

This paper builds a virtual disassembly teaching system that considers multiple difficulty gradients, enriches the difficulty gradients through guidance modes and interactive methods, and realizes the cooperative operation of complex equipment with two handles. The main contributions and conclusions include the following:

1) This paper designs a virtual disassembly system that considers multiple difficulty gradients, emphasizing the level of learning and interaction, and improving the operator's interest in learning.
2) This system is based on the constraint relationship of components, and is imported into the virtual disassembly system to realize the free disassembly mode.
3) Constructed and optimized objects and scene models, developed a virtual disassembly teaching system in the Unity 3D engine, including basic knowledge display module, mouse interaction module and HTC Vive interaction module. It realizes one-handed operation and two-handed cooperative operation in the process of HTC Vive handle interaction.

**Acknowledgment.** This work is supported by "the Fundamental Research Funds for the Central Universities (WUT: 202418001)".

## References

1. Pomares, J., Puente, S.T., Torres, F., et al.: Virtual disassembly of products based on geometric models. Comput. Ind. **55**(1), 1–14 (2004)
2. Fatima, S., Abdullah, S.: Improving teaching methodology in system analysis and design using problem based learning for ABET. Int. J. Modern Educ. Comput. Sci. **5**(7), 60–68 (2013)
3. Hao, W., Hong, Z., Jundong, Z., Yuefeng, G.: Modeling of virtual assembly and disassembly process based on selective disassembly. J. Graph. **36**(06), 903–908 (2015). (in Chinese)
4. Salleh, N.A.M., Kasolang, S., Mustakim, M.A., et al.: The study on optimization of streamlined process flow based on delmia quest simulation in an automotive production system. Proc. Comput. Sci. **105**, 191–196 (2017)
5. Yanfang, Y., Yang, P., Jia, L., Fan, Z., et al.: Research on virtual haptic disassembly platform considering disassembly process. Neurocomputing **348**, 74–81 (2018)
6. Wang, C.G., Mitrouchev, P., Li, G.Q., Lu, L.X.: Disassembly operations' efficiency evaluation in a virtual environment. Int. J. Comput. Integ. Manuf. **29**(3), 309–322 (2016). https://doi.org/10.1080/0951192X.2015.1033752
7. Hu, P., Wan, Z.: Research on virtual assembly sequence optimization based on genetic algorithm. In: International Conference on Advances in Image Processing, pp. 169–173 (2017)
8. Xie, Z., Zhou, Y., He, Y., Fu, X.: Design of virtual disassembly system of front pump based on unity-3D. In: International Conference on Virtual Reality and Intelligent Systems (ICVRIS), pp. 17–19 (2019)
9. Zhou, C., Wang, W., Tian, Y., Jiang, Z.: Research on virtual disassembly and assembly training and assessment system of engine. In: WCSE 2020: 10th International Workshop on Computer Science and Engineering, pp. 366–371 (2020)

10. Yang, P., Yanfang, Y.: Design and key technologies of the virtual disassembly system of portal cranes. Lifting Transp. Mach. **10**, 87–92 (2018). (in Chinese)
11. Chen, H., Zhang, G., Zhang, Q.: Virtual dismounting with physical effects based on Virtools platform. Modern Manuf. Eng. **2015**(01), 30–34 (2015). (in Chinese)
12. Jia, L., Wen, H., Wenwen, X., Yanfang, Y., Dingfang, C.: Research on the key technology of virtual disassembly system based on force feedback. J. Graph. **3**, 450–454 (2014). (in Chinese)
13. Xianmin, W.: Discovery and practice of EDA experimental teaching reform. Int. J. Educ. Manage. Eng. **1**(4), 41–45 (2011)
14. Jianjun, L.: Research on strategies and patterns of research teaching. Int. J. Educ. Manage. Eng. **2**(12), 67–72 (2012)

# Teaching Reform and Empirical Research of Logistics Professional English Based on Super Star Learning APP Platform

Jinling Xia(✉) and Jingjing Wu

Economics and Business Foreign Language School, Wuhan Technology and Business University, Wuhan 430074, China

**Abstract.** With the advancement of education informationization and technology, some colleges and universities have begun to encourage teachers and students to carry out flipped teaching to meet the needs of the information technology era. Our research group doubts whether flipped teaching is a more suitable way of English teaching for logistics majors compared with traditional face-to-face teaching. What problems will arise in implementing flipped teaching? Therefore, this research implemented flipped teaching experiment based on Super star learning APP platform in the logistics professional class, compared it with the traditional face-to-face teaching, and tested the effect of the two teaching modes through empirical research, so as to determine whether flipped teaching, as a new teaching mode, has better teaching effect. On the basis of the experiment, the result shows that flipped teaching based on Super star learning APP has better teaching effect than the traditional face-to-face teaching, and has more advantages than the traditional face-to-face teaching in terms of improving students' academic performance, optimizing students' learning efficiency and improving students' autonomous learning ability. In addition, the paper points out the problems that may arise when flipped teaching is carried out in our college, and puts forward corresponding solutions.

**Keywords:** Teaching reform · Empirical research · Logistics professional English · Super star learning APP platform

## 1 Introduction

Logistics is a practical industry. Graduates majoring in logistics management must have a certain level of English if they want to work in a large logistics company, such as some logistics companies that want to transport the goods to foreign countries [1]. How to use effective teaching methods to improve the effect of English teaching is the problem that English teachers pay great attention to [2]. Li Shanshan analyzes how to teach logistics English in colleges and universities based on the job demands [3]. Zhang Yu analyzed the problems existing in the course of Logistics English in higher vocational colleges in detail, put forward solutions, and explained how teachers should respond to the needs of enterprises in the teaching of professional English, so as to consolidate


Z. Hu et al. (Eds.): ICAILE 2021, LNDECT 82, pp. 152–160, 2021.
https://doi.org/10.1007/978-3-030-80475-6_15

students' professional English ability and continuously provide professional talents for social development [4]. As a new teaching mode, flipped teaching has attracted wide attention among the teachers [5–7]. In order to explore whether flipped teaching has better effect than traditional face-to-face teaching, this study takes college English teaching of our university as an example and adopts super star learning APP as the platform to implement flipped teaching, and tests the effect of the two teaching modes through empirical research [8].

## 2 Flip Teaching Based on Super Star Learning APP

From the perspective of teaching order, flipped teaching and traditional face-to-face teaching are the reverse of the order after class and in class. In the traditional face-to-face teaching, the emphasis is on knowledge transfer in class and knowledge internalization after class. The flipped teaching in this study is to reverse this order, which realizes knowledge transfer before class and knowledge internalization in class. In the implementation of flipped teaching between the two links, a suitable platform is essential.

In the context of the rapid development of mobile network and the heavy dependence of students on mobile phones, it is an inevitable trend to combine classroom learning with mobile learning. Super star learning APP is applied to mobile learning. The powerful functions, advanced teaching technology, advanced teaching concept and various teaching demands are organically integrated. It is a practical, comprehensive and easy to operate teaching software, integrating multiple teaching functions such as roll call, quick answer, interaction, query and questionnaire. Super star learning APP is used as a platform to implement flipped teaching. Not only can teachers use it to assist classes, but also students can use it to learn to charge, consult materials and share massive resources. Its powerful learning function and good user experience are the reasons why our research group chooses it as a flipped teaching platform. The flipped teaching design based on Super star learning APP in the study is shown in the figure below.

In the whole teaching process design, the research group deconstructs the traditional teaching materials and teaching ecological environment, takes Super star learning APP as a platform, and resets the teaching mode and content before class and in class.

Before class: Teachers upload teaching resources to the Super star learning APP. Teaching resources mainly include guided learning plan, PPT, audio and video, etc. Teachers set task points, assign learning tasks, publish before-class testing exercises or topic discussions. According to the teaching resources on the Super star learning APP, students learn independently under the guidance of learning plan and specific task points, and finish the homework assigned by teachers after forming their own understanding and internalization and absorption.

In class: The goal is to internalize knowledge in class. First of all, before classroom teaching, teachers summarize the problems existing in students' independent learning with the help of the learning access platform. Then, after the teacher completes the guided and framed lectures, the students, based on independent study before class, bring the summarized difficult problems to classroom discussion and make a presentation for the achievement exchange, and the teacher gives comments. Finally, teachers answer questions about common problems in independent learning and summarize key points of knowledge (Fig. 1).

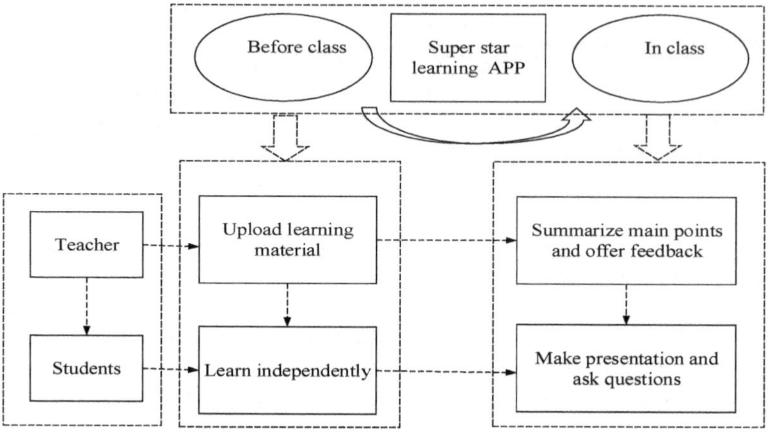


**Fig. 1.** Flip teaching design based on super star learning APP

# 3 Empirical Contrastive Study of Flip Teaching and Traditional Face-To-Face Teaching

## 3.1 Basic Information of Teaching Experiment

In this teaching experiment, freshmen who have just entered the university are selected as the experimental objects. In order to ensure the effectiveness and preciseness of the teaching experiment, the study investigated and studied students' scores of English subject in college entrance examination, self-learning ability and self-efficacy. SPSS software was used to analyze the data and strictly ensure the consistency of the selected classes in all aspects. According to the analysis results, the logistics major is determined to carry out flipped teaching, and the telecommunications major is determined to carry out traditional face-to-face teaching, with the same teacher to teach.

From October 2018 to January 2019, an empirical comparative study of flipped teaching and traditional face-to-face teaching was started in a 12-week research. There are four classes a week for College English 1, a total of 48 classes in a semester. The teacher used the first volume of the New Edition of Advanced College English Comprehensive Course, and the learning content was the first four units. Traditional English teaching was used in the control group, and flipped teaching was used in the experimental group. During the experiment period, the teaching content, teaching schedule and teaching teachers of the two classes were consistent.

Based on the reversal of the order between flipped teaching and traditional teaching, in the teaching experiment, the differences between the two classes in and out of class are embodied in the following aspects:

The experimental group of flipped teaching aims at knowledge transfer after class. Teachers upload teaching resources to the learning channel of the Super star learning APP, teaching resources mainly include learning guidance, PPT, audio and video, etc. set task points, assign learning tasks, and publish before-class testing exercises or topic discussions. According to the teaching resources on the Learning APP, students learn independently under the learning guidance and specific task points, and finish the homework assigned by teachers after forming their own understanding based on the internalization and absorption. The classroom session aims at internalizing knowledge. After the teacher completes the guided and structured lectures, the students, based on self-study before class, bring the summarized questions to class discussion, make presentations for the communication of results, make comments and answer questions, and then summarize the key points of knowledge.

The control group of traditional face-to-face teaching adopted multimedia teaching method. For the purpose of transferring knowledge in class, teachers carefully teach each key point of knowledge according to the rich teaching resources prepared, especially paying attention to students' understanding of the key points and difficult points. Questions, discussions, exchanges and other activities are interspersed in class. The major method of teaching is teacher speaking, students listening and taking notes. After class, students are tested through exercises, tests and other ways to promote the internalization of knowledge.

To sum up, the experimental group's flipped teaching focuses on students' autonomous learning, and students take the initiative to complete the construction and internalization of knowledge. The role of teachers is to guide students' learning ability of independent thinking and inquiry. In contrast, traditional face-to-face teaching in the control group is teacher-centered, in which teacher mainly indoctrinates students to passively accept knowledge, emphasizes the one-way transmission of knowledge, and pays less attention to the individual differences of students and cultivates their thinking ability.

### 3.2 Comparative Analysis of Teaching Effect

At the end of the semester, our research team collected research data from the following aspects: academic performance test, a questionnaire of self-efficacy in learning and a questionnaire of student's satisfaction of flipped teaching based on Super star learning APP.

#### 3.2.1 Academic Achievement Test

In order to verify the advantages of flip teaching in academic performance, the research group made statistical analysis of the final exam results. The final exam is based on the learning content of the semester. The full mark of the exam paper is 100, and the exam time is 100 min. The experimental class and the control class will test at the same time, and final exam paper will be evaluated by the same teacher. All the test results were analyzed by SPSS software, and Independent-Samples T test was conducted. The results are shown as following. In the table, 1 represents experimental class, 2 represents control class.

It can be seen from Table 1 that the average final score of the experimental class was 72.0250, the control class was 63.6905, and obviously the experimental class was higher than that of the control class, which was 8.33452, indicating that the results of the experimental class implementing the flipped teaching were higher than that of the control class implementing the traditional face-to-face teaching. Levene's Test for Equality of

**Table 1.** Independent-samples T Test (final exam)

| | Class | N | Mean | Std.Deviation | F | Sig. | T | Sig.(2-tailed) |
|---|---|---|---|---|---|---|---|---|
| Final Exam | 1.00 | 40 | 72.0250 | 16.42463 | .273 | .603 | 2.463 | 0.16 |
| | 2.00 | 42 | 63.6905 | 14.17915 | | | | |

Variances $p = 0.603 > 0.05$ indicates that the two groups have the same square difference. Check Sig.(2-tailed) $p = 0.016 < 0.05$, and 95% Confidence Interval of Difference does not contain 0, which fully indicates that there is a significant difference between the experimental class and the control class in the final examination results, and flipped teaching has obvious advantages over traditional teaching.

### 3.2.2 Questionnaire of Self-efficacy in Learning

At the end of the semester, the research team collected data from the English self-efficacy questionnaire for the two classes. The questionnaire includes four dimensions: the ability to complete the task, the confidence to achieve the goal, the ability to face setbacks, the ability to overcome difficulties. This questionnaire adopts 5-level Likert scale to score, 1–5 represents the progression from "completely inconsistent" to "very consistent". Each module is scored by adding up the scores for each question to get the average. The results are shown in the table below. In the table, 1 represents experimental class, 2 represents control class.

**Table 2.** Independent-samples T Test (Self-efficacy in English learning)

| Dimension | Class | N | Mean | Std.Deviation | t | p |
|---|---|---|---|---|---|---|
| First dimension | 1.00 | 40 | 22.0750 | 6.58432 | 2.112 | .038 |
| | 2.00 | 42 | 18.9048 | 6.99062 | | |
| Second dimension | 1.00 | 40 | 22.1750 | 3.52927 | 1.765 | .082 |
| | 2.00 | 42 | 20.5238 | 4.86520 | | |
| Third dimension | 1.00 | 40 | 10.0500 | 2.38532 | 1.995 | .049 |
| | 2.00 | 42 | 9.0000 | 2.37877 | | |
| Fourth dimension | 1.00 | 40 | 17.4750 | 3.01269 | 2.078 | .041 |
| | 2.00 | 42 | 15.8810 | 3.85849 | | |
| Total score | 1.00 | 40 | 71.7750 | 12.29444 | 2.507 | .014 |
| | 2.00 | 42 | 64.3095 | 14.51742 | | |

It can be seen from the Table 2 that there are significant differences between the experimental class and the control class in the three dimensions and the overall level of English self-efficacy at the end of the semester. In the first dimension of self-efficacy, we can see the sense of ability to complete tasks (T1, 4, 5, 10, 11, 13, 15, 17). It showed

Equal variances assumed ($P = 0.550$), which means there was significant difference between the two classes in the sense of ability to complete tasks ($P = 0.038 < 0.05$). In the second dimension of self-efficacy, the sense of confidence to achieve the goal (T2, 6, 8, 16, 19, 22), It showed Equal variances not assumed ($P = 0.041$), which means there was no significant difference between the two classes in the sense of confidence to achieve the goal ($P = 0.082 > 0.05$). In the third dimension of self-efficacy, the sense of ability to face setbacks (T9, 12, 21) showed Equal variances assumed ($P = 0.999$), and the two classes showed significant difference in the sense of ability to face setbacks ($P = 0.049 < 0.05$). In the fourth dimension of self-efficacy, the sense of ability to overcome difficulties (T3, 7, 14, 18, 20) showed Equal variances assumed($P = 0.157$), and the two classes showed significant difference in the sense of ability to face setbacks ($P = 0.041 < 0.05$).

To sum up, at the end of the semester, there were significant differences in English self-efficacy between the experimental class and the control class in the first, third and fourth dimensions, and the overall level of English self-efficacy was also significantly different ($P = 0.014 < 0.05$), and the average value of the experimental class was significantly higher than that of the control class. It can be found that compared with traditional teaching, flip teaching can improve students' sense of self-efficacy.

### 3.2.3 Questionnaire on Students' Satisfaction of Flipped Teaching

In this questionnaire, Likert scale at 5 levels were used, and 1–5 represented: totally disagree, relatively disagree, uncertain, relatively agree, and completely agree. 44 questionnaires were issued and 40 valid questionnaires were collected.

From the results of the questionnaire it is found that the vast majority of students believe that flipped teaching has more obvious advantages. For example, compared with the traditional teaching mode, the learning atmosphere of flipped teaching is better, it is easier to get help and improvement from teachers and classmates, and the assessment methods are more diversified and reasonable. The proportion of students who totally disagree and relatively disagree is relatively small.

To the students who were uncertain and disagreed with flipped teaching, the experiment teacher interviewed them and found that the students with uncertain attitude towards flipped teaching felt too unaccustomed or troublesome to use Super star learning APP for flipped learning for the first time, and generally preferred to use QQ. Because of it, the completion rate of homework is only 80% sometimes, and the submission rate is even lower for non-mandatory homework. For example, the total number of students in the experimental class is 44, 22 students submit the assigned composition to the Super star learning APP, and only 18 students participate in the discussion post. At the same time, this part of students think that getting familiar with the flipped teaching needs a learning process. If they are familiar with the Super star learning APP platform and have ability to complete the learning task, they are still willing to cooperate with the teacher in the flipped teaching. However, a very small number of students who totally disagree and relatively disagree with flipped teaching think that their English ability is poor, so they do not want to spend too much time on independent learning with difficulties and do not want to ask for help from teachers and other students, and they prefer to choose traditional face-to-face teaching, that is, teachers speak, they just listen, so as to save trouble. Despite this, the vast majority of students preferred the flipped teaching and still wanted to continue using the Super star learning APP as a platform (43.2% relatively agree, and completely agree13.6%) (Table 3).

**Table 3.** Questionnaire on the effect of flipped teaching

| Questions | Totally disagree | Relatively disagree | Uncertain | Relatively agree | Completely agree |
|---|---|---|---|---|---|
| Flipped teaching has more curriculum resources than the traditional classroom model | 2.30% | 4.50% | 31.80% | 50% | 11.40% |
| Flipped teaching has more diversified and reasonable curriculum assessment methods | 2.30% | 9.10% | 20.50% | 55% | 13.60% |
| Learning through learning Platform makes learning more free and open | 2.30% | 4.50% | 13.60% | 57% | 22.70% |
| Flipped teaching is more active and the classroom atmosphere is more active | 2.30% | 4.50% | 36.40% | 45% | 11.40% |
| Flipped teaching is more conducive to the improvement of self-study ability | 4.50% | 0.00% | 31.80% | 52% | 11.40% |
| Flipped teaching is more conducive to solving difficult problems | 4.50% | 0.00% | 38.60% | 46% | 11.40% |
| You prefer the flipped teaching model based on the super star learning APP | 2.30% | 11.40% | 29.50% | 43% | 13.60% |

## 4 Research Results

The results show that flipped teaching based on Super star learning APP has better teaching effect than the traditional face-to-face teaching, and has more advantages than the traditional face-to-face teaching in terms of improving students' academic performance, optimizing students' learning efficiency and improving students' autonomous learning ability. Compared with few students' recognition of traditional teaching, most students are more willing to choose the flipped teaching. Therefore, in college English teaching in private colleges and universities, we can try to introduce flipped teaching to compromise the traditional face-to-face teaching.

Although flipped teaching can obviously improve the teaching effect, the problems are not allowed to ignore. Firstly, The premise of the implementation of flipped teaching is that students must study autonomously, so that flipped teaching can be implemented. Secondly, the better teaching effect of flipped teaching is actually built on the basis of increasing teachers' burden. How to re-evaluate and assess teachers' teaching work is a practical problem that needs to be solved.

## 5 Conclusion

The logistics industry in China has been developing by leaps and bounds, which leads to a big gap in English talents. The logistics industry needs up to hundreds of thousands of English talents [9]. It is the responsibility of English teachers to improve the English teaching effect and train qualified graduates. Although flipped teaching is advocated [10–12], the mature use of flipped teaching is still a long process. The problems existing in flipped teaching can be solved as following.

First, flipped teaching can only be implemented on the premise of students' independent learning. However, the difficulty is how to ensure that all students can cooperate with the teacher's requirements and put themselves into self-study after class. Just as Milman said, flipped teaching mode meets the personalized needs and development of students and is a meaningful teaching activity jointly initiated by students and teachers. Under this teaching mode, students are responsible for their own learning [13]. Therefore, teachers should change the way of final examination for students. Only by adopting a diversified and long-acting evaluation system can the learning evaluation objectively and reasonably reflect the learning performance and learning effect of learners at different learning stages, can the students' initiative of independent learning be effectively enhanced and the flipped learning be put into practice. In addition, teachers should reasonably design flipped teaching links. In view of the nature of college English courses and the different English foundation of students in private colleges, teachers must take these characteristics into consideration to design flipped teaching. In particular, in the absence of teachers, students should avoid investing too much time and energy but learning is limited. Students with poor foundation should avoid psychological resistance such as heavy learning load and difficult implementation of independent learning.

Second, the better effect of flipped teaching is actually built on the basis of increasing teachers' burden. How to reevaluate and assess teachers' teaching work is a problem that cannot be ignored. In the view of the research group, it is necessary to change the traditional teachers' teaching evaluation standard to do well in flipped teaching. Flipped teaching seems to free teachers from the traditional cramming teaching, but it actually requires teachers to pay more energy in other aspects of teaching. Therefore, the teaching hours shown in the traditional teaching schedule can no longer be used as the evaluation standard of teaching workload, and it should be solved by multiplying the basic teaching hours by the coefficient. Secondly, the implementation of flipped teaching should also be considered as a factor in promotion, professional title assessment, annual assessment and other aspects.

**Acknowledgment.** This project is supported by 2017 Hubei University Provincial Teaching Reform Research Project (Project No.: 2017482).

## References

1. Weihua, G., Tingting, Z., Yuwei, Z.: On RFID application in the information system of rail logistics center. Int. J. Educ. Manage. Eng. **3**(2), 52–58 (2013)
2. Ling, J., Kai, S.: Application of task-based approach in college English teaching based on internet-assisted multimedia. Int. J. Educ. Manage. Eng. **2**(8), 58–64 (2012)
3. Shanshan, L.: Analysis of English teaching for logistics majors in higher vocational colleges under the requirements of vocational positions. Overseas English **5**, 53–54 (2020). (in Chinese)
4. Yu, Z.: Problem diagnosis in the teaching process of professional English courses in vocational colleges – a case study of logistics English. Logist. Eng. Manage. **5**, 174-175+165 (2020). (in Chinese)
5. Xiaoying, Y.: Research on flipped classroom teaching in vocational colleges under information environment. Overseas English **21**, 126–127 (2019). (in Chinese)
6. Shibo, Z.: Essence and element analysis of flipped classroom implementation. J. Ningbo Inst. Educ. **6**, 19–22 (2017). (in Chinese)
7. Sheng, W.: Construction and application of flipped classroom teaching model. J. Liaoning Vocat. College **12**, 27–31 (2017). (in Chinese)
8. Shunli, Z., Qingshuang, Y.: Research of network teaching platform based grid. Int. J. Educ. Manage. Eng. **2**(6), 41–46 (2012)
9. Lin, L.: Research on logistics English talent training under the economic background of Zhengzhou airport. Overseas English **12**, 46–47 (2020). (in Chinese)
10. Yayuan, Z., Chunlan, H.: Flipped classroom teaching evaluation based on student satisfaction. J. Chongqing Electric Power Jr. College **6**, 1–4 (2018). (in Chinese)
11. He, J., Huijun, D.: Design and practice of flipped classroom teaching based on micro lessons. Rural Educ. China **24**, 43 (2018). (in Chinese)
12. Bai, X., Bo , Q.: Application of formative assessment in flipped classroom. Chin. J. Explor. Med. Educ. **9**, 866–869 (2014). (in Chinese)
13. Milman, N.B.: The flipped classroom strategy: what is it and how can it best be used, vol. 62. Distance Learning, Greenwich (2012)

# Integrated Operation Mode of Warehousing and Distribution in the O2O Environment

Huiling Jian[1,2], Chuanlei Wang[1,2](✉), and Xiaocheng Jiang[1,2]

[1] School of Business, Anhui University, Hefei 230601, China
[2] Logistics and Supply Chain Research Center, Anhui University, Hefei 230601, China

**Abstract.** The O2O business model is increasingly impacting the traditional business model, bringing new opportunities for the transformation and upgrading of the warehousing industry. Combined with the O2O integration environment, Study the operating conditions of the integrated mode of warehousing and distribution from the aspects of warehousing and distribution center location layout, inventory structure design, cargo location management, business process organization, and distribution network planning. Four integrated warehousing and distribution modes with retail enterprise, warehousing enterprise, wholesale market and industry alliance as the core are respectively constructed, and compare them to promote the high-quality development of warehousing enterprises, toward warehousing and distribution integration.

**Keywords:** O2O model · Warehousing and distribution · Operating model

## 1 Introduction

With the rise of Internet information technologies such as cloud storage, cloud computing, Internet of Things, and mobile terminals, as well as the widespread use of automated, integrated, and intelligent logistics equipment such as RFID technology, video recognition technology and unmanned vehicles, Traditional warehousing enterprises are accelerating the transformation from a single-function warehousing center to a multi-functional and integrated logistics distribution center, and from a traditional warehousing type of commodity storage to a modern warehousing type of inventory control. In 2014, the "Medium and Long-term Plan for the Development of the Logistics Industry (2014–2020)" issued by the State Council proposed "to strengthen the planning of modern distribution centers around large and medium cities and manufacturing bases, build e-commerce logistics service platforms and distribution networks, and build a number of regional Warehousing and distribution base".

However, problems generally exist in the warehousing industry, such as single business function, low technical level, backward facilities and equipment, heavy pressure of transformation and upgrading, and weak integration of warehousing and distribution. In recent years, the development of Internet technology has promoted the combination of offline traditional industries and the Internet. O2O (Online to Offline) business has been promoted rapidly, which provides new development opportunities for retail, warehousing, wholesale and other industries, promotes industrial transformation and upgrading, and achieves the integrated development of real economy and network economy [1, 2].


Z. Hu et al. (Eds.): ICAILE 2021, LNDECT 82, pp. 161–170, 2021.
https://doi.org/10.1007/978-3-030-80475-6_16

## 2 O2O Business Model

With the rapid development of Internet technology, the Internet of Things will play an important role in people's life and work. Applying the concept of the Internet of Things to the retail industry can save enterprises' operating costs and improve their operating profits [3]. At the same time, the continuous innovation of Internet technology can bring huge business opportunities for O2O model. The O2O business model is an Internet business model that organically combines online channels and offline channels. It means that online promotion, transactions drive offline operations and consumption, that is, offline business opportunities are combined with Internet technology, so that the Internet becomes the front desk of offline transactions, with the characteristics of promotion effects can be checked and each transaction can be tracked [4]. Retail companies can promote their products online through O2O commerce, thereby stimulating offline consumption. Compared with O2O online banner adverts, digital coupons can not only promote online publicity, but also improve offline sales revenue [5]. Enterprises gain initiative in the highly competitive O2O market through an agent-based competitive O2O business model, in which suppliers and customers connect with each other through social networks and share their service experience [6]. Adopting O2O marketing strategy is conducive to expanding market demand, helping chain enterprises to obtain more profits, creating and maintaining sustainable competitive advantages [7].

The integration environment of the O2O business model is to integrate online and offline transaction channels to realize the sharing of online and offline resources. The online channel takes the online platform as the carrier to publicize and market the goods by displaying the pictures, prices, quantities and other information of the goods in the offline physical stores, so as to facilitate consumers to make purchasing decisions and complete online payment, and improve consumers' trust and satisfaction [8]. Offline channels take physical stores as the carrier and display a large number of goods through physical stores, so as to provide consumers with consumption experience and after-sales service.

Build an O2O operation system that integrates online trading platforms and offline physical stores to realize the free circulation of funds between online trading platforms and physical stores, as well as timely update and transmission of information, to achieve the purpose of online and offline resource sharing.

## 3 Integration of Warehousing and Distribution in the O2O Integration Environment

Logistics warehousing and distribution refers to the monitoring and management of warehouse entry and exit, storage locations and delivery picking process [9]. Integration of warehousing and distribution refers to the combination of warehousing and distribution business, to provide customers with integrated warehousing and distribution logistics services, and finally achieve supply chain management. The operation process of integrated warehousing and distribution in the O2O integrated environment is shown in Fig. 1.

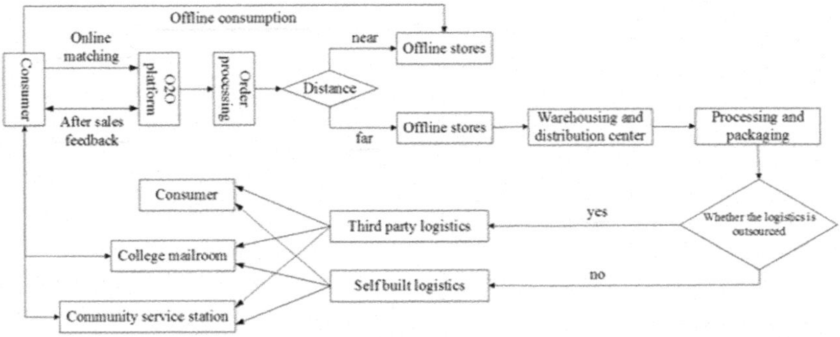


**Fig. 1.** Operation process of integrated warehousing and distribution in the O2O integrated environment

After consumers make purchase decisions on the O2O platform and complete online payment, the O2O platform automatically processes orders and sends order information to offline stores, If the location of consumers is close to the offline store, they can go to the nearest physical store for offline consumption, If it is far away, available the warehouse and distribution center of offline stores, and then through self-built logistics or third-party logistics to directly deliver the goods to the place designated by the consumer or to the community service stations or college mailrooms nearest to the consumer, After the goods are delivered to the community service stations and college mailrooms, the community service stations and college mailrooms will provide service experience for consumers.

# 4 The Operating Conditions of the Integrated Operation Mode of Warehousing and Distribution in the O2O Integrated Environment

## 4.1 Reasonable Warehousing and Distribution Center Location Layout

The layout of the warehousing distribution center should first comply with the overall urban planning and the requirements of commodities storage safety. The site selection and layout of the warehousing and distribution center should not only consider the land cost of the warehousing and distribution center construction, the water supply, power supply, communication, transportation and other infrastructure conditions around the warehouse area, the terrain and geological conditions, and the warehouse area environment from the perspective of reducing enterprise costs.

From the perspective of enhancing the location advantages of the warehousing and distribution center, consider the distribution scope of the warehousing and distribution center, target markets, industrial clusters, business development, warehousing and distribution networks, physical stores layout and cooperation with other enterprises.

### 4.2 Flexible Inventory Structure Design

Inventory management is a complex activity, including goods in and out of warehouse management, replenishment management, and related data management [10]. Maintaining an appropriate amount of inventory is the key to warehouse management [11]. According to the results of big data analysis, it is very necessary to predict the market demand of the inventory of goods and reserve urgently needed goods in advance to ensure the effective supply of goods.

By using the ABC inventory structure analysis method to serve the efficient management of the inventory structure, effectively adjust the inventory volume according to the sales volume and sales of A-type inventory commodities in different seasons and time periods. In terms of funds and marketing strategies, to promote the rapid circulation of A-type commodities and determine the minimum inventory to ensure the supply of best-selling commodities, and determine the maximum inventory for B and C commodities to ensure the effective use of funds and storage space resources.

### 4.3 Appropriate Cargo Location Management

Cargo location management refers to the optimal allocation of cargo location with appropriate storage methods, in order to achieve the minimum cost of storage operations management, and cargo location layout optimization of enterprise cargo location management activities [12]. Cargo location optimization refers to optimizing the storage location of the shelf part during the storage and exit process.

According to the variety, specification, volume, nature, seasonality and other characteristics of the commodities, different storage modes such as positioning storage and classification storage shall be adopted for the inventory commodities. Considering the sales volume, frequency of warehouse entry and exit, market demand of different commodities, store the commodities with large sales volume, high frequency of in and out of the warehouse, large market demand and waiting to be out of the warehouse in the place near the outlet, passage, delivery area or the middle and lower level of the shelf, place the commodities with small market demand, small sales volume and few times in and out of the warehouse far away from the outlet, the corner of an passage or upper shelf, so that the commodities can be allocated nearby and out of the warehouse.

### 4.4 Perfect Business Process Organization

The business of the warehousing distribution center can be divided into warehousing business and distribution business. The warehousing business mainly involves the warehousing management, in-warehouse management and out- warehouse management of goods, the distribution business involves goods distribution, processing, packaging, sorting, warehousing and other business. To realize the efficient operation of the warehousing and distribution center, it is necessary to optimize the business process of warehousing and distribution, effectively connect each link of the warehousing and distribution business, and form an efficient operation system to realize the rapid warehousing, storage, inventory, distribution, processing, packaging, sorting and delivery of goods.

### 4.5 Comprehensive Warehousing and Distribution Network Planning

Warehousing and distribution network planning is to rationally arrange the warehousing and distribution center layout, distribution network layout and distribution route selection.

Taking the warehousing and distribution center as the center, according to the distribution status, density and purchase frequency of the target market consumer groups, choose a place closer to the consumer group to deploy the distribution network, Shorten the distance between distribution network and warehousing and distribution centers, residential areas and physical stores, plan distribution routes, reduce distribution costs and improve distribution efficiency.

At the same time, strengthen the cooperation with logistics distribution enterprises, use the existing local distribution network, integrate distribution resources, improve the point, line and surface combination of warehousing and distribution network, promote the efficient operation of warehousing and distribution.

## 5 The Operation Mode of Integrated Warehousing and Distribution in the O2O Integrated Environment

### 5.1 Retail Enterprise as the Core of the Warehousing and Distribution Integration of the Operation Model

The retail enterprises in this model include department stores, chain supermarkets, large general supermarkets, convenience stores, specialty stores and warehouse supermarkets. Retail enterprises not only have mature offline chain stores, but also carry out online e-commerce transactions. They even have self-established warehousing and distribution centers or third-party warehousing and distribution centers in cooperation with them. Consumers make purchase decisions and complete online payment on the O2O platform of retail enterprises. The O2O platform automatically processes orders and transmits the order information to the offline physical stores of retail enterprises. Consumers can choose whether to go to the nearest physical store of retail enterprises for offline consumption according to the distance between their location and the offline physical stores of retail enterprises.

If it is far away from the physical stores of retail enterprise, consumers can use the third-party warehousing and distribution center or the self-built logistics of the self-built warehousing and distribution center or use the third-party logistics to directly deliver the commodities purchased by consumers to the places where consumers are located, or it can be delivered to the nearest community service stations, college mailrooms, or other distribution networks, and then delivered to the designated place by the consumer through these distribution networks.

### 5.2 Warehousing Enterprise as the Core of the Warehousing and Distribution Integration of the Operation Model

Warehousing enterprise in this model include warehousing and distribution enterprises and logistics warehousing enterprises independently operated by large enterprises with offline physical stores, such warehousing enterprises can not only independently operate their own warehousing and distribution business, but also strengthen cooperation with other enterprises to store and distribute the commodities of other enterprises, and carry out e-commerce transactions. Warehousing and distribution enterprises under the independent operation of large enterprises can use their own physical stores to conduct offline transactions, and then use online trading platforms to combine online and offline transactions.

After consumers make purchase decisions and complete online payment on the O2O platform, the O2O platform automatically processes orders and transmits the order information to the warehousing enterprises, the physical stores affiliated to the enterprises or the third-party physical stores cooperating with them. Consumers can choose whether to distribute or not. If they choose not to distribute, they can go to the nearest physical stores that has a cooperative relationship with the warehousing enterprises or the physical stores that belongs to the same enterprise as the warehousing enterprises for offline consumption. If consumers choose to make distribution, warehousing enterprises and physical stores can choose whether to make distribution through self-built logistics by warehousing enterprises or directly through self-built logistics by physical stores according to the distance between the location and the location of consumers.

## 5.3 Wholesale Market as the Core of the Warehousing and Distribution Integration of the Operation Model

In this model, the wholesale market, on the one hand, operates the traditional business model of the entity wholesale market, and strengthens the cooperation with the third-party physical stores or warehousing and distribution enterprises [13]. On the other hand, they can also independently or strengthen the cooperation with other wholesalers to create an online trading platform.

The wholesale market can strengthen cooperation with physical stores such as chain supermarkets, warehousing and distribution enterprises, and establish long-term strategic partnerships. They build a complete online information trading platform among the wholesale market, physical stores, warehousing and distribution enterprises and other entities by setting up sales outlets in the leasing booths of large physical stores or renting warehouses of warehousing and distribution enterprises, so as to extend the business chain of the wholesale market and realize resource sharing.

After consumers, individual purchasers, and group purchasers make purchase decisions on the O2O platform and complete online payment, they can choose to pick up the goods at the nearest wholesale market, third-party physical stores that have a cooperative relationship with the wholesale market, and warehousing and distribution enterprises, or use wholesale market, third-party physical stores that have a cooperative relationship with the wholesale market, self-built logistics of warehousing and distribution enterprises, or third-party logistics to deliver the purchased commodities to the locations of consumers, individual purchasers, and group purchasers.

### 5.4 Industry Alliance as the Core of the Warehousing and Distribution Integration of the Operation Model

The industry alliance in this model refers to the retail industry, warehousing industry, wholesale industry, commerce industry and other industries that combine offline transactions with online transactions, the main content is to strengthen business cooperation and build strategic alliances by signing contracts or agreements. The partners in the industry alliance carry out online and offline transactions and have frequent business contacts and common business contacts, they implement joint distribution.

After consumers make purchase decisions on the O2O platform of each partner and complete online payment, the O2O platform of each partner automatically processes the order and transmits the order information to the joint distribution public information service platform, the joint distribution public information service platform conducts order summary and classification, formulates distribution plans, selects the optimal distribution route, and makes joint distribution management decisions, then the joint distribution public information service platform will communicate the decision of joint distribution management to the joint distribution center or third-party logistics enterprise warehousing and distribution center.

### 5.5 A Comparison of the Four Models

In the O2O integrated environment, the integrated operation modes of warehousing and distribution centered on retail enterprise, warehousing enterprise, wholesale market and industry alliance have their own advantages and disadvantages, and their application conditions are also different. The enterprises should choose the most suitable operating mode according to their actual operating situation. The comparison results between the four modes are shown in Table 1.

As shown in Table 1, the integrated operation modes of warehousing and distribution centered on retail enterprise, warehousing enterprise, wholesale market and industry alliance can reduce the distribution cost. The difference lies in the following three modes can centralized distribution of commodity, forming the distribution scale. In terms of shortcomings, the four modes have their own shortcomings. The disadvantages of the first mode are mainly the high operating cost of self-established logistics and low distribution efficiency. The second and third modes are mainly limited by the low level of informationization of distribution. The deficiency of the fourth mode lies in the high cost of communication and coordination and the difficulty of resource integration. In terms of application conditions, the four modes all emphasize cooperation with third-party logistics. As different enterprises have different business modes, their application conditions are also different. The specific application conditions of the four modes are described in Table 1. The four modes are widely used by many enterprises, SUNING is the typical representative of using the first mode, The Motion Optimization has actively explored the second mode, Xin Fa Di Wholesale Market of Agricultural Products Center and RRS Logistics are typical enterprises using the third mode and the fourth mode respectively.

**Table 1.** Four modes

| Name | The advantages of the integrated operation model of warehousing and distribution in the O2O integrated environment | The disadvantages of the integrated operation mode of warehousing and distribution in the O2O integrated environment | The operating conditions of the integrated operation mode of warehousing and distribution in the O2O integrated environment | Typical cases |
|---|---|---|---|---|
| Retail enterprise | There are many physical stores, which facilitate the offline consumption experience of customers and reduce the number of deliveries. Strengthening cooperation with third-party logistics and distribution enterprises can greatly reduce distribution costs | Self-built logistics has high operating costs and low distribution efficiency. The frequency of commodity distribution is high, the batch is small, and the distributed commodities are too scattered, which makes it difficult to form an effective distribution scale | It is necessary to shift from a single physical store operation model to an O2O business model combining online and offline. Establish close strategic partnerships with third-party logistics and distribution enterprises to ensure the timeliness and safety of distribution | SUNING |
| Warehousing enterprise | The existing warehouses, distribution facilities, and high-efficiency distribution teams of warehousing enterprises can be used, the distribution of commodities are concentrated, and it is easy to form a scale of goods distribution, with low distribution costs and high efficiency | The O2O business model of warehousing enterprise is still in its infancy, the level of distribution informatization is low, and the cooperation with physical stores is still being explored, an integrated distribution system of warehousing and distribution has not been formed in the O2O environment, and it is difficult for consumers to experience offline | Carry out O2O e-commerce transactions, strengthen cooperation with third-party physical stores and third-party logistics, and build an O2O e-commerce transaction platform for warehousing enterprises and physical stores | The motion optimization |

(*continued*)

**Table 1.** (*continued*)

| Name | The advantages of the integrated operation model of warehousing and distribution in the O2O integrated environment | The disadvantages of the integrated operation mode of warehousing and distribution in the O2O integrated environment | The operating conditions of the integrated operation mode of warehousing and distribution in the O2O integrated environment | Typical cases |
|---|---|---|---|---|
| Wholesale market | Commodities with low distribution frequency and large batches can be distributed in a centralized manner. The scale of commodity distribution is significant, with low distribution costs and high efficiency | The distribution system of the wholesale market is not yet perfect, and the level of distribution informatization is low. The cost of communication and coordination with third-party physical stores and warehousing and distribution enterprises is high | Expand O2O e-commerce business and strengthen cooperation with third-party logistics distribution enterprises or self-built logistics for distribution | Xin Fa Di wholesale market of agricultural products center |
| Industry alliance | Distributed commodities with high distribution frequency and small batches can be concentrated for joint distribution, which improves the utilization efficiency of warehousing and distribution related resources, reduces distribution costs, improves distribution efficiency, and easily creates large-scale economic benefits | The cost of communication and coordination between industries within the industry alliance is high, resource integration is difficult, distribution routes are complex, and the development of joint distribution mode is immature | Establishing their own joint distribution centers or renting warehouse and distribution centers of third-party logistics enterprises, building public information service platforms for joint distribution, and improving the information exchange mechanism within the industry alliance | RRS Logistics |

## 6 Conclusion

In order to promote the transformation and upgrading of retail, warehousing, wholesale and other industries, improve their quality and efficiency, and promote the high-quality development of integrated warehousing and distribution. In the O2O integrated environment, this paper considers the operation process of integrated warehousing and distribution, analyzes the operation conditions of the integrated warehousing and distribution mode, and compares the differences among the integrated operation modes of warehousing and distribution centered on retail enterprise, warehousing enterprise, wholesale market and industry alliance.

The results showed that: (1) The integrated operation process of warehousing and distribution in the O2O integrated environment is a process of warehousing and distribution linkage, coordination and cooperation between online and offline; (2) The efficient operation of the integrated warehousing and distribution mode in the O2O integrated environment is inseparable from the reasonable warehousing and distribution center location layout, flexible inventory structure design, appropriate cargo location management, perfect business process organization, comprehensive warehousing and distribution network planning; (3) The integrated operation mode of warehousing and distribution centered on retail enterprise, warehousing enterprise, wholesale market and industry alliance has differences in advantages, disadvantages and application conditions, but these four modes can effectively reduce the distribution cost. The enterprises should proceed from the reality and choose the integrated operation mode of warehousing and distribution which is consistent with their own business conditions.

## References


1. Li, P., Zhong, Y.: Research on adoption behavior of offline merchants in mobile O2O business. Contemp. Finan. Econ. **9**, 75–87 (2014). (in Chinese)
2. Nie, L.: The characteristics and trend of e-commerce development in China. China Bus. Mark. **6**, 97–101 (2014). (in Chinese)
3. Syed, K.A.S., Waqas, M.: Smart home automation using IOT and its low cost implementation. Int. J. Eng. Manuf. **5**, 28–36 (2020)
4. Wu, Z.: A brief analysis of O2O e-commerce model. J. Chongqing Univ. Sci. Technol. (Soc. Sci. Ed.) **13**, 73–74 (2012). (in Chinese)
5. Phang, C.W., Tan, C., Sutanto, J., et al.: Leveraging O2O commerce for product promotion: an empirical investigation in mainland China. IEEE Trans. Eng. Manag. **61**(4), 623–632 (2014)
6. He, Z., Cheng, T.C.E., Dong, J., et al.: Evolutionary location and pricing strategies for service merchants in competitive O2O markets. Eur. J. Oper. Res. **254**(2), 595–609 (2016)
7. Noori-Daryan, M., Taleizadeh, A.A., Rabbani, M.: Advance booking pricing in O2O commerce with demand leakage using game theory for tourism supply chains. Int. J. Prod. Res. **2**, 1–36 (2019)
8. Jairak, R., Praneetpolgrang, P., Chirawichitchai, N.: A roadmap for establishing trust management strategy in e-commerce services using quality based assessment. Int. J. Inf. Eng. Electron. Bus. **6**(5), 1–9 (2014)
9. Xu, Q.: Logistics warehousing and distribution optimization and its visual operation management based on radio frequency identification. China Bus. Mark. **1**, 26–30 (2011). (in Chinese)
10. Mohamed, M.: Smart warehouse management using hybrid architecture of neural network with barcode reader 1D/2D vision technology. Int. J. Intell. Syst. Appl. **11**(11), 16–24 (2019)
11. Chang, S.: Analysis on warehouse management of publishing enterprises. Sci.-Technol. Publ. **7**, 71–72 (2013). (in Chinese)
12. Zeng, Z.: The existing problems and countermeasures of warehouse cargo location management management. J. Liaoning Econ. Vocat. Tech. Coll. **3**, 25–26 (2010). (in Chinese)
13. Wang, X., Zhang, Q.: Research on agricultural product circulation mode based on online and offline integration – O2O framework and trend of agricultural products. J. Beijing Technol. Bus. Univ. (Soc. Sci. Ed.) **29**(3), 18–25 (2014). (in Chinese)

# A Comparison of Machine Learning Algorithms for Prediction Higher Education Institution's Entrants Admissions

Pavlo Zhezhnych, Khrystyna Zub(✉), Oleksandr Berezko, and Anna Shilinh

Lviv Polytechnic National University, S. Bandera str., 12, Lviv 79013, Ukraine
khrystyna.v.zub@lpnu.ua

**Abstract.** A significant influence on the choice of both specialty and specific educational institution made by each entrant during the admission campaign is the assessment of the applicant's chances of admission. One possible way to solve this task is to build specialized web services using machine learning methods based on the processing of entry data for previous years. In this paper, research and experimental analysis of existing machine learning methods to solve this task were made. Experimental evaluation of Linear Regression, Stochastic Gradient Descent (SGD), Tree, AdaBoost, Random Forest, Support Vector Machine (SVM) and k-nearest neighbors' algorithm (k-NN) was performed on a real data set. It was found that the highest prediction accuracy based on all accuracy indicators was obtained using methods based on Linear Regression and SGD. Since the latter method is the basis of a number of algorithms for training artificial neural networks, in the future it is advisable to evaluate and improve existing and develops, ANN-based approaches to improve the accuracy of the task in this study.

**Keywords:** Higher education institution · Machine learning · Prediction · Graduate admissions · Enrollment

## 1 Introduction

Right choice of a specialty to obtain an educational qualification level is a highly challenging task for both the HEI and those interested in enrollment. The choice of university and field of knowledge is not always the result of a considered personal, professional, and educational self-determination of applicants. Entrants might spend a considerable amount of money and time to get advice from consultancy organizations (such organization are common in the USA) to help them identify their admission chances. Another existing solution that aims to support entrants' decision is commercial software solutions, which mostly used by educational institutions and only a small part offers prediction mechanisms. However, most decisions are focused on the very process of accepting and processing applications from applicants. In addition, there remains the problem of privacy and data security. They provide broad functionality, but the high cost of purchase and support process becomes a difficulty for many universities [1].


Z. Hu et al. (Eds.): ICAILE 2021, LNDECT 82, pp. 171–179, 2021.
https://doi.org/10.1007/978-3-030-80475-6_17

The way a candidate makes course selection will directly influence the quantitative and qualitative set of entrants, that is the key to the high rating of the educational institution. In addition, a rational choice of future students is the key to their successful learning outcomes and at the same time a significant factor in shaping the university popularity [2, 3]. With today's rapid development of information technologies, the one of the most priorities tasks of HEIs is to find and implement the most effective technological solutions to support the entrant's decision making.

Machine learning provides the capability to automatically learn and improve from experience without being explicitly programmed. It is a natural extension of traditional statistical approaches focusing primarily on predictions, automatically identifying patterns within data, and performing tasks. Using the knowledge gained in the process allows one to adjust their future actions of institutions and entrants accordingly.

The HEI administration and the researchers have the possibility to apply prediction on various tasks with a different level of complexity [19, 21]. Using prediction for student enrollment primarily aids the generate information could influence future strategy and resource decisions. Understanding patterns, association, changes, significant structures, and anomaly detection are some of the benefits perceived after knowledge extraction [4, 20, 22]. However, the selection of the correct prediction methods for a particular problem is still a quest since data mining algorithms to be used are dependent on the availability of the HEI data.

The aim of the work is to experimentally evaluate and compare the effectiveness of machine learning algorithms in solving the task of predicting the chances of admission to the university. The choice of the most accurate algorithm can become the basis of the component of the intelligent decision support system for the admission of entrants to the HEI, including Ukraine.

## 2 Related Works

The task set in the work is not new. There are a lot of scientific works with methods and means to solve it. It is advisable to consider a number of the most common approaches that provide the high accuracy of the result.

There are different approaches to evaluating potential university entrants. Each university, in accordance with its admission rules, forms an appropriate list of criteria for admission. In particular, many criteria have been identified, which can be formulated into six groups: academic background, grades, cognitive ability, research background, personality and personal competencies, motivation factors [5].

In the other research, the accuracy of the Linear Regression method to solve the problem of predicting the probability of admission to the university was investigated. The author used a well-known data set [6, 7]. Preliminary correlation analysis has identified a number of features that have the greatest impact on the model. They were included for modeling. The coefficient of determination (R2) served as an indicator of accuracy. Other authors used the decision tree algorithm to solve the problem [8].

Despite the high accuracy of the obtained result, the authors do not present the results of comparison of the applied model with the existing ones.

In the next research the authors from studied three known machine learning algorithms: linear regression, decision tree, and logistic regression model, for ranking potential universities for admission to the applicant according to his indicators [9]. It is experimentally established that logistic regression provides the highest prediction accuracy. However, the peculiarity of this work is that the studied algorithms were considered as part of a large system of decision support for admission to universities.

The study and experimental analysis of six known classification methods were performed by Koutina and Kermanidis, where authors included C4.5, k-nearest neighbor's classifiers, Naïve-Bayes classifier, RIPPER, Random Forest and SVM [10]. The simulation was performed on the author's data set, which contained 117 instances using the WEKA software environment. The authors found that the highest accuracy of solving the problem was obtained using Naïve Bayes and 1-NN methods.

Other authors developed a new regression method using the boosting approach [11]. The authors called it the Gradient boosting regressor model. The simulation of his work was based on the use of a known data set [7]. The results of the prediction were evaluated using R2, MAE, MSE and RMSE indicators. The authors established a high accuracy of the developed model in comparison with existing regressors.

The effectiveness of solving the problem using four known machine learning methods: Linear Regression, SVM, Decision tree and Random Forest was made by Acharya and other [12]. Evaluation was performed using R2, Mean Squared Log Error, MSE and RMSE. The results of modeling based on a sample from dataset source with the optimal values of all parameters of the studied algorithms showed that the highest prediction accuracy is provided by Linear Regression among all studied methods [7].

The main contribution of this paper is the experimental investigation of the seven machine-learning algorithms with the selection of their optimal parameters. In addition, a numerical evaluation and comparison was performed using a well-known dataset using four different accuracy indicators.

## 3 Materials and Methods

An experimental study was conducted to determine the effectiveness of a number of existing regression methods based on machine learning to predict the success of entrants to the HEI. Experimental comparison of their work was performed using the online platform Orange Data Mining - a component-based visual programming software package for data visualization, machine learning, data mining, and data analysis [13].

### 3.1 Dataset Descriptions

The solution of the task of predicting the chances of an entrant entering the university was done using a real data set [7]. The authors of the mentioned research collected 500 instances when applying for a master's degree [12]. Each of them contained seven independent attributes that significantly affect the results of admission to the university. The name, explanation of the content of each attribute, as well as the minimum, maximum and average value of each of them are given in Table 1. The initial dependent attribute is an estimate of the probability of entering the university. It should be noted that data used in this research contains two data sets: initial (smaller) and supplemented (larger). For the research conducted in this work, we used a supplemented data set (Admission_Predict_Ver1.1.csv). Details of the collection and processing of this data set can be found in [12].

**Table 1.** Dataset attributes and its main characteristics

| Parameter title | Parameter's explanation | MIN value | MAX value | MEAN value |
|---|---|---|---|---|
| GRE score | General test scores | 290 | 340 | 316.472 |
| TOEFL score | Test of English as a foreign language score | 92 | 120 | 107.192 |
| University rating | University rating value | 1 | 5 | 3.114 |
| SOP | Statement of purpose and letter of recommendation strength | 1 | 5 | 3.374 |
| LOR | | 1 | 5 | 3.484 |
| CGPA | Undergraduate grade/mark average | 6.8 | 9.92 | 8.576 |
| Research | Research experience | 0 | 1 | 0.56 |
| chance of admit | Output value | 0.34 | 0.97 | 0.722 |

### 3.2 Accuracy Indicators

The evaluation of the studied methods was performed using the following indicators: Mean Square Error (MSE), Root Mean Square Error (RMSE), Mean Absolute Error (MAE), Coefficient of determination (R2). The use of all indicators will summarize the results obtained on the effectiveness of each of the studied methods [18].

### 3.3 Investigated Methods

To solve this task, several most used regression methods based on machine learning were used: SVM, SGD, Linear regression, k-NN, Tree, AdaBoost and Random Forest. Consider them in more detail.

SVM is a machine learning method used to solve various classification and regression tasks [14]. It converts the input data of set task into a space of higher dimension to build a separating hypersurface. The main idea is to build an optimal hypersurface with the maximum gap in the new space [15].

Linear regression is the most widely used method of regression analysis in various fields of science. It establishes a relationship between independent and dependent variables by finding the appropriate coefficients. In Orange, this method is accompanied by the possibility of using different regularization options: (L1 (LASSO), L2 (ridge) or L1L2 (elastic net)) [15].

The SGD machine learning method involves performing a stochastic approximation of gradient descent to minimize the objective function. Although it is an iterative method, it is one of the fastest. Among the main parameters of the implementation of this method in Orange should be noted the possibility of using different variants of loss function [15].

k-NN is a non-parametric machine learning method that does not require a special learning procedure and uses all available data for learning when solving classification or regression problems. This algorithm predicts new data points based on the degree of proximity of their features to the points in the existing training data set. It is obvious that the choice of the number of neighboring points that will participate in the calculations will significantly affect the overall accuracy of the method [15].

Among the advantages of machine learning methods based on decision trees, such as Tree and Random Forest, is the ease of interpretation of the results of their work. A significant difference between these two methods is that the Tree method in Orange is a simple method of dividing observations into classes, while Random Forest is an ensemble method, which in most cases provides high accuracy in solving various problems [16]. The main parameter of the Tree method is a minimum number of instances in leaves, and Random Forest is the number of trees that form an ensemble [15].

Unlike previous methods, AdaBoost builds a committee using a booting approach. According to the idea, the method increases the accuracy of the prediction or classification at each subsequent level, by taking into account the errors of weak classifiers at the previous level. Among the disadvantages of the method should be noted the sensitivity to the dimensionality of the training sample as well as to noise in the data [17]. Among the main parameters of his work is the number of estimators [15].

According to the purpose of the paper, all these methods were used to solve the task of predicting the chances of an applicant to enter the university in order to choose the best of the methods based on the accuracy of his work.

The training procedure of each method was performed on a randomly generated training sample of 80% of the entire data sample. The remaining 20% of observations were used to test the performance of each of the studied methods. This procedure was repeated 10 times using the additional settings of the Random Sampling option of the Test and Score widget. The result of each method was formed after performing the above procedures. Repetition of the training procedure of each method, 10 times on each, randomly formed subsets of a given data set provided an increase in the level of reliability of the results. Each method was run 10 times. This, in turn, increases the reliability of the results obtained, for the application of a particular method to solve a specific application problem.

An important stage in the application of machine learning methods is the selection of their optimal parameters [18]. This significantly affects the operation of each method. The selection of optimal parameters of each of the studied methods is carried out in the work. The results are summarized in Table 2. They were used during the application of the studied methods of machine learning to solve the task.

**Table 2.** Optimal parameters of the investigated methods

| Method | Parameter | Parameter's value |
|---|---|---|
| Linear Regression | – | – |
| SGD | Loss function | "Huber" |
| k-NN | Metric, number of neighbors | "Manhattan", 15 |
| SVM | Kernel, iteration limit | "rbf", 200 |
| Tree | Min. number of instances in leaves | 14 |
| AdaBoost | Number of estimators | 99 |
| Random Forest | Number of trees | 300 |

## 4 Results and Discussion

In this paper, the following indexes are used to evaluate: Estimates of accuracy of work of all investigated methods with optimum values of parameters of each of methods (Table 2), based on the accuracy indicators (1)–(4) are summarized in Table 3.

**Table 3.** Results (MSE, RMSE, MAE and R2) for all investigated methods

| Method | MSE | RMSE | MAE | R2 |
|---|---|---|---|---|
| Linear Regression | 0.004 | 0.066 | 0.047 | 0.798 |
| SGD | 0.004 | 0.066 | 0.047 | 0.798 |
| k-NN | 0.006 | 0.075 | 0.054 | 0.739 |
| v-SVM | 0.006 | 0.075 | 0.055 | 0.742 |
| Tree | 0.005 | 0.073 | 0.052 | 0.755 |
| AdaBoost | 0.006 | 0.075 | 0.051 | 0.743 |
| Random Forest | 0.005 | 0.071 | 0.050 | 0.770 |

As can be seen from Table 3, the lowest accuracy for all indicators is shown by classifiers based on SVM, k-NN and Tree. Slightly better results were obtained when using the ensemble methods: AdaBoost and Random Forest. Moreover, the boosting method in comparison with Random Forest shows a significantly lower error in this task based on both R2 and RMSE indicators. The highest accuracy based on all four indicators (1)–(4) was demonstrated by two methods: Linear Regression and SGD.

To visually assess the work of all studied methods, Fig. 1 presents the results of their work in the form of Scatter Plots. Ox-axis shows the real (true) values of the chance of admit, oy-axis shows the predicted values by a – k-NN, b – SVM, c – AdaBoost, d – Tree, e – Random Forest, f – Linear Regression, g – SGD.

Visual analysis of the obtained results confirms the accuracy estimates from Table 3. In particular, the correlation fields of the k-NN, SVM and AdaBoost methods are quite

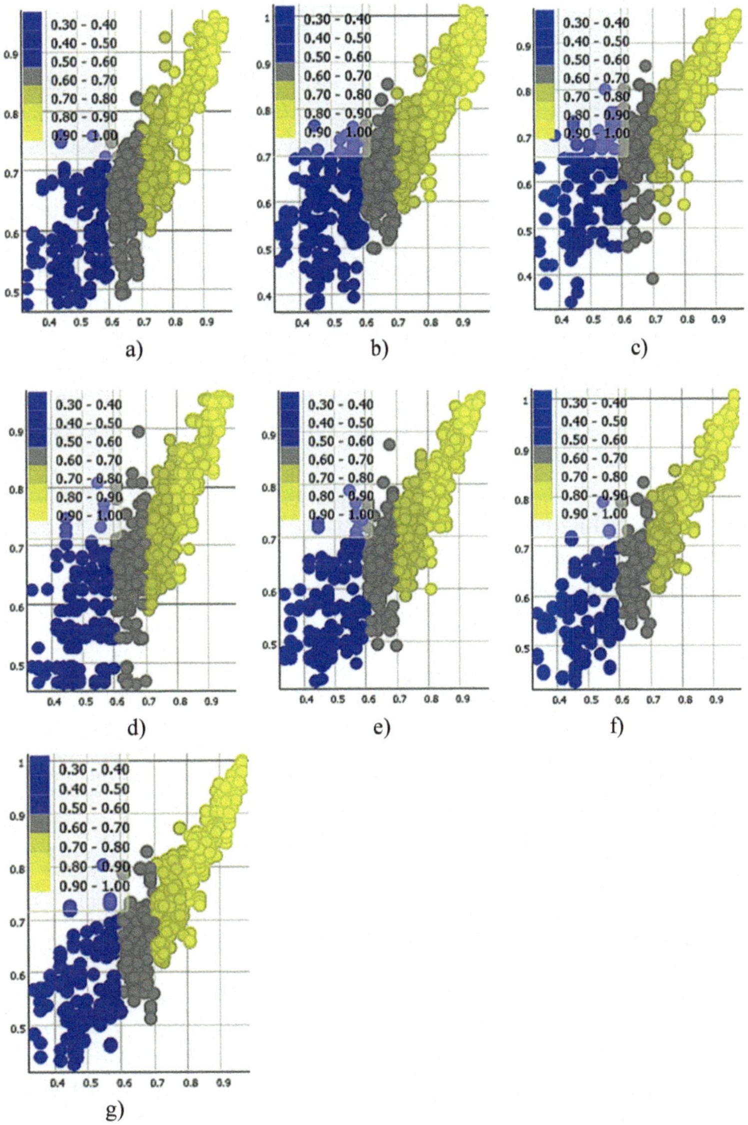


**Fig. 1.** Scatter plots for all investigated methods

wide, which indicates the low accuracy of their work. Somewhat better results were obtained when using the methods of Fig. 1d, and e. The narrowest correlation fields, the highest accuracy, are demonstrated by two methods among the subjects: Linear regression and SGD.

Since the latter method is the basis of a number of learning algorithms for different topologies of artificial neural networks, in the future it seems appropriate to conduct research to evaluate such means of artificial intelligence to solve this task.

## 5 Conclusion

The paper presents the results of solving the task of predicting the chance of admission by entrants to HEI. The practical value of the work is to give entrants the opportunity to assess their chances of entering a particular university. Effective solution of the task will allow entrants to make a short list of universities (specialties) for admission, which is relevant in the realities of the relevant rules of admission HEI of Ukraine.

The authors made an experimental comparison of a number of known machine learning methods in terms of the accuracy of their work. The simulation was performed using a real data set containing 500 vectors with seven independent attributes in each. The optimal parameters of each studied method are selected. The obtained results testify to the highest accuracy of the prediction using methods based on linear regression, as well as stochastic gradient descent among all considered. However, it provides not high enough results to build information technology for the support of decision-making by HEI's entrants.

Further research will be based on the application and evaluation of different topologies of artificial neural networks to solve this task. It is also advisable to build stacking ensembles based on the studied machine learning methods, which should increase the accuracy of the prediction results.

## References


1. Zhezhnych, P., Berezko, O., Zub, K., Demydov, I.: Analysis of features and abilities of online systems and tools meeting information needs of HEIs' entrants. In: Proceedings of the 2nd IWCOAPSN, vol. 2616, pp. 76–85 (2020)
2. Zhezhnych, P., Demydov, I., Berezko, O., Shilinh, A.: Corporate culture influence on the HEI's information image on the internet. In: Proceedings of the 2nd IWCOAPSN, vol. 2392, pp. 286–296 (2020)
3. Korzh, R., Peleschyshyn, A., Holub, Z.: Analysis of integrity and coverage completeness of the informational image of a higher education institution. In: Proceedings of 13th ICTCSET, pp. 825–827 (2016)
4. Cruz, A.P.D., et al.: Higher education institution (HEI) enrollment forecasting using data mining technique. Int. J. Adv. Trends Comput. Sci. Eng. **9**(2), 2060–2064 (2020)
5. Kurysheva, H.V., Rijen, M.V., Dilaver, G.: How do admission committees select? Do applicants know how they select? Selection criteria and transparency at a Dutch University'. Int. J. Tert. Educ. Manag. **25**(4), 367–388 (2019)
6. Sujay, S.: Supervised machine learning modelling & analysis for graduate admission prediction. Int. J. Trend Res. Dev. **7**(4), 5–7 (2020)
7. Graduate Admission 2. https://kaggle.com/mohansacharya/graduate-admissions. Accessed 26 Nov 2020
8. Janani, P., Hema, P., Monisha, P.: Prediction of MS graduate admissions using decision tree algorithm. Int. J. Sci. Res. **9**(3), 492–495 (2018)
9. AlGhamdi, A., Barsheed, H., AlMshjary, H., AlGhamdi, H.: A machine learning approach for graduate admission prediction. In: Proceedings of 2nd ICIVSP, pp. 155–158 (2020)
10. Koutina, M., Kermanidis, K.L.: Predicting postgraduate students' performance using machine learning techniques. In: Proceedings of ICAIAI, pp. 159–168 (2011)

11. Chakrabarty, N., Chowdhury, S., Rana, S.: A statistical approach to graduate admissions' chance prediction', in innovations in computer science and engineering. In: Proceedings of 7th ICICSE, pp. 333–340 (2020)
12. Acharya, M.S., Armaan, A., Antony, A.S.: A comparison of regression models for prediction of graduate admissions', in computational intelligence in data science. In: Proceedings of ICCIDS, pp. 1–5 (2019)
13. Demšar, J., et al.: Orange: data mining toolbox in Python. J. Mach. Learn. Res. **14**, 2349–2353 (2013)
14. Izonin, A., et al.: The combined use of the wiener polynomial and SVM for material classification task in medical implants production. Int. J. Intell. Syst. Appl. **10**(9), 40–47 (2018)
15. Orange Visual Programming 3 documentation'. https://orange3.readthedocs.io/projects/orange-visual-programming/en/latest/widgets/model/tree.html. Accessed 27 Nov 2020
16. Tkachenko, R., et al.: Development of machine learning method of titanium alloy properties identification in additive technologies. East.-Eur. J. Enterp. Technol. **3**(12), 23–31 (2018)
17. Izonin, K.N., Tkachenko, R., Zub, K.: An approach towards missing data recovery within IoT smart system. Procedia Comput. Sci. **155**, 11–18 (2019)
18. Fedushko, S., Ustyianovych, T.: Predicting pupil's successfulness factors using machine learning algorithms and mathematical modelling methods. In: Proceedings of ICCSEEA, pp. 625–636 (2020)
19. Raghavendra, C.K., Srikantaiah, K.C., Venugopal, K.R.: Personalized recommendation systems (PRES): a comprehensive study and research issues. Int. J. Mod. Educ. Comput. Sci. **10**(10), 11–21 (2018)
20. Qingyun, L.: A quantitative analysis of postgraduate programs admission of the Hong Kong institute of education. Int. J. Mod. Educ. Comput. Sci. **7**(3), 1–11 (2015)
21. Muhammad, U.F., Sani, M.I.: Data mining to prediction student achievement based on motivation, learning and emotional intelligence in man 1 ketapang. Int. J. Mod. Educ. Comput. Sci. **10**(6), 53–60 (2018)
22. Abdulkadir, S., et al.: Assessment of students' academic performance using admission entry requirements under the computer-based test and paper-pencil-based test in Kaduna State University, Kaduna – Nigeria. Int. J. Mod. Educ. Comput. Sci. **11**(8), 48–60 (2019)

# Ocean Carrier Price Competition with Bulk Cargo Transportation and Empty Container Repositioning

Mingzhu Yu[1] and Hanzhe Liu[2](✉)

[1] College of Management, Shenzhen University, Shenzhen 518060, Guangdong, China
[2] College of Civil and Transportation Engineering, Shenzhen University, Shenzhen 518060, Guangdong, China

**Abstract.** The price competition game between two ocean carriers providing container transportation service and empty container repositioning between two ports is studied. Two kinds of cargos are considered: containerized cargos and bulk cargos. The optimal pricing strategies are derived and sensitivity analysis is conducted to analyze the impacts of ocean carriers' heterogeneity and the potential demand imbalance. The findings suggest that efficient and powerful hinterland transport systems are not necessarily a good thing for ocean carriers. The transportation prices of two carriers increase with hinterland transportation cost. The price of two-way transport by ocean carriers increases with the increase of the laden container operation cost. The potential imbalance demand for containerized and bulk cargoes will result in greater losses for developing carriers.

**Keywords:** Empty container repositioning · Containerized cargos · Bulk cargos

## 1 Instruction

Maritime transportation has become the main mode of international trade with its great advantages [1]. Ocean carriers are competing in different dimensions in the market. The service quality and transportation price provided by the ocean carriers are the two main factors that affect their competition advantages. The liner company will take measures in terms of price and service to attract the goods. Zhou and Lee studied the competitive pricing game problem when the empty container repositioning cost was considered by two transportation enterprises [2]. Chen et al. divided marine transport products into two categories: cargos and waste, and established a monopoly and duopoly model to find the optimal pricing strategy of shipping companies [3]. Martin and Chris used average cost pricing method and marginal cost pricing method to analyze the shipping price under specific demand [4].

Nowadays, containerized maritime transportation has become an indispensable way for goods to flow around the world, accelerating the process of global economic integration and greatly promoting the development of the world economy. However, due to the trade imbalance between different ports, the problem of empty containers has been


Z. Hu et al. (Eds.): ICAILE 2021, LNDECT 82, pp. 180–189, 2021.
https://doi.org/10.1007/978-3-030-80475-6_18

an inevitable issue [5]. Scholars have also done some research on this issue. Lai et al. proposed an empty container repositioning strategy that could save the actual operation cost and opportunity cost by using simulation and heuristic methods [6]. Shintani et al. studied the optimization of empty container transportation in the hinterland by establishing an integer programming model [7]. Zhou et al. established a two-stage stochastic programming model for random empty container repositioning problem [8]. Xu and Liu considered the empty container repositioning of the folding container [9]. Lei et al. found that sea-land combined transportation can greatly reduce the cost of empty container transportation [10]. Legros et al. found that managing containers through a time-based threshold policy can save costs [11]. Yu and Zhi discussed the impact of different levels of port service on the decision to allocate empty containers [12].

Although containerized liner transportation is the mainstream mode, some goods, such as coal, grain, oil, wheat and other dry bulk cargos, are generally transported by bulk ships. Dry bulk shipping provides the most cost-effective way to transport large quantities of low value goods for seaborne trade. One way to recover some of the empty container repositioning costs is to fill them with bulk cargos.

In this paper, bulk cargo transportation and empty container repositioning are considered. A price game model is established to deduce the equilibrium solutions of two carriers under homogeneous and heterogeneous carrier cases [13]. Numerical studies are used to fully demonstrate the impact of various parameters on ocean carriers. Our paper is of great relevance to Zhou and Lee [2]. However, their research is limited to homogeneous transport enterprises, and the linear demand function in their model is not quite suitable for marine transport systems. The Hotelling model characterizes the demand and considers shipping bulk cargos in empty containers.

## 2 Model Basics

In this paper, only two ports (port A and port B) are considered. There are two ocean container carriers, who transport containerized cargos and bulk cargos (which are put in the containers to transport) between two ports. The transportation from port A to port B is delineated as "direction AB", and naturally the "direction BA" refers to the transportation from port B to port A.

Carriers attract customers in two directions separately. The realized demands are affected by the carriers' price decisions. The realized demands are divided into two kinds, the bulk cargo demand and the containerized cargo demand. Containers can be utilized to transport bulk cargos. Here, let $D_A$ and $D_B$ represent the potential demand for containerized cargos in AB and BA directions, respectively, and $D_A^{'}$ and $D_B^{'}$ be the potential demand for bulk cargos in AB and BA directions, respectively. Without loss of generality, we assume that the potential demand for containerized cargos and bulk cargos in the direction AB is greater than that in the direction of BA. Namely, $D_A \geq D_B$, $D_A^{'} \geq D_B^{'}$. For example, direction AB is from China to America and direction BA is from America to China.

Different models are used to describe the demands of these two cargos. In a certain period of time, the total demand of containerized cargo transportation in a port area is determined by the external factors, such as global economy, national import and export strategic policies, etc., and not affected by the prices in the system. Therefore, the Hotelling model is applicable to containerized cargos. While, the realized demand for total bulk cargos in the port area is affected by the prices in the system. The increase of a bulk cargo shipping price may cause the bulk cargo demands flow to traditional bulk ships. Therefore, the classical linear model is suitable to describe the demand of bulk cargos.

Following the characteristic of Hotelling model, the demand for containerized cargos in the direction of AB (or BA) can be derived,

$$d_{i,A}^{c}\left(p_{i,A}, p_{j,A}\right) = D_A \times \frac{p_{j,A} - p_{i,A} + t}{2t} \tag{1}$$

$$d_{i,B}^{c}\left(p_{i,B}, p_{j,B}\right) = D_B \times \frac{p_{j,B} - p_{i,B} + t}{2t} \tag{2}$$

Here, $t$ is the hinterland unit transportation cost, $i, j \in \{1, 2\}$, $i \neq j$. The two carriers' demands of the bulk cargos in direction AB (or BA) are, $d_{i,A}^{b} = D_A^{'} + \alpha p_{j,A} - \beta p_{i,A}$ and $d_{i,B}^{b} = D_B^{'} + \alpha p_{j,B} - \beta p_{i,B}$. Here, $\beta > \alpha$. Therefore, the total realized demand for the transportation services of two ocean carriers in direction AB and BA can be expressed as $d_{i,A} = d_{i,A}^{c} + d_{i,A}^{b}$, $d_{i,B} = d_{i,B}^{c} + d_{i,B}^{b}$.

When a transportation service occurs, two types of costs are generated, the unit laden container transportation cost $v_i$ and unit empty container repositioning cost $e_i$. The laden container transportation cost is generally higher than the empty container repositioning cost. It is set that,$e_i = av_i$, $(0 \leq a \leq 1)$.

Define $x^{+} = \max(x, 0)$. The following equation characterizes the profit function of carrier $i$ $(i, j \in \{1, 2\}, i \neq j)$:

$$\pi_i\left(p_{i,A}, p_{i,B}, p_{j,A}, p_{j,B}\right) = \left(p_{i,A} - v_i\right)d_{i,A} + \left(p_{i,B} - v_i\right)d_{i,B} - e_i\left(d_{i,B} - d_{i,A}\right)^{+} - e_i\left(d_{i,A} - d_{i,B}\right)^{+} \tag{3}$$

The carrier's profits of transporting laden containers in both directions are represented by the first two terms of Eq. (3). The latter two terms refer to the empty containers repositioning costs.

## 3 The Homogeneous Carrier Situation

The homogeneous situation signifies that the operating costs of the two ocean carriers are the same when they provide basic container transportation services. Both carriers have the identical unit transportation cost of laden and empty containers, i.e. $v_i = v_j = v$, $e_i = e_j = e$. The profit function of the homogeneous ocean carriers $i$ $(i, j \in \{1, 2\}, i \neq j)$ are as follows.Here, "$\sim$"stands for the homogeneous carrier case.

$$\tilde{\pi}_i\left(p_{i,A}, p_{i,B}, p_{j,A}, p_{j,B}\right) = \left(p_{i,A} - v\right)d_{i,A} + \left(p_{i,B} - v\right)d_{i,B} - e\left(d_{i,A} - d_{i,B}\right) \tag{4}$$

The optimal pricing strategies of two homogeneous ocean carriers are as follows:

$$\left(\tilde{p}_{i,A}^{*}, \tilde{p}_{i,B}^{*}\right) = \left(\frac{t\left(D_A + 2D_A^{'}\right) + (v + e)(D_A + 2t\beta)}{D_A + 2t(2\beta - \alpha)}, \frac{t\left(D_B + 2D_B^{'}\right) + (v - e)(D_B + 2t\beta)}{D_B + 2t(2\beta - \alpha)}\right) \tag{5}$$

Accordingly, equilibrium demands and equilibrium profits are:

$$\tilde{d}^*_{i,A} = \frac{(D_A + 2t\beta)\left[\left(D_A + 2D'_A\right) + 2(\alpha - \beta)(v + e)\right]}{2[D_A + 2t(2\beta - \alpha)]},\ \tilde{d}^*_{i,B} = \frac{(D_B + 2t\beta)\left[\left(D_B + 2D'_B\right) + 2(\alpha - \beta)(v - e)\right]}{2[D_B + 2t(2\beta - \alpha)]},\ \pi_i$$
$$= \frac{t\left[\left(D_A + 2D'_A\right) + 2(v + e)(\alpha - \beta)\right]^2 (D_A + 2t\beta)}{2[D_A + 2t(2\beta - \alpha)]^2} + \frac{t\left[\left(D_B + 2D'_B\right) + 2(v - e)(\alpha - \beta)\right]^2 (D_B + 2t\beta)}{2[D_B + 2t(2\beta - \alpha)]^2}.(i \in \{1, 2\}). \quad (6)$$

It is easy to find that if $\beta - \alpha \leq \frac{D_A + 2D'_A}{2(v+e)}$ and $\beta - \alpha \leq \frac{D_B + 2D'_B}{2(v-e)}$, the transportation prices of two carriers increase with hinterland transportation cost $t$. If the operation costs of the ocean carriers ($v$) are small, especially when the hinterland transportation system is also relatively inefficient (i.e. high $t$ value), the carrier will raise the prices in two directions. Therefore, an inefficient and underdeveloped hinterland transportation system may suffer the shippers both in the hinterland transportation cost and ocean transportation cost. For ocean carriers, an efficient and powerful hinterland transportation system may not be preferred. The improvement of hinterland transportation system reduces the hinterland transportation cost. A lower hinterland transportation cost is a blessing for shippers but a curse for ocean carriers.

The ocean carriers' prices in both directions increase with the cost of transporting laden containers. The changes in equilibrium prices caused by the impact of laden container transportation cost are consistent in both directions. However, the increase of empty container repositioning cost in direction AB will cause the ocean carriers to raise the prices in this direction. In direction BA, if the empty container repositioning cost increases, the carriers will reduce the prices. By doing this, there will be more demand in direction BA, which will reduce the total empty container repositioning cost.

## 4 The Heterogeneous Carrier Situation

Two heterogeneous ocean carriers are analyzed, whose laden container transportation costs are $v_i$, $v_j$, and the empty container repositioning costs are $e_i$, $e_j$. We assume the operation costs of carrier j are higher, which indicates that carrier $i$ is more developed than carrier j. Namely, $V = v_j - v_i \geq 0$, $E = e_j - e_i \geq 0$.Then, $V - E = (1 - a)V \geq 0$. The equilibrium solutions and their related properties are derived in the following theorem.

There is a unique Nash equilibrium solution in the pricing game of two duopoly heterogeneous ocean carriers. The NE prices of two heterogeneous ocean carriers are asymmetric and the equilibrium solutions are listed in Table 1.

Here, $G_{A_1} = \frac{D_A}{2t} + \alpha$, $G_{B_1} = \frac{D_B}{2t} + \alpha$, $G_{A_2} = \frac{D_A}{2t} + \beta$, $G_{B_2} = \frac{D_B}{2t} + \beta$, $H_A = \frac{2G_{A_2}^2}{(2G_{A_2} - G_{A_1})(2G_{A_2} + G_{A_1})}$, $H_B = \frac{2G_{B_2}^2}{(2G_{B_2} - G_{B_1})(2G_{B_2} + G_{B_1})}$, $K_A = \frac{D_A + 2D'_A}{2(2G_{A_2} - G_{A_1})}$, $K_B = \frac{D_B + 2D'_B}{2(2G_{B_2} - G_{B_1})}$, $Q_A = \frac{G_{A_2} G_{A_1}}{(2G_{A_2} + G_{A_1})(2G_{A_2} - G_{A_1})}$, $Q_B = \frac{G_{B_2} G_{B_1}}{(2G_{B_2} + G_{B_1})(2G_{B_2} - G_{B_1})}$, $N_A = \frac{G_{A_2}\left(G_{A_1}^2 - 2G_{A_2}^2\right)}{(2G_{A_2} + G_{A_1})(2G_{A_2} - G_{A_1})}$, $N_B = \frac{G_{B_2}\left(G_{B_1}^2 - 2G_{B_2}^2\right)}{(2G_{B_2} + G_{B_1})(2G_{B_2} - G_{B_1})}$, $M_A = \frac{G_{A_1} G_{A_2}^2}{(2G_{A_2} + G_{A_1})(2G_{A_2} - G_{A_1})}$, $M_B = \frac{G_{B_1} G_{B_2}^2}{(2G_{B_2} + G_{B_1})(2G_{B_2} - G_{B_1})}$, $N_A - N_B = R_N^-$, $N_A + N_B = R_N^+ (R_N^+ < 0)$, $M_A - M_B = R_M^-$, $M_A + M_B = R_M^+$ $(R_M^+ > 0)$. $(R_M^- - R_N^- > 0,\ R_M^+ - R_N^+ > 0,\ R_M^+ + R_N^+ < 0)$,

**Table 1.** The equilibrium solutions of two heterogeneous ocean carriers

| Case | | Optimum solutions |
|---|---|---|
| $e_iR_N^+ + e_jR_M^+ \geq 0$ | $\eta_2 \leq v_j - v_i \leq \eta_1$ and $S_4 \leq -W < S_1$ | if $\bar{\pi}_i(\bar{p}_{i,A}, \bar{p}_{i,B}) \geq \underline{\pi}_i(\underline{p}_{i,A}, \underline{p}_{i,B})$, $(p^*_{i,A}, p^*_{i,B}) = (\bar{p}_{i,A}, \bar{p}_{i,B})$, $(p^*_{j,A}, p^*_{j,B}) = (\underline{p}_{j,A}, \underline{p}_{j,B})$ if $\bar{\pi}_i(\bar{p}_{i,A}, \bar{p}_{i,B}) \leq \underline{\pi}_i(\underline{p}_{i,A}, \underline{p}_{i,B})$, $(p^*_{i,A}, p^*_{i,B}) = (\underline{p}_{i,A}, \underline{p}_{i,B})$,$(p^*_{j,A}, p^*_{j,B}) = (\underline{p}_{j,A}, \underline{p}_{j,B})$ |
| | $v_j - v_i \geq \eta_1$ and $S_3 \leq -W < S_1$ | |
| | $\eta_2 \leq v_j - v_i \leq \eta_1$ and $S_3 \leq -W < S_4$ | if $\bar{\pi}_i(\bar{p}_{i,A}, \bar{p}_{i,B}) \geq \underline{\pi}_i(\underline{p}_{i,A}, \underline{p}_{i,B})$, $(p^*_{i,A}, p^*_{i,B}) = (\bar{p}_{i,A}, \bar{p}_{i,B})$, $(p^*_{j,A}, p^*_{j,B}) = (\hat{p}_{j,A}, \hat{p}_{j,B})$ if $\bar{\pi}_i(\bar{p}_{i,A}, \bar{p}_{i,B}) \leq \underline{\pi}_i(\underline{p}_{i,A}, \underline{p}_{i,B})$, $(p^*_{i,A}, p^*_{i,B}) =$ $(\underline{p}_{i,A}, \underline{p}_{i,B})$,$(p^*_{j,A}, p^*_{j,B}) = \left(\check{p}_{j,A}, \check{p}_{j,B}\right)$ |
| | $v_j - v_i \leq \eta_2$ and $S_3 \leq -W < S_1$ | |
| | $\eta_2 \leq v_j - v_i \leq \eta_1$ and $S_2 \leq -W < S_3$ | $(p^*_{i,A}, p^*_{i,B}) = (\bar{p}_{i,A}, \bar{p}_{i,B})$ $(p^*_{j,A}, p^*_{j,B}) = (\hat{p}_{j,A}, \hat{p}_{j,B})$ |
| | $v_j - v_i \geq \eta_1$ and $S_2 \leq -W < S_4$ | |
| | $v_j - v_i \leq \eta_2$ and $S_2 \leq -W < S_3$ | |
| $e_iR_N^+ + e_jR_M^+ \leq 0$ | $v_j - v_i \geq \eta_2$ and $S_2 \leq -W < S_4$ | |
| | $v_j - v_i \leq \eta_1$ and $S_2 \leq -W < S_1$ | |
| | $\eta_1 \leq v_j - v_i \leq \eta_2$ and $S_2 \leq -W < S_1$ | |
| $e_iR_N^+ + e_jR_M^+ \geq 0$ | $\eta_2 \leq v_j - v_i \leq \eta_1$ and $-W \geq S_1$ | $(p^*_{i,A}, p^*_{i,B}) = (\underline{p}_{i,A}, \underline{p}_{i,B})$ $(p^*_{j,A}, p^*_{j,B}) = (\underline{p}_{j,A}, \underline{p}_{j,B})$ |
| | $v_j - v_i \geq \eta_1$ and $-W \geq S_1$ | |
| | $v_j - v_i \leq \eta_2$ and $-W \geq S_4$ | |
| $e_iR_N^+ + e_jR_M^+ \leq 0$ | $v_j - v_i \geq \eta_2$ and $-W \geq S_3$ | |
| | $v_j - v_i \leq \eta_1$ and $-W \geq S_4$ | |
| | $\eta_1 \leq v_j - v_i \leq \eta_2$ and $-W \geq S_3$ | |
| $e_iR_N^+ + e_jR_M^+ \geq 0$ | $\eta_2 \leq v_j - v_i \leq \eta_1$ and $-W < S_2$ | $(p^*_{i,A}, p^*_{i,B}) = (\bar{p}_{i,A}, \bar{p}_{i,B})$ $(p^*_{j,A}, p^*_{j,B}) = (\bar{p}_{j,A}, \bar{p}_{j,B})$ |
| | $v_j - v_i \geq \eta_1$ and $-W < S_2$ | |

(*continued*)

**Table 1.** (*continued*)

| Case | | Optimum solutions |
|---|---|---|
| | $v_j - v_i \le \eta_2$ and $-W < S_2$ | |
| $e_iR_N^+ + e_jR_M^+ \le 0$ | $v_j - v_i \ge \eta_2$ and $-W < S_2$ | |
| | $v_j - v_i \le \eta_1$ and $-W < S_2$ | |
| | $\eta_1 \le v_j - v_i \le \eta_2$ and $-W < S_2$ | |
| $e_iR_N^+ + e_jR_M^+ \ge 0$ | $v_j - v_i \ge \eta_1$ and $S_4 \le -W < S_3$ | $\left(p_{i,A}^*, p_{i,B}^*\right) = \left(\bar{p}_{i,A}, \bar{p}_{i,B}\right)$<br>$\left(p_{j,A}^*, p_{j,B}^*\right) = \left(\underline{p}_{j,A}, \underline{p}_{j,B}\right)$ |
| $e_iR_N^+ + e_jR_M^+ \le 0$ | $v_j - v_i \ge \eta_2$ and $S_4 \le -W < S_1$ | |
| $e_iR_N^+ + e_jR_M^+ \ge 0$ | $v_j - v_i \le \eta_2$ and $S_1 \le -W < S_4$ | $\left(p_{i,A}^*, p_{i,B}^*\right) = \left(\underline{p}_{i,A}, \underline{p}_{i,B}\right)$<br>$\left(p_{j,A}^*, p_{j,B}^*\right) = \left(\check{p}_{j,A}, \check{p}_{j,B}\right)$ |
| $e_iR_N^+ + e_jR_M^+ \le 0$ | $v_j - v_i \le \eta_1$ and $S_3 \le -W < S_4$ | |
| $e_iR_N^+ + e_jR_M^+ \le 0$ | $v_j - v_i \ge \eta_2$ and $S_1 \le -W < S_3$ | $\left(p_{i,A}^*, p_{i,B}^*\right) = \left(\check{p}_{i,A}, \check{p}_{i,B}\right)$<br>$\left(p_{j,A}^*, p_{j,B}^*\right) = \left(\underline{p}_{j,A}, \underline{p}_{j,B}\right)$ |
| | $\eta_1 \le v_j - v_i \le \eta_2$ and $S_4 \le -W < S_3$ | |
| | $v_j - v_i \le \eta_1$ and $S_1 \le -W < S_3$ | if $\hat{\pi}_i\left(\hat{p}_{i,A}, \hat{p}_{i,B}\right) \ge \check{\pi}_i\left(\check{p}_{i,A}, \check{p}_{i,B}\right), \left(p_{i,A}^*, p_{i,B}^*\right) = \left(\hat{p}_{i,A}, \hat{p}_{i,B}\right)$<br>if $\hat{\pi}_i\left(\hat{p}_{i,A}, \hat{p}_{i,B}\right) \le \check{\pi}_i\left(\check{p}_{i,A}, \check{p}_{i,B}\right), \left(p_{i,A}^*, p_{i,B}^*\right) = \left(\check{p}_{i,A}, \check{p}_{i,B}\right)$<br>if $\hat{\pi}_j\left(\hat{p}_{j,A}, \hat{p}_{j,B}\right) \ge \check{\pi}_j\left(\check{p}_{j,A}, \check{p}_{j,B}\right), \left(p_{j,A}^*, p_{j,B}^*\right) = \left(\hat{p}_{j,A}, \hat{p}_{j,B}\right)$<br>if $\hat{\pi}_j\left(\hat{p}_{j,A}, \hat{p}_{j,B}\right) \le \check{\pi}_j\left(\check{p}_{j,A}, \check{p}_{j,B}\right)$, $\left(p_{j,A}^*, p_{j,B}^*\right) = \left(\check{p}_{j,A}, \check{p}_{j,B}\right)$ |

$S_1 = v_i R_N^- + v_j R_M^- + e_i R_N^+ + e_j R_M^+$, $S_2 = v_i R_M^- + v_j R_N^- + e_i R_M^+ + e_j R_N^+$, $S_3 = v_i R_N^- + v_j R_M^- - e_i R_N^+ - e_j R_M^+$, $S_4 = v_i R_M^- + v_j R_N^- - e_i R_M^+ - e_j R_N^+$, $W = \frac{1}{2}(D_A - D_B) + \left(D_A^{'} - D_B^{'}\right) + (\alpha - \beta)(K_A - K_B)$, $\eta_1 = \frac{(R_M^+ - R_N^+)(e_j - e_i)}{R_M^- - R_N^-}$, $\eta_2 = \frac{-(R_M^+ + R_N^+)(e_j + e_i)}{R_M^- - R_N^-}$. $(i, j \in \{1, 2\}, i \neq j)$, $\frac{1}{4G_{A_2}(G_{B_2} + G_{A_2})} = F_A$, $\frac{1}{4G_{B_2}(G_{A_2} + G_{B_2})} = F_B$.

$$\bar{p}_{i,A} = K_A + H_A(v_i + e_i) + Q_A(v_j + e_j), \underline{p}_{i,A} = K_A + H_A(v_i - e_i) + Q_A(v_j - e_j).$$

$$\bar{p}_{i,B} = K_B + H_B(v_i - e_i) + Q_B(v_j - e_j), \underline{p}_{j,B} = K_B + H_B(v_i + e_i) + Q_B(v_j + e_j).$$

$$\bar{p}_{j,A} = K_A + Q_A(v_i + e_i) + H_A(v_j + e_j), \underline{p}_{j,A} = K_A + Q_A(v_i - e_i) + H_A(v_j - e_j).$$

$$\bar{p}_{j,B} = K_B + Q_B(v_i - e_i) + H_B(v_j - e_j), \underline{p}_{j,B} = K_B + Q_B(v_i + e_i) + H_B(v_j + e_j).$$

$$\begin{matrix} 2G_{A_1}(G_{B_2} + 2G_{A_2})(K_A + Q_A v_i + H_A v_j) - 2G_{A_2} G_{B_1}(K_B + Q_B v_i + H_B v_j) \\ + (G_{B_2} + 2G_{A_2})(D_A + 2D_A^{'}) - G_{A_2}(D_B + 2D_B^{'}) + 4G_{A_2} G_{B_2} v_i \end{matrix} = J_{i,A}.$$

$$\begin{matrix} 2G_{B_1}(G_{A_2} + 2G_{B_2})(K_B + Q_B v_i + H_B v_j) - 2G_{B_2} G_{A_1}(K_A + Q_A v_i + H_A v_j) \\ + (G_{A_2} + 2G_{B_2})(D_B + 2D_B^{'}) - G_{B_2}(D_A + 2D_A^{'}) + 4G_{B_2} G_{A_2} v_i \end{matrix} = J_{i,B}.$$

$$\frac{Q_A G_{A_1}(G_{B_2} + 2G_{A_2}) + Q_B G_{B_1} G_{A_2}}{2G_{A_2}(G_{B_2} + G_{A_2})} = \theta_1, \quad \frac{Q_B G_{B_1}(G_{A_2} + 2G_{B_2}) + Q_A G_{A_1} G_{B_2}}{2G_{B_2}(G_{A_2} + G_{B_2})} = \theta_2.$$

$$\frac{H_A G_{A_1}(G_{B_2} + 2G_{A_2}) + H_B G_{B_1} G_{A_2}}{2G_{A_2}(G_{B_2} + G_{A_2})} = \mu_1, \quad \frac{H_B G_{B_1}(G_{A_2} + 2G_{B_2}) + H_A G_{A_1} G_{B_2}}{2G_{B_2}(G_{A_2} + G_{B_2})} = \mu_2.$$

$$\hat{p}_{i,A} = F_A J_{1,A} - \theta_1 e_i + \mu_1 e_j, \overset{\vee}{p}_{i,A} = F_A J_{1,A} - \theta_1 e_i - \mu_1 e_j.$$

$$\hat{p}_{j,A} = F_A J_{2,A} + \mu_1 e_i + \theta_1 e_j, \overset{\vee}{p}_{j,A} = F_A J_{2,A} - \mu_1 e_i - \theta_1 e_j.$$

$$\hat{p}_{i,B} = F_B J_{1,B} - \theta_2 e_i - \mu_2 e_j, \overset{\vee}{p}_{i,B} = F_B J_{1,B} + \theta_2 e_i + \mu_2 e_j.$$

$$\hat{p}_{j,B} = F_B J_{2,B} - \mu_2 e_i - \theta_2 e_j, \overset{\vee}{p}_{j,B} = F_B J_{2,B} + \mu_2 e_i + \theta_2 e_j.$$

## 5 Computational Studies

A series of numerical analysis are conducted in the heterogeneous case to further explore the managerial inspirations. In the numerical experiments, let $D_A \geq D_B$, $D_A^{'} \geq D_B^{'}$. Also, $v_1 \leq v_2$, which means that carrier 1 is a developed carrier and carrier 2 is a relatively developing carrier.

### 5.1 The Impact of the Operation Cost Parameters

In this subsection, the impact of operation cost parameters on two carriers' profit are analyzed. First, for carrier 1, we change the value of $v_1$ while keeping other parameters fixed. We set the laden container operation cost $v_1$ of carrier 1 to vary from 10 to 20. Values of other parameters are as follows: $D_A = 600$, $D_B = 100$, $D_A^{'} = 400$, $D_B^{'} = 150$, $v_2 = 20$, $a = 0.6$, $\alpha = 1.5$, $\beta = 6.5$, $t = 15$.

As shown in Fig. 1, with the increase of carrier 1's operation costs ($v_1$ and $e_1$), carrier 1's profit is decreasing, but his rival's profit is increasing. When $v_1$ reaches 20 (according to our previous parameter setting, the value of $v_2$ is 20), it can be seen that the profits of two heterogeneous carriers are the same. This conclusion coincides with the results in the previous homogeneous ocean carrier case.

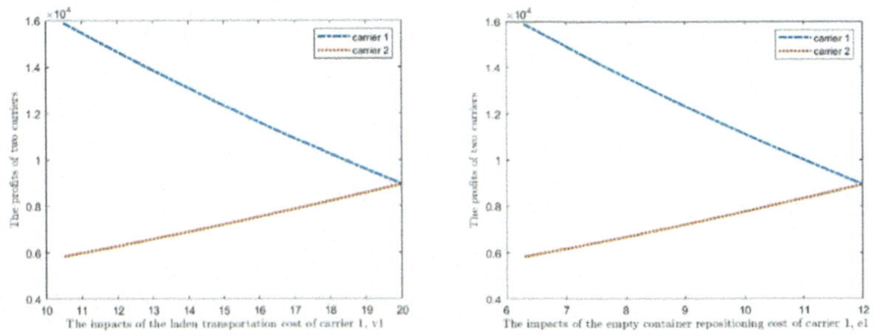


**Fig. 1.** The impact of carrier 1's cost parameters ($v_1$ and $e_1$) on two carriers' profits

## 5.2 The Effects of the Potential Demand Imbalance

The effects of potential demand imbalance for bulk cargos and containerized cargos between the two ports area are studied. Firstly, for traditional containerized cargos, we set the value range of $D_A$ as (300, 500), and the value range of $D_B$ as (100, 500). Therefore, the degree of potential demand imbalance for containerized cargos can be described as $D_A - D_B$, which varies from 0 to 200. For bulk cargos, we let $D_A^{'}$ to raise in (150,300), and $D_B^{'}$ to increase in (50, 300). The degree of potential demand imbalance of bulk cargos which is portrayed by $D_A^{'} - D_B^{'}$ is changed from 0 to 100. Values of other parameters are as follows:$v_1 = 10, v_2 = 15, a = 0.6, e_1 = 6, e_2 = 9, t = 15, \alpha = 1.5, \beta = 6.5, D_A = 600, D_B = 100, D_A^{'} = 400, D_B^{'} = 150$.

As shown in Fig. 2, the profits of the two carriers decreases with the potential demand imbalance of containerized cargos. At the same time, the profit gap between the two carriers decreases with the growth of the potential containerized cargos demand imbalance. Moreover, the developed carrier 1 has higher profit than the developing carrier 2. The increase of demand imbalance will lead to higher empty container repositioning which harms the carriers' profits. It can be seen that the potential containerized cargos demand

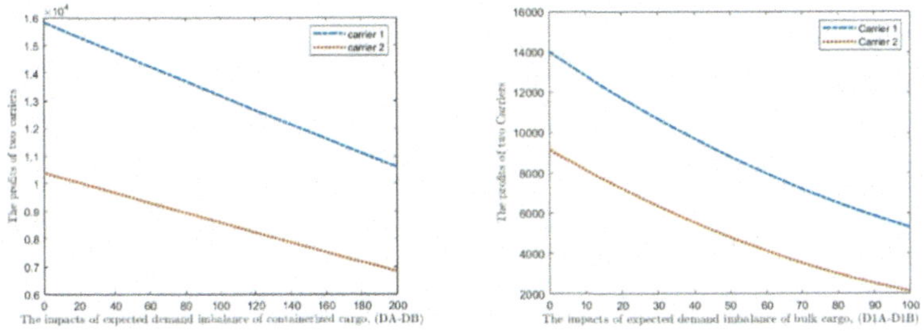


(a) (b)

**Fig. 2.** The impact of expected demand imbalance of containerized cargos and bulk cargos on two carriers' profits imbalance will bring more harm to the profit of the developing carrier 2. Similar conclusions can be drawn from the potential bulk cargos demand imbalance (as shown in Fig. 2).

## 6 Conclusion

Our paper studies the pricing game model of two ocean carriers providing container transportation services for shippers between two ports. The integration of bulk cargo transportation and traditional ocean container transportation is emphasized while considering the problem of empty container repositioning. Several suggestive conclusions can be drawn. An underdeveloped hinterland transport system is a boon for shippers but a bane for ocean carriers. Demand imbalance of containerized cargos and bulk cargos hurts carriers' profits, and the developing carrier suffers more. Our paper simplify the relevant settings of the model. For example, we study the shipping system with only two ports. The shipping system with port group may be explored in the future. The situation of asymmetric information obtained by two homogeneous or heterogeneous ocean carriers could also be a focus of future research.

## References

1. Kumar, V.K.N., Srinivasan, B.: Implementation and performance effect on electronic procurement and its ship management companies. Int. J. Inf. Eng. Electron. Bus. **5**(5), 10–16 (2013)
2. Zhou, W., Lee, C.: Pricing and competition in a transportation market with empty equipment repositioning. Transp. Res. Part B: Methodol. **43**(6), 677–691 (2009)
3. Chen, R., Dong, J., Lee, C.: Pricing and competition in a shipping market with waste shipments and empty container repositioning. Transp. Res. Part B: Methodol. **85**, 32–55 (2016)
4. Martin, W., Chris, H.: Transportation Investment and Pricing Principles: An Introduction for Engineers, Planners and Economists. Wiley, New York (1984)
5. Yu, M., Zhang, Y.: Multi-agent-based fuzzy dispatching for trucks at container terminal. Int. J. Intell. Syst. Appl. **2**(2), 41–47 (2010)
6. Lai, K.K., Lam, K., Chan, W.K.: Shipping container logistics and allocation. J. Oper. Res. Soc. **46**(6), 687–697 (1995)
7. Shintani, K., Konings, R., Imai, A.: The impact of foldable containers on container fleet management costs in hinterland transport. Transp. Res. Part E: Logist. Transp. Rev. **46**(5), 750–763 (2010)
8. Shaorui, Z., Xiaopo, Z., Zhiming, C., et al.: A new separable piecewise linear learning algorithm for the stochastic empty container repositioning problem. Math. Probl. Eng. **03**, 1–16 (2020)
9. Xu, T., Liu, H.: An empty container reposition study with foldable containers. Int. Core J. Eng. **6**(5), 258–268 (2020)
10. Xing, L., Yang, Z., Xu, Q., et al.: Optimizing empty container dispatching of China Railway Express under sea-land intermodal transport. J. Dalian Maritime Univ. **45**(2), 1–8 (2019). (in Chinese)
11. Legros, B., Bouchery, Y., Fransoo, J.C.: A time-based policy for empty container management by consignees. Prod. Oper. Manag. **28**(6), 1503–1527 (2019)

12. Jiang, Y., Yao, Z.: Collaborative empty container allocation between liner companies in random environment. Pack. Eng. **40**(5), 162–168 (2019). (in Chinese)
13. Beltadze, G.N., Giorgobiani, J.A.: The stability of equilibrium situation in lexicographic strategic games. Int. J. Mod. Educ. Comput. Sci. **8**(12), 38–45 (2016)

# Practical Personnel Training Mode of Logistics Engineering Specialty Based on CDIO Concept Under the Background of Emerging Engineering Education

Yanhui Liu[1], Xiaoguang Zhou[2](✉), Ruiceng Meng[1], and Liang Fang[1]

[1] Century College of Beijing University of Posts and Telecommunications, Beijing 102101, China
[2] Beijing University of Posts and Telecommunications, Beijing 100876, China
zxg@bupt.edu.cn

**Abstract.** Under the background of Emerging Engineering Education construction, taking the cultivation of application-oriented talents as the goal, combined with the concept of CDIO Engineering Education, to meet the needs of social development for compound logistics talents, this paper puts forward the CDIO-P & CG Practical talents training mode of logistics engineering specialty, and constructs the "integrated" curriculum system around the core ability of "intelligent logistics and supply chain management and optimization". A three-level project construction is carried out to realize the horizontal connection between projects and the vertical integration between courses. From the implementation of teaching, the achievements of teachers and students can be well transformed into scientific research and technological innovation, with outstanding achievement and remarkable effects. Based on CDIO-P & CG, the training mode of practical logistics engineering specialty is proposed, which has great reference value for the establishment of application-oriented talents training system of related majors in other colleges and universities.

**Keywords:** Logistics engineering · Application-oriented · CDIO · Talent training mode

## 1 Introduction

Since February 2017, the Ministry of education has actively promoted the construction of Emerging Engineering Education called 3E, formed the "Fudan Consensus", "Tianda Action" and "Beijing Guide" successively, and issued the notice on carrying out new engineering research and practice and the notice on promoting 3E research and practice projects. The Ministry of Education has fully developed the formation of a Chinese model and Chinese experience leading the global engineering education, so as to help higher education become stronger National construction [1].


Z. Hu et al. (Eds.): ICAILE 2021, LNDECT 82, pp. 190–199, 2021.
https://doi.org/10.1007/978-3-030-80475-6_19

The construction of 3E courses is a major action plan to continuously deepen the reform of Engineering Education from the perspective of serving the national strategy, meeting the needs of industry and facing the future development in response to the challenges of the new economy.

In recent years, CDIO (Conceiving, Designing, Implementing, Operating) education concept has become the new direction and trend of engineering education reform, and is an important way to cultivate professional talents [2–5]. It takes the life cycle from product development to product operation as the carrier, combines basic theory teaching, course experiment, comprehensive training and engineering practice together with each other, so that students can learn engineering in an active, practical and organic way, reflecting the characteristics of systematic and comprehensive learning [6, 7].

In 2016, our school officially launched the professional comprehensive teaching reform based on the CDIO Engineering Education Concept, and was approved to join the "CDIO Engineering Education Alliance" in 2019. Based on years of teaching research and practice of logistics engineering major in our university, the group investigated and analyzed the talent training objectives and core courses of logistics engineering major in 80 colleges and universities in China, found that the talent training characteristics of various colleges and universities are not prominent, the overlapping of subjects is obvious, and the theoretical teaching and practical training are lack of systematic thinking [8]. This paper puts forward the talent training direction of "intelligent logistics and supply chain management and optimization". After four years of teaching reform and practice, we proposed the application-oriented talents training mode of Logistics Engineering Specialty in line with the characteristics of our students, and carried out the construction and implementation. Facts have proved that such reform is important and necessary [9].

## 2 Construction of Applied Talents Training Mode of Logistics Engineering Specialty Based on CDIO-P & CG

Under the background of 3E construction, the comprehensive teaching reform of Logistics Engineering Specialty in our university has introduced CDIO-P&CG (Project teaching and Course Group) applied talents training mode which is constructed and implemented from the following aspects: training requirements, curriculum system, construction process, achievement transformation, platform construction and teaching inspection, shown in Fig. 1.

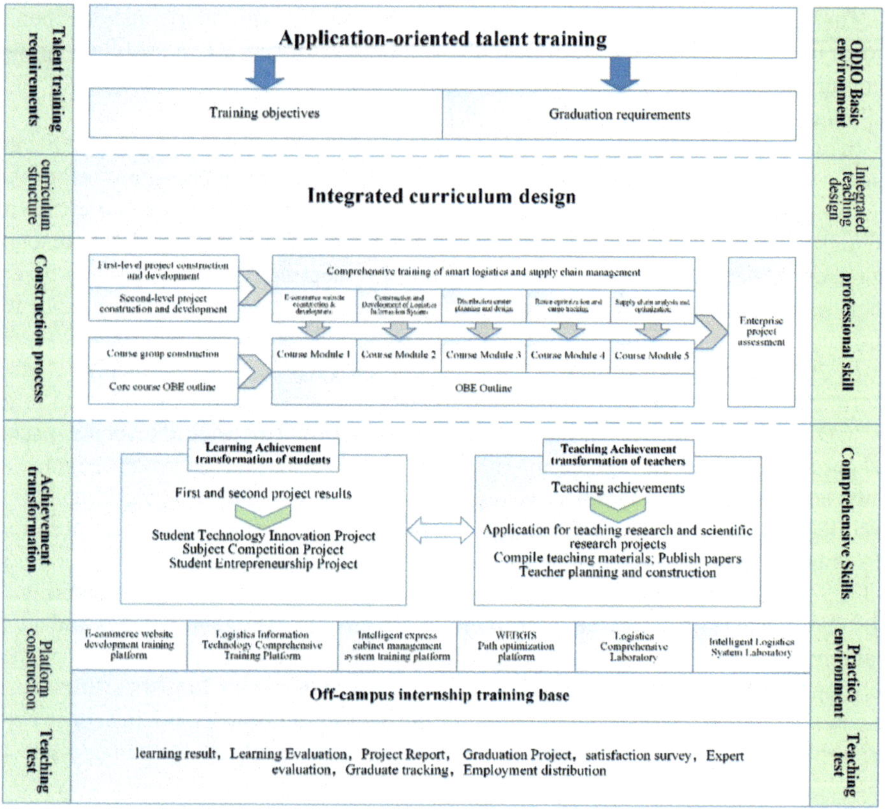


**Fig. 1.** Training mode of applied talents in logistics engineering specialty based on CDIO-P&CG

## 3 The Construction of Professional Integrated Curriculum System

According to the training objectives and graduation requirements, the logistics engineering specialty of our university highlights the comprehensive ability training of "intelligent logistics and supply chain management and optimization" [10]. From theoretical teaching to practical training, from basic knowledge to professional skills, from single skill to comprehensive ability, the integrated curriculum system of logistics engineering specialty of our University is proposed, presented in Fig. 2.

From the course system, it can be seen that "intelligent logistics and supply chain management and optimization" belongs to the first level project; "e-commerce website construction and development", "logistics information system design and development", "distribution center planning and design", "route optimization and goods tracking", "supply chain management and optimization" and "innovation and entrepreneurship" pertain to the second level project, and the course group established by the core courses subordinates to the third level project. The construction mode of the first, second and third level project system enables us to clearly analyze the corresponding relationship between the project and the course, and gradually realize the horizontal connection between the

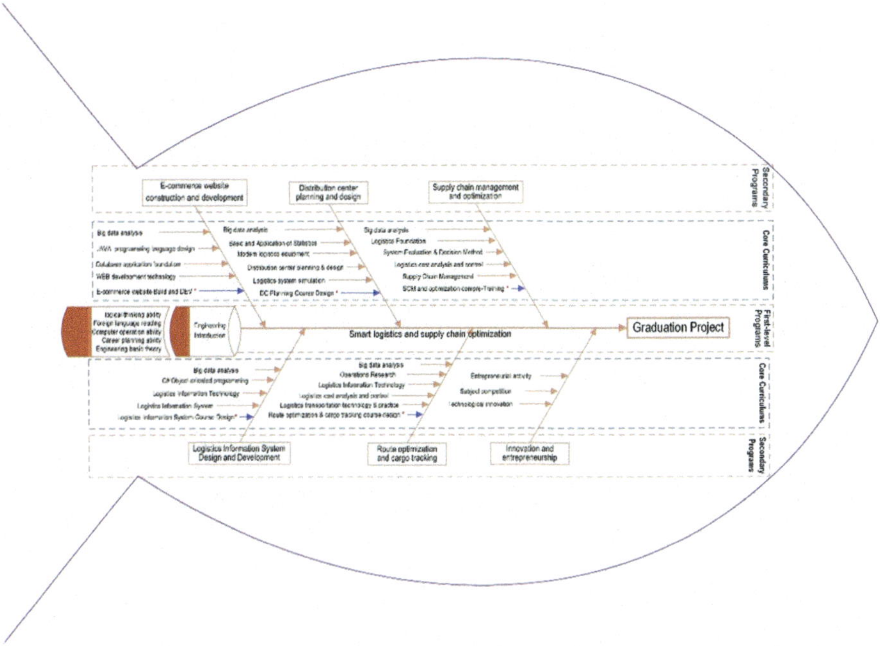


**Fig. 2.** Fishbone of integrated curriculum system of Logistics Engineering Specialty

project and the vertical integration between the courses, showing the systematic and integrated teaching design idea.

# 4 Construction and Implementation of Three Level Projects

## 4.1 Building and Implementation of First Level Project

The effect of project construction depends on a perfect assessment system [11, 12]. The establishment of "first level project" is not only the direction of talent training of Logistics Engineering Specialty, but also the embodiment of students' comprehensive ability training. It plays a closing role in the integration and promotion of secondary projects.

According to the operation procedure of the whole process of e-commerce logistics supply chain, we designed the first-class project of "smart logistics and supply chain management and optimization", which mainly involves e-commerce website construction, warehouse management system development, order review, data analysis, sorting and unloading, distribution path optimization and GIS based visualization, as well as express delivery and supply chain performance evaluation. In order to ensure the smooth development of the first level project, the evaluation index (as shown in Fig. 3) and the implementation plan (shown in Table 1) are formulated.

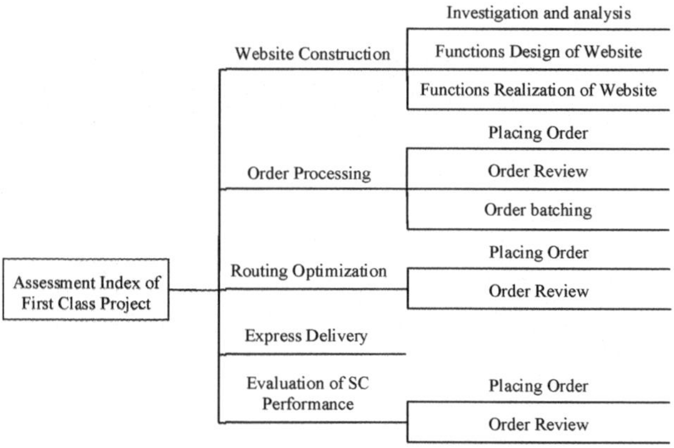


**Fig. 3.** First-level of project assessment indicators

### 4.2 Building and Implementation of Second Level Project

The second level project is carried out around the first level project, which focuses on the cultivation of students' professional core competence. It mainly includes five second level projects: "e-commerce website construction and development", "logistics information system design and development", "distribution center planning and design", "route optimization and goods tracking" and "supply chain management and optimization". In the process of construction, it is necessary to introduce the concepts of "project driven" and "task driven", and from the actual project of the enterprise, formulate the secondary project assessment standard. This paper takes "logistics information system design and development" as an example to illustrate the process of secondary project construction, as well as the ideas and assessment indicators are shown in Figs. 4 and 5.

In the implementation process of the second level project, the phased assessment method is well used to transform the teaching content of the second level project to the knowledge points of the course, realizing the integration of the course and the second level project.

### 4.3 Building and Implementation of Third Level Projects

The construction of third-level project is not only the foundation but also the key to the comprehensive teaching reform. In order to break the barriers between the courses of logistics engineering and eliminate the overlapping of course contents, the concept of curriculum group is introduced, and the knowledge structure of each course is sorted out under the guidance of OBE (Outcomes Based Education).

Taking the knowledge structure of "logistics information system" course as an example, the theoretical part includes eight aspects: logistics system overview, system development method, topic selection discussion, demand analysis, system analysis, system design and system implementation, and periodic review is timely introduced; while, in the practice part, there are 6 design experiment projects which are: logistics information system preliminary knowledge, demand analysis, project cost estimation, Axure prototype design, system analysis and system design, and stage assessment is also set up to

**Table 1.** Implementation plan of the first level project

| No. | Project teaching process | Student activities | | Teachers activities | Class hours |
|---|---|---|---|---|---|
| 1 | Project introduction | A. Understand the overall content of the project and establish the actual concept of the project in the workplace;<br>B. Understand the learning objectives of the project;<br>C. Observe and understand the procedures, methods and quality requirements for the completion of attempted tasks;<br>D. By trying to complete the task, accurately understand the specific tasks in the project that you want to complete, and enter the work role | | A. Demonstration project example;<br>B. Describe the project content, result form and quality requirements;<br>C. Explain the learning objectives of the project;<br>D. Describe the content, quality requirements and working methods of the trial task;<br>E. Demonstrate the completion process and operation method of the trial task;<br>F. Make sure that all students understand the task and enter the role of worker | 2 class hours |
| 2 | Project implementation | Task 1: E-commerce website design | A. Select background analysis, there is a field investigation process | A. Guide the students to complete the trial task one by one, and judge the quality of the task. Correct the existing mistakes strictly;<br>B. Inductive explanation of the common problems in the process of trying to complete the task | 4 weeks |
| | | | B. Website function design, improve shop- ping cart and order query function | | |

(*continued*)

**Table 1.** (*continued*)

| No. | Project teaching process | Student activities | | Teachers activities | Class hours |
|---|---|---|---|---|---|
| | | | C. Website function realization, reasonable realization shopping cart and order inquiry function | | |
| | | Task 2: order processing | A. Order data is standard and reasonable | | |
| | | | B. Generate the order picking list according to the order batching strategy | | |
| | | Task 3: path optimization | A. Use operational research to establish the optimization model | | |
| | | | B. Use MATLAB to solve and use reasonable algorithms | | |
| | | Task 4: express delivery | Using express intelligent cabinet for delivery can be completed quickly and accurately | | |
| 3 | Project acceptance | A. Complete the exhibition<br>B. Summarize the learning results obtained through the project<br>C. Supply chain operation management and evaluation | | Comprehensive evaluation examines the overall performance of students and their performance in the work process | 4 class hours |

lay a solid foundation for the smooth development of the follow-up project teaching. As shown in Fig. 6, the construction of the course highlights the connotation of individual difference and ability standard emphasized in OBE.

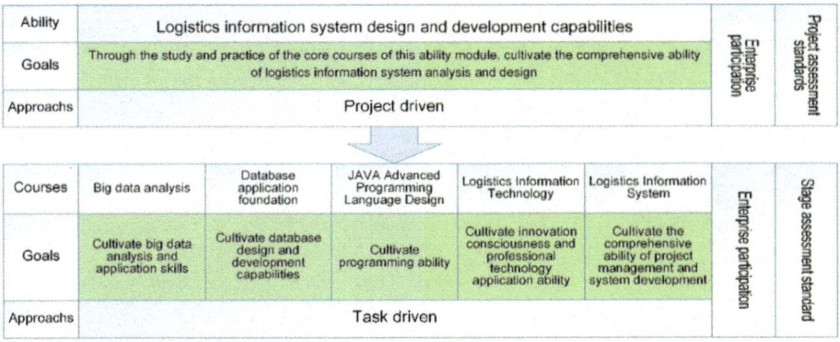


**Fig. 4.** Second level of project construction

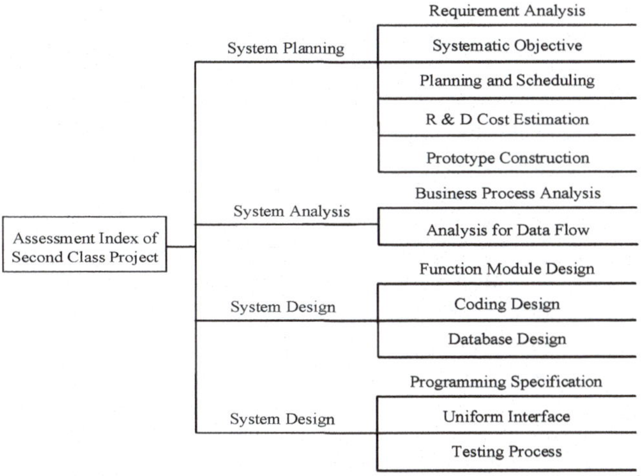


**Fig. 5.** Evaluation index system for secondary projects

# 5 Teaching Implementation

## 5.1 Curriculum Organization Mode of Active Participation and Interaction Between Teachers and Students

In the process of organizing teaching, considering the characteristics of our students, the heuristic teaching mode is adopted to guide the students to explore and study independently and stimulate their innovative thinking. In the process of teaching implementation, several project teams are formed with 2–3 students, including one leader separately. Under the guidance of the tutor group, the members of the team experience all aspects of the comprehensive reform and practice of CDIO specialty, and constantly improve the conception and design of the system, so as to lay a solid foundation for the future professional conception, design, implementation and operation.

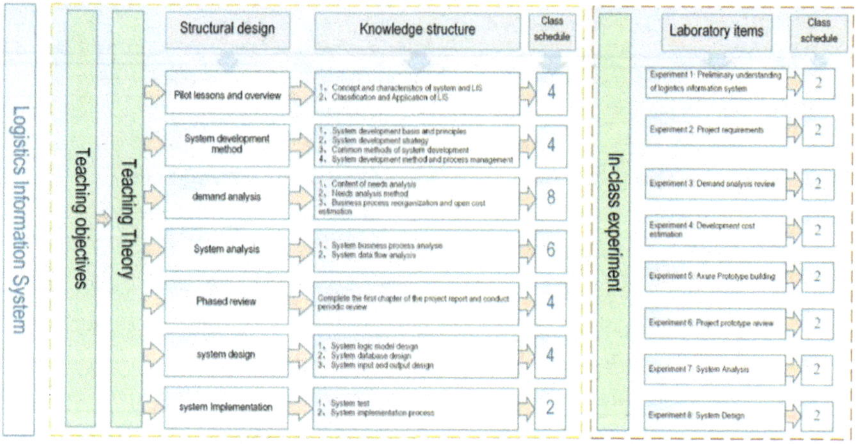


**Fig. 6.** Knowledge structure of the course logistics information system

### 5.2 Interactive Mode of Curriculum Learning and Innovation, Subject Competition and Graduation Design

Through systematic theoretical study and comprehensive practical training, the students have laid a firm foundation for the ability of intelligent logistics and supply chain management and optimization; through the transformation and redesign of projects in the comprehensive teaching reform, it provides rich research topics for scientific and technological innovation, discipline competition and graduation design, and enhances the confidence of students to participate in scientific and technological innovation and discipline competition, and students' participation is also improved.

## 6 Conclusion

The talent training mode of Logistics Engineering Specialty Based on CDIO-P&CG engineering education concept proposed in this paper meets the needs of application-oriented talents training in our university; the fish bone diagram of integrated course shows the systematic design of curriculum teaching; the combination of curriculum assessment and project evaluation can transform the appraising standard of enterprise actual project into teaching, and implement the stage assessment in the practice teaching with the results as the guidance, which presents the connotation of OBE education concept;

The positive participation and interaction between teachers and students has achieved good results. It has brought the students' subjective initiative into play, enhanced their self-confidence, stimulated their learning interest and learning potential, greatly improved the quality of personnel training and employment level, and received high evaluation from employers, reflecting the effectiveness of comprehensive professional teaching reform. At the same time, this achievement has been recognized by the relevant professional teachers and experts in the same kind of colleges and universities, and has also opened up an innovative way for the cultivation of applied talents, which has a high reference value for the construction of related majors in similar colleges and universities.

**Acknowledgment.** This paper is supported by the teaching research project of Century College of Beijing University of Posts and Telecommunications "Research and practice of Logistics Engineering Specialty Construction Based on CDIO Engineering Education Mode", (2016JKY-02).

## References

1. Wu, Y.: Emerging engineering education-the future of higher engineering education: strategic thinking on the future of higher education. High. Eng. Educ. Res. **6**, 7–9 (2018). (in Chinese)
2. Hua, K., Han, Y.: Integrated CDIO talents training plans design and realization. J. Civil Aliat. Univ. China **30**(5), 34–37 (2012)
3. Liu, Y., Zhou, X., Fang, L., Lin, G.: Applied logistics engineering talent training mode based on CDIO concept. China Mark. **15**, 37–38 (2016). (in Chinese)
4. Chen, L., Wu, Q.: Optimization of talent training of application-oriented undergraduate MICE economy and management specialty under CDIO. Logist. Eng. Manag **41**(1), 174–176 (2019). (in Chinese)
5. Liu, J., Li, P., Geng, Y.: Curriculum architecture design of logistics engineering based on CDIO mode. Logist. Technol. **35**(9), 177–180 (2016). (in Chinese)
6. Wei, Z.: Design of capstone projects under CDIO mode. In: International Conference on Computational Science and Computational Intelligence (CSCI), pp. 849–852 (2019)
7. Khan, R., Ercan, M., Kristian, N., et al.: Engineering and design: an integrated course with real world projects. In: Proceedings of the 11th International CDIO Conference, no. 35, pp. 1–15 (2015)
8. Yu, L.: Research on the "cooperative education" model cultivating in higher vocational education. Int. J. Educ. Manag. Eng. **2**(1), 521–524 (2012)
9. Wang, X., Cao, Z.: Discussion reform of forestry panorama course teaching. Int. J. Mod. Educ. Comput. Sci. **12**(29), 41–45 (2012)
10. Xu, Y.: A new education mode of engineering specialty graduate students—co-founding by college and enterprises, uniting of production, learning and researching. Int. J. Educ. Manag. Eng. **2**(12), 590–593 (2012)
11. Zhang, S., Zhou, Z.: Talent training of software testing based on university-enterprise cooperation. Softw. Eng. Knowl. Eng. Theory Pract. **2**, 315–320 (2012)
12. Zhao, C., Wang, Y., Ma, J., Yu, A.: Study on the examination pattern of the engineering specialized courses. Int. J. Mod. Educ. Comput. Sci. **2**(29), 59–65 (2012)

# Teaching Reform of Sports Dance Course in Ethnic Areas Under the Background of "Intelligent Teaching"

Qunqun Sun[1] and Qi Shan[2](✉)

[1] Institute of Physical Education, Qiannan Normal University for Nationalities, Duyun 55800, China

[2] Zhongshan College, University of Electronic Science and Technology, Zhongshan 528400, China

**Abstract.** The teaching reform of sports dance course in ethnic areas is a new mode of curriculum teaching reform which is an exploration and innovation of higher education reform under the new situation. Intelligent teaching has become an irresistible development trend. In the face of the problems existing in the traditional way of sports dance teaching, this paper discusses the teaching mode of sports dance in class and extracurricular clubs as two wings, and actively integrates extracurricular sports dance activities into the physical education curriculum system. With the help of the digital Internet platform of intelligent teaching, the teaching contents, teaching organization forms, exercise time, use of venues and equipment and teaching evaluation are made, so as to provide new ideas for the reform and development of sports dance curriculum in ethnic areas.

**Keywords:** Intelligent teaching · Ethnic areas · Sports dance course · Teaching reform

## 1 Introduction

Physical education is an important part of the teaching system in colleges and universities [1]. Not only in China, but also in other countries in the world, the teaching of physical education is highly valued [2–4].

The world is now in an era of information technology, and the rise of intelligent teaching has led to the emergence of many new things, but also promoted the integration of other industries [5]. The reform of the education industry has become increasingly rapid in the times of intelligence, and the education industry is facing a new round of education revolution, in which, the reform and innovation of sports dance course conform to the trend of education reform.

With the development of information, communication and Internet technology, mobile phone has become an indispensable "small partner" in students' study and gradually entered the classroom. Various sports APPs have been constantly developed, which have different effects on the teaching of different courses [6]. It must be admitted that


Z. Hu et al. (Eds.): ICAILE 2021, LNDECT 82, pp. 200–208, 2021.
https://doi.org/10.1007/978-3-030-80475-6_20

the intelligent convenience of the Internet in the new era gives more vitality to the education industry. Compared with the traditional physical education teaching mode, the emergence of new things "intelligence + sports" not only brings freshness, but also stimulates students' initiative, enthusiasm and participation in learning [7]. It can also return time and subject status to students, and teachers are no longer "crammed", only need to fulfill the responsibility of a mentor [8]. This kind of "intelligence + physical education" classroom teaching mode is gradually favored by teaching reformers [9].

It is an exploration and innovation of higher education reform under the new situation. For a long time, sports dance courses in colleges and universities in ethnic areas have basically followed the traditional mode, ignoring the particularity of sports dance in ethnic areas. Teaching methods are outdated and teaching effects are not obvious [10]. Therefore, it is necessary to explore a sports dance teaching mode that conforms to the development of the times and conforms to the characteristics and laws of ethnic areas, and give full play to the sports dance education in ethnic areas. The role of talent training has become an important issue in the reform of higher education [11].

Due to the backward economy in ethnic areas, there is also a conservative trend in education concepts. Many advanced teaching methods and intelligent teaching modes cannot be spread out in an all-round way, which leads to many college students unable to integrate into the society well after graduation. How to reform the sports courses in ethnic areas on those intelligent times, so that students can feel in the intelligent era to the convenience brought by intelligence, it can better promote the teaching reform of sports dance in ethnic areas and narrow the gap between the teaching level in those areas and that in developed areas. This is a new topic that teachers of sports dance course in ethnic areas should seriously study.

## 2 The Object and Method of the Research

### 2.1 Research Object

In this paper, based on the background of "intelligent teaching", the teaching reform of sports dance course in ethnic areas is the research object. It mainly studies the teaching reform of sports dance students, teaching content and methods by using intelligent methods.

### 2.2 Research Methods

This project uses a variety of methods to carry out the research, including literature search, interview, logical analysis and experimental method.

#### 2.2.1 Documentation Study

Through CNKI, Wanfang, VIP database and school e-library and other network channels, access to the relevant intelligent teaching reform of physical education curriculum related master's and doctoral papers, journals, newspapers, and through the collection, collation, analysis of relevant literature, the basic grasp of the theoretical points of the study, laid a theoretical foundation for this study.

#### 2.2.2 Interviewing

According to the requirements of the relevant interview outline, focusing on the relevant content of intelligent teaching reform in ethnic areas, it is obliged to conduct interviews on the relevant issues of school physical education teachers and leaders, and carry out investigation and analysis on the content of the interview, at the same time, sort out the relevant data, and get the corresponding results.

#### 2.2.3 Logical Analysis

Using the theory of physical education, logic, statistics and other related theoretical knowledge, this paper analyzes the data obtained from mathematical statistics, and then arranges the data. On this basis, it makes logical analysis, probes the problems existing in the teaching of sports dance course in colleges and universities in ethnic areas at this stage, and puts forward effective reform strategies by using logical thinking according to the existing problems.

#### 2.2.4 Experimental Research

According to the research content and purpose of this subject, using the requirements of sports scientific research methods and experimental ways, the experimental group and the control group were designed for experimental control, and then the differences between them were obtained.

1) *Purpose of the experiment*
   In the sports dance course, the experimental group implemented the interactive teaching in and out of class under the network platform, while the control group still used the traditional teaching method. Through the comparison of the relevant indicators before and after the experiment between the two groups, the influence of the interactive teaching in and out of class supported by the network platform on the learning effect was verified.
2) *Subjects and time of the experiment*
   In this study, class 1 of Physical Education (10 girls) was selected as the subject and class 2 of physical education (11 girls) taken as the control class in Qiannan Normal University for Nationalities. The experimental site was the mirror wall dance room of the university.
   The experiment lasted for one semester from April 2, 2019 to June 20, 2019, with a total of 16 weeks and 32 class hours.
3) *Experimental content*
   As the experimental subjects are at the entry level, the main content of this semester's teaching is defined as the combination of sports and dance "bronze routine", which has a low degree of difficulty and is in line with the characteristics of learners.
4) *Experimental variables*
   The independent variable is the interactive teaching and traditional teaching under the network platform; the dependent variable is the students' basic physical quality, sports dance skills, sports dance learning interest and mental health level; before the experiment, the initial ability diagnosis of the two groups of students is carried

out to ensure that there is no significant difference in the relevant indicators. The syllabus was the same and the levels of teachers were identical. The experiment was designed with single blind.

5) *Experimental effect index*
The results showed that there were significant differences between the two groups which mainly in four aspects, i.e. Basic physical fitness ($4 \times 10$ m turn back run, standing long jump, sit three-dimensional forward bend, 1 min sit up); Sports dance technical ability (breathing and balance, technical quality, movement and music, arrangement and performance); Sports dance learning interest (questionnaire); and Symptom checklist SCL90.
6) *Experimental process and operation*
On the premise of ensuring the same initial state of the two groups, the experimental group implemented interactive teaching in and out of class: it was composed of SPOC teaching mode in class and extracurricular club activities; the control group was taught according to the traditional process. The differences between the two groups before and after the experiment were considered and evaluated.

## 3 Literature Review and Analysis

### 3.1 Relevant Theories of Intelligent Teaching

Luckin R, Holmes pointed out that educational artificial intelligence is a new research field formed by the combination of artificial intelligence and learning science.

The goal of educational artificial intelligence is to create conditions for students' learning by observing and understanding the occurrence process of learning [12];

Timms MJ proposed that educational artificial intelligence mainly enables students to better participate in teaching through computers and teaching platforms, and helps teachers teach more effectively. He also emphasized that in the future, it is necessary to integrate the power of artificial intelligence with the progress in the field of robotics, and much more sensors are used to monitor the learning environment and learning behavior [13].

Some other scholars have discussed the problems in the application of artificial intelligence in teaching, the challenges that the application of AI in education is facing or will face, and the future development trend of educational AI [14].

Li Zhaohan and others discussed the combination of artificial intelligence and education, including four directions: automatic homework marking, online question answering by taking photos, intelligent evaluation and personalized learning, and discussed the innovative changes of artificial intelligence in three aspects: learner model, teaching model and specific learning field [15];

Wang Haifang and Li Feng studied six application fields of artificial intelligence widely used in education, and discussed the latest progress of its application in Education [16]; Yu Minghua et al. have found that the application of machine learning in education mainly focuses on six aspects: student modeling, student behavior modeling, prediction of learning behavior, early warning of dropout risk, learning support and evaluation, and Resource Recommendation [17];

Chen Ying studies the application of artificial intelligence in middle school English teaching, aiming to optimize the teaching methods and effects of middle school English [18];

On the basis of discussing the support of artificial intelligence to stem education, Tang Yiwei et al constructed the stem interdisciplinary integration mode based on artificial intelligence, in order to better promote the integration of stem interdisciplinary [19].

To sum up, intelligent education has become a new favorite of current education. How to carry out intelligent optimization teaching in sports dance class, so that students in ethnic areas can feel the charm of intelligent teaching, and let sports dance teachers feel the benefits of intelligent teaching, has become a new proposition that should be seriously considered.

### 3.2 Teaching Mode Innovation of Sports Dance Course

The most dynamic factor is teaching method. Under the influence of contemporary educational thoughts, teaching theories and the development of modern educational technology, we must attach importance to the reform and exploration of learning methods and highlight the subjectivity of students' learning; we must attach importance to the unity of teaching methods and learning methods, and highlight the interaction between teachers and students; we must attach importance to the integration of emotional teaching and highlight the life of teaching activities; we must attach importance to the development of teaching technology and highlight teaching The modernity of the way.

The reform of teaching methods needs to learn from other people's experience and lessons, but also combined with the characteristics of this course, to find more suitable methods and ways.

Special attention should be paid to the following points when learning from others' teaching methods.

7) *Teaching reform should be based on the change of educational ideas*
   To learn excellent experience is to transplant the ideas, to be guided by scientific theory, to form their own teaching personality. In this sense, any teaching method reform is not a simple "method" change, but a change in educational ideas.
8) *The reform of teaching methods should be combined with the curriculum practice*
   There is a method in teaching, but there is no fixed method. It is important to get the method. No matter how good the teaching methods summed up by others, we can't mechanically copy them mechanically. We should make appropriate adaptations according to our own specific situation and the specific situation of students.

In a word, the improvement of teaching methods needs to be analyzed from the perspective of innovation, and reasonable reform programs should be designed and implemented seriously.

## 4 Practice of Teaching Reform of Sport Dance Course

Under the background of intelligent teaching, this paper discusses the innovative research on the teaching mode of college sports dance course, which is based on the teaching mode of physical education class and extracurricular club, actively brings extracurricular sports activities into the physical education curriculum system.

With the help of the intelligent digital Internet platform, the two sides can achieve positive interaction in teaching content, teaching organization form, exercise time, field equipment use and teaching evaluation.

## 4.1 Interactive Teaching Mode of Sports Dance Course Under the Background of Intelligent Teaching

After investigating the teaching mode of sports dance course in several colleges and universities in ethnic areas, it is found that no university has applied this teaching mode. Therefore, combining with my own teaching practice in colleges and universities in ethnic areas, the author attempts to carry out reform and research on college sports dance from the above aspects, and use the "intelligent + " mode to promote the interactive teaching mode of sports dance inside and outside the classroom.

Through two semesters of experimental teaching research, it is indicated that "intelligent + " has brought great changes to the teaching mode of physical education teachers, but also put forward more challenges. With the help of online teaching APP, Pan-Ya network teaching platform, excellent resource sharing course and SPOC mode of combination of online and offline are implemented.

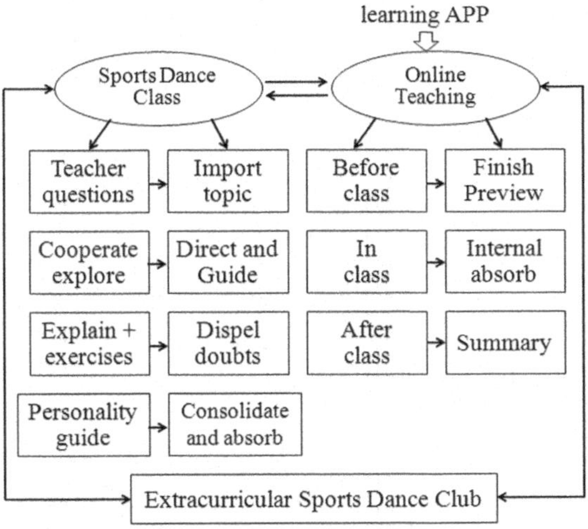


**Fig. 1.** "Intelligent + course of sports dance" interactive teaching mode inside and outside class

Figure 1 shows the interactive teaching mode of "intelligent + " sports dance course.

From the perspective of classroom teaching, there are four main links, which are: to introduce the theme through the teacher's questions; to give guidance in the process of cooperative exploration; to solve doubts through explanation and communication; finally, through personalized guidance, students can consolidate and absorb the knowledge and skills they have learned.

Another important content is online teaching. Through the intelligent technical means, complete the preview before class, absorption in class, as well as the summary and carding after class.

Both classroom teaching and online teaching are complementary.

Considering the particularity of the course "Sports Dance", there is also an "extracurricular sports dance club" which is also an important place for practice and exercise.

### 4.2 Differences of Students' Performance of Sports Skills Under the Background of Intelligent Teaching of Sports Dance

Motor skill refers to the ability of human body to master and effectively complete special movements in sports.

Before the experiment, a group of simple movements were taught and tested, and the results were tested by independent sample T test. The P values were all greater than 0.05 (see Table 1). It is concluded that the two groups can be used as equivalent experimental objects.

After 16 weeks of sports dance skills learning, the sports skills of the two groups were improved in varying degrees, but the difference was significant, P value was less than 0.05 (see Table 2). This shows that the effectiveness of intelligent and interactive teaching in sports dance skills is better than the traditional teaching.

The reasons are as follows: first of all, SPOC teaching provides students with space to set their own pace before class, which creates the possibility of repeated learning for students with weak learning ability. The degree of polarization is reduced and the average score of motor skills is improved;

Secondly, the active atmosphere of student-student interaction and teacher-student interaction enhances students' ability to accept special skills. In a free and relaxed environment, learning efficiency is improved.

**Table 1.** Comparison of sports dance performance between the experimental group and the control group before the experiment

| | Experience group | Control group | T | P |
|---|---|---|---|---|
| | $X \pm S$ | $X \pm S$ | | |
| Breathing and balance | $7.36 \pm 0.54$ | $7.37 \pm 0.55$ | 1.143 | >0.05 |
| Action quality | $8.12 \pm 0.45$ | $8.09 \pm 0.53$ | 1.252 | >0.05 |
| Action and music | $7.10 \pm 0.68$ | $7.11 \pm 0.66$ | 0.254 | >0.05 |
| Arrangement and performance | $7.61 \pm 0.60$ | $7.59 \pm 0.66$ | 1.012 | >0.05 |
| General comments | $7.61 \pm 0.53$ | $7.61 \pm 0.66$ | −1.029 | >0.05 |

**Table 2.** Comparison of sports dance performance between the experimental group and the control group after the experiment

| | Experience group | Control group | T | P |
|---|---|---|---|---|
| | $X \pm S$ | $X \pm S$ | | |
| Breathing and balance | 8.88 ± 0.61 | 8.13 ± 0.49 | 1.82 | <0.05 |
| Action quality | 9.34 ± 0.38 | 8.84 ± 0.33 | 1.58 | <0.05 |
| Action and music | 9.08 ± 0.47 | 8.73 ± 0.39 | 1.35 | <0.05 |
| Arrangement and performance | 9.37 ± 0.54 | 8.93 ± 0.98 | 2.15 | <0.05 |
| General comments | 9.29 ± 0.34 | 8.75 ± 0.48 | 1.04 | <0.05 |

## 5 Conclusion

The research conclusions of this paper are induced as follows:

1) Practice has proved that the intelligent and interactive sports dance teaching in and out of class has a significant effect on strengthening students' physique, promoting the formation and improvement of sports skills, stimulating students' interest in exercise, and improving students' mental health level. At the same time, it solves the practical problems of effective integration of online and offline, in class and out of class, which has certain research and promotion value.
2) Sports dance course teaching in ethnic minority areas has its particularity and independence. It should be differentiated and innovated from other colleges and universities to find a suitable model for the local sports dance course teaching. The emergence of modern information technology and intelligent teaching methods makes the university sports dance course teaching more advanced, Physical education teachers should use intelligent teaching methods to integrate the excellent national dance movements, dance costumes, dance music and other national cultural elements into the sports dance classroom teaching, and use modern intelligent technology to further reform and innovation.
3) Using modern intelligent information technology in teaching content, teaching organization form, exercise time, field equipment uses and teaching evaluation to realize the positive interaction mode between students and teachers, inside and outside the classroom, should be promoted in more colleges and universities in ethnic areas. Definitely, in the experimental promotion stage, it should be paid attention to the actual situation of the school and the situation of school teachers, students as well as the teaching environment.

**Acknowledgment.** This paper is supported by the teaching reform project of Qiannan Normal University for Nationalities (2019xjg0510).

## References


1. Sun, J.: Strategic development of physical education discipline in colleges and universities of China. Educ. Teach. Forum **3**, 43–44 (2015). (in Chinese)
2. Schulenkorf, N., Sherry, E., Rowe, K.: Sport for development: an integrated literature review. J. Sport Manag. **30**(1), 22–39 (2016)
3. Zhou, P.: Development trend of college physical education from the perspective of comparative sports between China and the west. J. Guangzhou Sports Univ. **34**(02), 109–112 (2014). (in Chinese)
4. Zhang, J., Wang, J., Wang, T.: School sports development patterns and enlightenment of some developed countries. J. Wuhan Sports Univ. **49**(10), 5–20 (2015). (in Chinese)
5. Xu, Y.: A new education mode of engineering specialty graduate students—co-founding by college and enterprises, uniting of production, learning and researching. Int. J. Educ. Manag. Eng. **2**(12), 590–593 (2012)
6. Yu, L.: Research on the "cooperative education" model cultivating in higher vocational education. Int. J. Educ. Manag. Eng. **2**(1), 521–524 (2012)
7. Zhang, L., Wei, D., Huang, L.: The research on the program of educational technology ability building for vocational education teachers. Int. J. Educ. Manag. Eng. **2**(11), 46–51 (2012)
8. Li, C., Wang, C.K.J., Pyun, D.Y.: Talent development environmental factors in sport: a review and taxonomic classification. Quest **66**(4), 433–447 (2014). (in Chinese).
9. Wei, X.: Discovery and practice of EDA experimental teaching reform. Int. J. Educ. Manag. Eng. **1**(4), 41–45 (2011)
10. Liu, J.: Research on strategies and patterns of research teaching. Int. J. Educ. Manag. Eng. **2**(12), 67–72 (2012)
11. Boschman, F., McKenney, S., Voogt, J.: Understanding decision making in teachers' curriculum design approaches. Educ. Tech. Res. Dev. **62**(4), 393–416 (2014)
12. Luckin, R., Holmes, W.: Intelligence unleashed: an argument for AI in Education [EB/OL]. https://www.pearson.com/news/blogs/CompanyBlog/2016/11/intelligence-unleashed-an-argument-for-ai-in-education.html. Accessed 24 Nov 2016
13. Timms, M.J.: Letting artificial intelligence in education out of the box: educational cobots and smart classrooms. Int. J. Artif. Intell. Educ. **2**, 701–712 (2016)
14. Mc Arthur, D., Lewis, M., Bishary, M.: The roles of artificial intelligence in education: current progress and future prospects. J. Educ. Technol. **4**, 42–80 (2005)
15. Li, Z., Jin, H., Liu, Y.: Artificial intelligence opens the new mode of "Internet plus education." Telecommun. Netw. Technol. **12**, 6–10 (2016). (in Chinese)
16. Wang, H., Li, F.: Application of artificial intelligence in education. Mod. Educ. Technol. **11**, 18–20 (2008). (in Chinese)
17. Yu, M., Feng, X., Zhu, Z.: Educational application and innovative exploration of machine learning from the perspective of artificial intelligence. Distance Learn. **3**, 11–21 (2017). (in Chinese)
18. Chen, Y.: Application of artificial intelligence in middle school English teaching. Nanchang University, Nanchang (2013). (in Chinese)
19. Tang, Y., et al.: Stem interdisciplinary integration model supported by educational artificial intelligence. China Electron. Educ. **8**, 46–52 (2017). (in Chinese)

# Design and Implementation of "Task Driven" Teaching Mode Based on Flipped Classroom in Logistics Management Course

Qinxian Chen[1] and Saipeng Xing[2](✉)

[1] School of Economics and Management, Wuchang Shouyi University, Wuhan 430070, China
[2] School of Management, Wuhan Technology and Business University, Wuhan 430065, China

**Abstract.** Logistics management course has the characteristics of wide range of professional knowledge, many contents, strong practicality and strong comprehensiveness. The traditional teaching mode is difficult to meet the teaching requirements. Flipped classroom is a new teaching mode which is guided by information technology and can significantly improve the teaching effect. Taking the course reform of logistics management as an example, this paper expounds the design and implementation process of "task driven" teaching mode based on flipped classroom, This paper discusses the time arrangement, implementation process and evaluation method of task driven teaching method in logistics teaching, and provides feasible operation scheme for teachers to learn from and apply the method, so as to provide reference for deepening the curriculum reform of logistics specialty.

**Keywords:** Flipped classroom · Task driven logistics management · Curriculum reform

## 1 Instruction

Task driven teaching method is a kind of teaching method based on constructivism learning theory [1]. Its main feature is that teachers include the teaching objectives in the learning task list in the teaching design, and take clear tasks as the driving force in the teaching process, so that students can actively construct knowledge, improve their ability and achieve teaching objectives in the process of completing learning tasks independently [2]. Flipped classroom is a new teaching mode which is guided by information technology and can significantly improve the teaching effect. In logistics management teaching, the task driven teaching method based on flipped classroom is to drive students' learning motivation with clear, clear and essential tasks in the teaching process, stimulate students' interest in learning, and let students experience the learning process by displaying and testing the results of tasks, so as to improve students' learning state and promote students' learning Deep learning can effectively cultivate students' autonomous learning ability and logistics management comprehensive ability [3, 4].


Z. Hu et al. (Eds.): ICAILE 2021, LNDECT 82, pp. 209–218, 2021.
https://doi.org/10.1007/978-3-030-80475-6_21

## 2 Design and Operation Elements of Task Driven Flipped Classroom

### 2.1 Elements of Task Driven Teaching Method

Task driven teaching method has three elements: task as the main line, students as the main body, and teachers as the leading [5]. First, the task as the main line, the teaching process students around the task to learn, emphasizing learning by doing. Secondly, students are the main body, and the design, implementation and evaluation of tasks are all around students [6]. Third, the teacher is the leading, the teacher carries on the preliminary analysis work, the classroom instruction and the instruction student, after class carries on the teaching appraisal [7].

### 2.2 Task Driven Teaching Method is Mainly Divided into Three Part Spreliminary

Analysis, teaching process and teaching evaluation are important contents [8]. The preliminary analysis includes the analysis of teaching objectives, teaching contents and teaching objects, which is the premise of task driven teaching method [9]. The teaching process includes six steps: situation creation, task design, task arrangement, task implementation, task evaluation and task extension [10]. The last link is teaching evaluation, in which the teacher carries out process evaluation and summative evaluation according to the students' situation (as shown in Fig. 1).

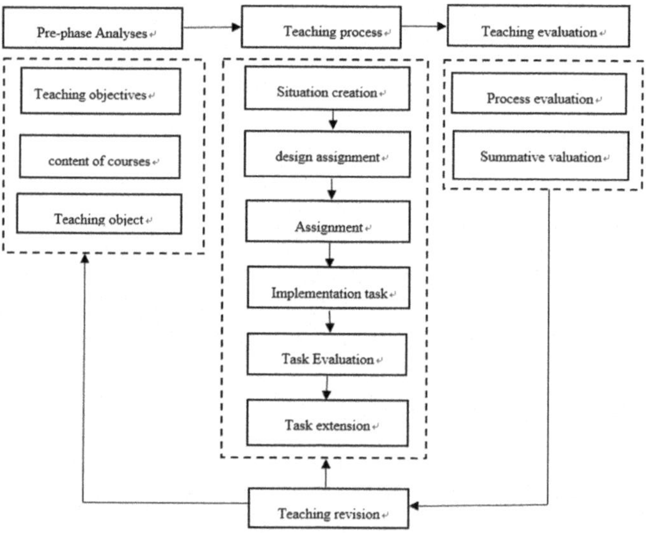


**Fig. 1.** Curriculum design framework of task driven teaching method

### 2.3 Design of Task Driven Flipped Classroom

The concept of "task driven" is to teach knowledge, which advocates solving problems and completing one task after another [11]. It has the characteristics of multi-dimensional interaction. Flipped classroom will return the decision-making power to students by optimizing the time arrangement inside and outside the classroom. Based on this, in the normal classroom time, students can be more focused on professional knowledge learning. Flipped classroom can even be said to be the subversion of the traditional teaching form, leading to the reform of management form, curriculum and even the role of teachers [12, 13]. "Task driven" is a bold attempt in the field of teaching to combine theory with practice, and to cultivate students' professional skills.

## 3 Task Driven Flipped Classroom Teaching Implementation

### 3.1 Design of Task Driven Flipped Classroom

Finish the task by self-study before class. The main purpose of this stage is to enable students to understand the learning content in advance through self-study before class, and find teaching difficulties by completing tasks. Teachers can prepare classroom teaching activities and assign new tasks by analyzing the completion of students' tasks. With the help of the network, teachers can release teaching resources and arrange learning tasks in advance according to the teaching contents, and choose to push learning resources including pictures and texts, videos and learning links that meet the teaching objectives and teaching contents of the course [14].

Students can independently learn pre class teaching resources and complete pre class tasks through mobile terminals such as tablet, computer and mobile phone, combined with books. In the process of self-study, if you encounter any problems, you can seek help from classmates or teachers through "we chat" or "QQ" and other communication tools [15]. At the same time, students submit tasks in time, teachers analyze the learning tasks, view the learning situation of each student and the overall situation of the whole class, and according to the feedback results of the task, the teaching activities and teaching are carried out Tasks are developed or modified [16].

### 3.2 In Class Internalization Promotion Stage

The main purpose of this stage is to help students solve problems, internalize knowledge, comprehensively use language knowledge and improve their learning ability through the links of teachers' answering questions, collaborative inquiry, testing and application. The teaching process of this stage mainly includes five links: answering questions and solving puzzles - new knowledge supplementary teaching - task completion - group presentation - question answering and puzzle solving.

Firstly, the teacher answers the questions found in the pre class stage and the questions raised by the students, and then explains the teaching knowledge points according to the teaching arrangement.

Then, the students carry out cooperative exploration, discuss and study, complete the task according to the classroom teaching task arranged by the teacher, and finally carry out the group display [17, 18]. The teacher answers questions and solves doubts. Each student evaluates himself or herself according to the group, and the teacher summarizes Sexual evaluation is used to test the learning effect. The teacher gives guidance to the students who have the ability to learn, and to the students who have not mastered the knowledge content Individual guidance, through repeated practice to achieve teaching objectives.

### 3.3 Consolidation and Transfer Stage after Class

The purpose of this stage is to consolidate students' knowledge, stimulate students' interest in learning, and expand students' breadth and depth of knowledge. According to the students' learning situation before and in class, teachers should give timely feedback and evaluation, arrange learning tasks, encourage students to actively complete the tasks, and test the mastery of knowledge, so as to maximize the learning effect [18].

## 4 Application of Practical Task Driven Flipped Classroom in Logistics Management Course

### 4.1 Task Design

The first work is the task design. Firstly, the task driven teaching mode needs to design the corresponding teaching tasks according to the characteristics of the course and teaching objectives, combined with the principles and methods of the task driven mode. Secondly, the course of logistics management is an introductory course in the undergraduate teaching of logistics management major. The teaching objective is to enable students

**Table 1.** Task list of logistics management course

| Nine tasks | Teaching hours | Completion mode |
|---|---|---|
| (1) Logistics around you | 4 | Practice, theory |
| (2) Understanding modern packaging | 4 | Practice, theory and extracurricular activities |
| (3) Understanding loading and unloading and handling | 4 | Practice, theory |
| (4) Design an effective supply chain | 4 | Practice, theory |
| (5) Design a storage plan | 6 | Practice, theory |
| (6) Rationalization of a transportation plan | 6 | Practice, theory |
| (7) Design a distribution plan | 6 | Practice, theory |
| (8) Introduction of circulation processing | 4 | Practice, theory |
| (9) Understanding logistics information system | 4 | Practice, theory and extracurricular activities |

to understand the concept of modern logistics and the basic concepts and methods of various elements of logistics. Lastly, in the practice of task driven teaching mode, nine tasks are designed according to the knowledge structure of the course (as shown in Table 1).

### 4.2 Interest Leads People

The use of guiding cases, video materials, etc., combined with the needs of the professional, to create a situation, through the association of students, observation, appreciation, communication, to stimulate students' interest in learning and hands-on initiative. At this point, you should know what you need to accomplish the task Knowledge structure to guide and prompt, guide students to find the corresponding knowledge, to avoid students unnecessary links waste too much time.

### 4.3 Subtask Design – Taking Task 9 as an Example

After the students' interest in learning is stimulated, they should put forward the task timely. According to the teaching content, combined with the actual situation of students, around the big task of teaching objectives, through demonstration of examples, we analyze and put forward small tasks together with students. In small tasks, old knowledge and new knowledge are included, which are promoted layer by layer. Finally, the overall teaching goal is realized, so that students can gradually learn new knowledge and skills in the completion of small tasks.

For example, in the task of "logistics information system", students are required to understand various logistics information technologies according to the given environment (as shown in Table 2). Driven by this task, if students want to complete the task, they must actively understand the relevant technology and application of logistics

**Table 2.** Sub task composition of logistics information system

| Task: logistics information system | | Requirement |
|---|---|---|
| Subtask 1 | RFID Technology | Understanding, application and practice |
| Subtask 2 | Sensing technology | Understanding, application and practice |
| Subtask 3 | barcode technology | Understanding, application and practice |
| Subtask 4 | GPS Technology | Understanding, application |
| Subtask 5 | Internet of things technology | Understanding, application and practice |
| Subtask 6 | EDI technology | Understanding, application |
| Subtask 7 | GIS technology | Understanding, application |

information system under the guidance of teachers. Table 2 Understanding sub tasks of logistics information system.

Give an example: Sensing technology.

The dynamic characteristic of the sensor refers to the response characteristics of the test system to the input signal when the input changes with time.

Description method of dynamic characteristics

1) *Time domain differential equation*

$$a_n\frac{d^n y(t)}{dt^n} + a_{n-1}\frac{d^{n-1}y(t)}{dt^{n-1}} + \ldots + a_1\frac{dy(t)}{dt} + a_0 y(t) \tag{1}$$

2) *transfer function*

On the premise of zero initial condition, the transfer function is defined

$$H(s) = \frac{Y(s)}{X(s)} = \frac{b_m s^m + b_{m-1}s^{m-1} + \cdots + b_1 s + b_0}{a_n s^n + a_{n-1}s^{n-1} + \cdots + a_1 s + a_0} \tag{2}$$

Where s is a complex variable, $s = \alpha + j\omega$

The transfer function of the first order system is obtained

$$\tau\frac{dy(t)}{dt} + y(t) = x(t) \tag{3}$$

Physical meaning: describes the frequency characteristics of the system.

$$H(\omega) = \frac{Y(\omega)}{X(\omega)} = \frac{|Y(\omega)|e^{j\phi_y}}{|X(\omega)|e^{j\phi_x}} = \frac{|Y(\omega)|}{|X(\omega)|}e^{j(\phi_y-\phi_x)} = A(\omega)e^{j\phi(\omega)} \tag{4}$$

$A(\omega)$ Amplitude frequency characteristics of the system

$\phi(\omega)$ Phase frequency characteristics of the system

$H(\omega)$ The relationship between the harmonic input and the steady-state output of the system is described, and the transient response information is not included.

## 4.4 Send Relevant Information and Tasks

The components of pre-class teaching resources are presented in Table 3.

**Table 3.** Components of pre-class teaching resources

| Name | Specific content | Release method |
|---|---|---|
| Teaching video | Recorded by the teacher according to students' learning objectives, including the teaching content and teaching key and difficult points | Video |
| Learning links | Students who have spare power to learn can click the learning link to expand learning | Website |
| Textbook learning page range | Show the page range of the textbook to the students in advance and prompt them to learn the textbook by themselves | Page number text |
| Task list | Learning task list refers to the content of teaching objectives and learning tasks of this class, which requires students to self-study and actively complete the task according to learning resources before class, so that teachers can prepare classroom teaching content pertinently | Task list text |

### 4.5 Exchange Experience and Improve Learning

After trying to practice and analyze, students are required to exchange experience. In view of the problems in the process of trying to practice, teachers should guide students to analyze and share around the task. At the same time, teachers should guide students to learn operation skills through explanation, demonstration and demonstration, and learn the key points and difficulties in operation, and teach students some difficult knowledge points or skills, so as to promote students' awareness [19].

### 4.6 Operation Practice

In view of the problems exposed in practice, after the exchange and sharing, students are required to further improve their own programs, deepen their understanding and application, and consolidate their learning achievements. In the operation practice, we can ask the students who try to finish the task better in practice to guide the students who are not proficient in practice.

### 4.7 Summary of Evaluation

After the completion of the second operation, the teacher should analyze and evaluate the improved scheme, clearly point out the shortcomings, and students can also show and evaluate it. At this stage, we should pay special attention to the objective and pertinent evaluation of the students' performance, and protect the enthusiasm of students to complete the task. At the same time, according to the students' situation of completing the task, the knowledge and skills required by the syllabus are sorted out, and the omissions are found.

# 5 Analysis and Summary of Teaching Effect

Through practice, it is found that the flipped classroom teaching mode based on task driving has certain effect on promoting students' knowledge mastery, learning enthusiasm and learning ability. The conclusions of this study are as follows:

## 5.1 The Task Driven Teaching Mode Based on Task Driven in Colleges and Universities is Conducive to Promoting Students' Knowledge Mastery

Starting from the construction of a new teaching mode, gradually study the current teaching mode, classroom situation and students' learning situation, introduce the task driven hybrid teaching mode based on flipped classroom, and apply it to the university curriculum is an effective means for the Ministry of education to call for the construction of first-class colleges and universities. Before class, the learning resources will be displayed on the Internet to cooperate with the task driven to stimulate students' learning initiative Mobility and the knowledge necessary to complete the task. "Task driven" and "flipped classroom teaching" will become an important method and means to promote students to master knowledge actively in the future [20].

## 5.2 Enhance the Interaction and Cooperation between Students

Task driven teaching method emphasizes the cooperation between groups, completes tasks in the form of groups, and enhances students' sense of cooperation. At the same time, students conduct group mutual evaluation in the task evaluation stage, so that students can learn from the experience of other groups and enhance the interaction between students. Task driven teaching method emphasizes student-centered, focusing on students in the early stage of analysis, teaching process and teaching evaluation, giving full play to students' learning initiative and enthusiasm. Students in the process of completing the task, master the knowledge and skills of logistics management [21].

## 5.3 Teachers Should Play a Leading Role in the Evaluation of the New Model

Starting from the construction of a new teaching mode, gradually study the current teaching mode, classroom situation and students' learning situation, introduce the task driven hybrid teaching mode based on flipped classroom, and apply it to the university curriculum is an effective means for the Ministry of education to call for the construction of first-class colleges and universities. Before class, the learning resources will be displayed on the Internet to cooperate with the task driven to stimulate students' learning initiative Mobility and the knowledge necessary to complete the task [22]. "Task driven" and "flipped classroom teaching" will become an important method and means to promote students to master knowledge actively in the future.

## 6 Conclusion

Based on the flipped classroom task driven teaching mode, teachers should have a better understanding of the talent training program and syllabus, correctly understand the teaching content, master the practical skills, grasp the students' psychology in time, and guide the classroom atmosphere timely. At the same time, there are still many problems worthy of discussion in the practice of task driven teaching mode in economics and management courses. For example, how to master the teaching time, how to effectively combine with practical training and practical courses, how to realize the effective coverage of tasks on teaching objectives? However, as an effective means to guide students to "learn by doing, do in learning", and implement the quality education policy, The practice of task driven teaching mode based on flipped classroom in the undergraduate course of logistics management is a beneficial exploration.

**Acknowledgment.** This project is supported by Key Projects of Wuchang Shouyi University "Research on the application of "project introduction task driven teaching method" in flipped classroom of marketing" (2020Z08); Key Projects of Education Science Planning in Hubei Province "Research on the management mode and operation mechanism of Teacher Development Center in Applied Undergraduate Colleges" (2019GA064).

## References

1. Xu, Y.: A new education mode of engineering specialty graduate students—co-founding by college and enterprises, uniting of production, learning and researching. Int. J. Educ. Manag. Eng. **2**(12), 590–593 (2012)
2. Zeng, Y., Li, M., Lin, H., et al.: Application of mixed teaching in organic chemistry experiment teaching. Univ. Chem. **3**, 47–52 (2020). (in Chinese)
3. Li, X.: Research on the application of hybrid teaching mode in basic accounting teaching under the background of internet. In: Proceedings of the Annual Meeting of the 2019 Academic Forum on Educational Informatization and Educational Technology Innovation, Chongqing, no. 11, pp. 220–224 (2019)
4. Li, H., Wang, W.: Barrage video: a new orientation of online video interactive learning. Mod. Educ. Technol. **6**, 12–17 (2015)
5. Yu, L.: Research on the "Cooperative Education" model cultivating in higher vocational education. Int. J. Educ. Manag. Eng. **2**(1), 521–524 (2012)
6. Beranuy, M., Carbonell, X., Griffiths, M.D.: A qualitative analysis of online gaming addicts in treatment. Int. J. Health Addict. **4**, 149–161 (2012)
7. Tian, X., Zhang, S., Wu, Y.: Status quo and outlook of the studies of entrepreneurship education in China: statistics and analysis based on papers indexed in CSSCI (2004–2013). Chin. Educ. Soc. **3**, 217–227 (2016). (in Chinese)
8. Zhang, L., Wei, D., Huang, L.: The research on the program of educational technology ability building for vocational education teachers. Int. J. Educ. Manag. Eng. **2**(11), 46–51 (2012)
9. Wei, X.: Discovery and practice of EDA experimental teaching reform. Int. J. Educ. Manag. Eng. **1**(4), 41–45 (2011)
10. Veselinovska, S.S.: The effect of teaching methods on1 cognitive achievementin biology studying. Procedia Soc. Behav. Sci. **12**, 221–227 (2011)

11. Liu, J.: Research on strategies and patterns of research teaching. Int. J. Educ. Manag. Eng. **2**(12), 67–72 (2012)
12. Kocadere, S., Ozgen, D.: Assessment of basic design course in terms of constructivist learning theory. Procedia – Soc. Behav. Sci. **51**, 115–119 (2012). https://doi.org/10.1016/j.sbspro.2012.08.128
13. Puteh, M., Ibrahim, M.H., Adnan, M., Che'Ahmad, C.N., Noh, N.M.: Learning theory. Procedia – Soc. Behav. Sci. (2), 165–175 (2012)
14. Willis, J.: A Framework for Task-Based Learning, pp. 87–98. Addison Wesley Publishing Company (1996)
15. Li, Y.: Research on project driven basic accounting teaching mode. Sci. Technol. Wind (21), 34+40 (2018). (in Chinese)
16. Huang, L.: On the application of "task driven teaching method" in accounting course teaching. Strait Sci. **3**, 86–87 (2013). (in Chinese)
17. Li, W., Zhang, P., Dai, Y.: "Application of task driven" teaching method in "project management" course. China Constr. Educ. **6**, 50–53 (2019). (in Chinese)
18. Zhao, W., Feng, Q., Zhang, Y.: The study of "Internet plus" blended teaching theory based on constructive learning. Heilongjiang Educ. (High. Educ. Res. Eval.) **4**, 24–26 (2019). (in Chinese)
19. Zhao, P., Jin, Y., Jiang, Z.: Influencing factors of students' learning effect quality education and promotion strategy. Qual. Educ. Western China **15**, 208 (2019). (in Chinese)
20. Yang, L.: Research and practice on task-driven teaching based on blended learning. In: International Conference on Economics, Management Engineering and Marketing (EMEM 2017) (2017)
21. Guo, S.: Connotation of task driven teaching method. Audio Vis. Educ. China **7**, 57–59 (2006). (in Chinese)
22. Inan, F.A., Lowther, D.L.: Factors affecting technology integration in K-12 classrooms: a path model. Educ. Tech. Res. Dev. **2**, 157–165 (2010)

# The Operation Mode and Vehicle Scheduling Problem of Urban Co-distribution

Mingwang Dong, Yi Qiu(✉), and Yong Gu

Logistics Management, Wuhan University of Technology, Wuhan 430000, China

**Abstract.** The urban population's living standard have been improved in recent years, the continuous upgrading consumption concepts and consumption methods of those people are followed by the increasing demand for urban distribution. Urban co-distribution has been planned by many local governments into priority development of urban services, which will surely play an increasingly important role in urban distribution services. The article analyzes the development status of urban co-distribution from the perspective of national support development and enterprise operation, integrates two different types of urban co-distribution development methods, propose a platform operation model combining business and public welfare, which is mainly operated by enterprises with the assistance of the government. Based on this background, the multi-vehicle vehicle scheduling problem in the city's peripheral distribution center is studied, and a mathematical model with the minimum total distribution cost as the objective function is constructed. Genetic algorithm is used to solve it in MATLAB. The results of the example can verify the model and algorithm rationality.

**Keywords:** Urban distribution platform · Vehicle scheduling problem · Urban distribution

## 1 Instruction

China is rapidly developing towards a comprehensive urbanization, population structure is changing significantly, up grading the industrial layout has cropped up. Consumer demand also from the direction of change from the original few varieties, high-volume, low frequency to more personalized, precision-oriented, which makes logistics service model of urban distribution activities was shifted to high frequency, low-volume, multi-species [1]. As an effective advanced distribution method, urban joint distribution has the benefits of economies of scale. It can not only meet the needs of urban residents' production activities, but also meet the normal operation of cities and promote the development of emerging industries. In the rapidly changing social environment, the definition of modern logistics services is constantly being expanded and its influence is also increasing [2]. In the urban distribution problem, backward distribution methods and inefficient distribution services cannot meet the increasing demand for logistics distribution [3]. The convenience brought by the city's joint distribution will become more and more obvious. In order to improve operating efficiency and reduce operating costs, building a


Z. Hu et al. (Eds.): ICAILE 2021, LNDECT 82, pp. 219–227, 2021.
https://doi.org/10.1007/978-3-030-80475-6_22

good urban common distribution platform is an essential part of urban distribution. The city's joint distribution has been included in the city's priority development projects by many governments [4].

Therefore, this article takes the city's joint distribution operation mode as the research object, and discusses how to accelerate the implementation of the platform. The discussion proceeded from the following aspects: by collecting academic papers on urban distribution published by core journals at home and abroad, analyzing and summarizing the urban distribution problems and solutions, and discussing the economic and environmental protection of urban joint distribution, the operation mode of urban joint distribution platform, Research on issues such as the optimization of the city's joint delivery vehicle route.

## 2 Contents and Methods

The contents and methods of the research are explained as follows.

### 2.1 The Model of the City's Common Distribution Platform

Aiming at some of the problems exposed in the logistics distribution process of cities that have carried out urban joint distribution projects in China, combined with the current status of urban development, find out the reasons why the development of urban joint distribution cannot be implemented. For small and medium-sized goods companies with relatively insufficient capital, analyze their development models and obstacles. Study whether there is a way to speed up the construction of a city's common distribution platform, how to use the platform to help it match supply and capacity, and build a model city for a city's common distribution platform. Finally, the economic and environmental protection analysis is carried out from the perspectives of society, merchants, and freight companies to analyze and compare the impact of different urban logistics and distribution modes on the urban environment and economy, sum up experience, and achieve green and sustainable development.

### 2.2 Vehicle Route Optimization Plan for Urban Common Distribution Platform

Based on the proposed operation mode of the urban common distribution platform, the distribution scheduling problem is discussed, the multi-model vehicle path optimization problem is described, and the mathematical model based on the minimum distribution cost is established. Based on the real data of the same-city distribution company L, MATLAB is used. Carry out algorithm design and simulation test [5].

## 3 The Model of Urban Co-distribution Platform

### 3.1 Analysis of the Operation Mode of the Existing Urban Co-distribution Platform

As early as 2012, the State Council issued the policy, in 22 cities across the country to carry out urban co-distribution pilot cities. These cities have also summed up different experiences in optimizing the distribution model, which can be summarized in the following 4 ways: Common distribution model in business district, Merchants' joint distribution model in professional market, End resource sharing mode, and, Information platform integration resource model [6]. Moreover, by comparing the volume of road cargo transportation from 2012 to 2019, it can be concluded that this data increase in pilot cities can reach a maximum of 184%. It can be seen that the promotion of policies has provided good development opportunities for urban co-distribution, and the continuous improvement of the mode has also uncovered a large number of potential distribution services (Table 1).

**Table 1.** Comparison of road cargo transportation volume from 2012 to 2019

| Comparison of road cargo traffic (Ten thousand tons) | | | | |
|---|---|---|---|---|
| | 2012 | 2016 | 2019 | Increase in 8 years |
| Tianjin | 27735 | 32000 | 36711 | 132% |
| Hebei | 195530 | 190390 | 226809 | 116% |
| Fujian | 59431 | 85822 | 109200 | 184% |
| Shandong | 296754 | 250000 | 325969 | 110% |
| Henan | 251772 | 183121 | 253457 | 101% |
| Hubei | 97136 | 122000 | 176402 | 184% |
| Sichuan | 158396 | 145830 | 188464 | 119% |
| Shanxi | 104593 | 113283 | 136184 | 130% |

The demand for co-distribution in cities continues to expand. Moreover, the continuous improvement of urban residents' living standards and purchasing power will continue to stimulate the rapid growth of urban commodity consumption. Under this influence, it has provided a good soil for the development of urban distribution services in Xiamen, and has spawned a large number of distribution services. In addition to the logistics between industrial and trade enterprises, the demand for door-to-door distribution is also growing, especially in the central city with outstanding benefits.

## 3.2 Analysis on the Operation Mode of Freight Transport Enterprises

Since the development of intra-city freight companies, many functions have been derived, such as establishing an information matching platform to become a trading hub for both supply and demand; responsible for the monitoring of industry operations and management of service evaluation systems; reducing information asymmetry on both sides of resources, enhancing customer experience, etc., But there are still many problems in the enterprise [7].

1) *It is difficult for the industry to establish "user stickiness".*
   People with frequent freight needs often have their own fleets or fixed cooperative logistics companies. Only people with temporary freight needs will seek help from

the freight platform. Freight is still very different from daily life services such as food delivery and express delivery. Freight has the characteristics of low frequency of demand. Unlike other industries that win on scale, they can occupy a high ground in user habits. The freight industry is unable to continue because it is difficult to build user stickiness. The dilemma of downward penetration is not to be ignored [8].

2) *The size of the city determines the number of users.*
Unlike service platforms such as food delivery and taxi rides, which have potential development scales in cities at all levels, the demand for intra-city freight in first-tier cities and second-tier cities is greater than that in other cities, and it is easier to generate freight service demand. The demand for intra-city freight in tier cities is obviously insufficient. Due to the large population and fast pace of life in big cities, the basic characteristics of the large number of target users in these cities are determined. The characteristics of the fragmentation of the life time of these urban residents are relatively obvious, and they are more proficient in the use of Internet-related platforms. The same-city freight transportation in the form of "Internet+" service is more willing to try and accept.
3) *It is difficult to have both scale and high efficiency.*
Urban joint distribution is an industry that should be operated in a refined manner. It needs to ensure the quality of business through new technologies such as big data, cloud computing, and artificial intelligence [9]. However, in the initial stage, the platform needs to rapidly expand in scale to seize the blank market, and the rapid expansion of scale will inevitably bring about large initial capital investment, and lack of management in the medium term will lead to low operating efficiency of the platform [10].

### 3.3 Chapter Summary

This chapter studies the development model of the city's common delivery platform by comparing and analyzing the development status of various domestic cities and enterprises. First, by comparing the operating modes of 22 pilot cities carrying out urban joint distribution, four platform operating modes were summarized, and their development status and development obstacles were compared and analyzed. Secondly, by studying the mode of intra-city distribution companies, analyze the feasibility of the commercialization of the city's common distribution platform.

## 4 Vehicle Scheduling for Urban Co-distribution Platform

Facing the fast-growing LTL market and fierce market competition, how to reasonably plan the loading plan of delivery vehicles to reduce the empty load rate of vehicles and improve the efficiency of enterprises to reduce the cost of business outlets is a problem for the construction of a common urban delivery platform [11].

### 4.1 Problem Description and Model Establishment

#### 4.1.1 Multi-model Vehicle Path Planning Problem Description

Urban distribution platform within a period of accumulated LTL orders centrally, the purpose is to work reasonable selection of transport vehicles to match orders, the overall path of the vehicle in the case of meet the transport needs of the shortest in order to achieve the minimum distribution costs. Aiming at the problem of how to match the supply and demand relationship of urban freight, this article considers the loading restrictions of different vehicles and the weight of each order, and establishes a 0–1 planning model. The following assumptions are now given [12]:

1) The distribution center is the start and end of the journey.
2) A car serves one route in a cycle, and each customer is visited by one car only once and each customer must be served.
3) In the process of platform operation, the supply of vehicles is slightly larger than the demand.
4) Regardless of vehicle transportation, the weight of each order is less than the load of the transportation vehicle.
5) The customer's order requirements are known and determined.

### 4.1.2 Minimal Model Construction of Total Delivery Cost

It is known that the capacity of the distribution center L is unlimited, and there are a total of k models for service, where $k = \{1, 2, 3, \ldots, m\}$, $k \geqslant 2$. The corresponding load capacity of each vehicle type is $W_k = \{W_1, W_2, ..., W_m\}$, and the starting price and traveling price of the K vehicle type are $S_k = \{S_1, S_2, ..., S_m\}$ and $F_k = \{F_1, F_2, ..., F_m\}$. The order quantity is n, where $n = \{1, 2, 3, ..., n\}$, the weight of each order $Q = \{Q_1, Q_2, ..., Q_n\}$. It is known that $d_{ij}$ is the distance from node (distribution center and customer point) $i$ to node $j$, where $i = 0, 1, ..., n;\ j = 0, 1, ...n$. The service route number corresponding to each vehicle is corresponding to r, $r = \{1, 2, 3..., m\}$, the starting point of route r is $G_r$, the end point is $H_r$, and the material demand on each route is $Q_r$. The model for the minimum distribution cost problem can be established as follows:

$$MIN\ C = \sum_{k=0}^{m}\sum_{i=0}^{n}\sum_{j=0}^{n} x_{ijk} d_{ij} F_k + \sum_{k=0}^{m}\sum_{i=0}^{n}\sum_{j=0}^{n} x_{ijk} S_k$$
$$Q_r \leq W_r, r = 1, 2, ..., m; k = 1, 2, ..., m \tag{1}$$

$$G_r = H_r, r = 1, 2, ..., m \tag{2}$$

$$\sum_{i=0}^{n} x_{ijk} = 1, j = 1, 2, ..., n; k = 1, 2, ..., m \tag{3}$$

$$x_{ijk} \begin{cases} 1, \mathit{Vehicle\ k\ from\ i\ to\ node\ j} \\ 0, \mathit{Others} \end{cases} \tag{4}$$

In the above model, formula (1) means that the vehicle cannot be overloaded, formula (2) means that the starting and ending points of the vehicle are both distribution centers, and formula (3) means that all customers are served, and one customer is only served by one vehicle, (4) is a 0–1 variable.

## 4.2 Algorithm Design

This paper selects genetic algorithm to solve the multi-vehicle optimization problem. The characteristic of genetic algorithm can search for global optimization. It is a kind of random search algorithm. Due to the powerful global search function and adaptability to parallel processing, it is widely used to solve vehicle scheduling problems.

## 4.3 Case Analysis

### 4.3.1 Parameter Design

Based on the background of a logistics company, after conducting a field investigation on the actual operation of the company, it is found that there are two types of vehicles used for freight in the same city. Different types of vehicles have different load capacities and charging standards. See the table for specific vehicle information (Table 2).

**Table 2.** Vehicle information form

| Cars | Deadweight (kg) | Starting price | Excess price |
|---|---|---|---|
| A | 350 | 20 | 5 |
| B | 500 | 40 | 5 |

Through on-site investigations, the company's business conditions on a certain day in April 2020 were obtained. There were 5 Type A vehicles and 2 Type B vehicles in the logistics park that day, and it was necessary to deliver goods to 40 customers in the city.

### 4.3.2 Calculation Example Results

The genetic algorithm parameters are set as follows: population size NP = 100, crossover probability PC = 0. 8, mutation probability PM = 0. 2, and the algorithm termination criterion is set to 1000 iterations. The optimal path results obtained are shown in Table 3.

**Table 3.** Optimal scheduling plan

| Route | Cars | Vehicle path | Weight | TC | Loading rate | Total cost |
|---|---|---|---|---|---|---|
| 1 | A | 0 → 11 → 5 → 24 → 38 → 36 → 19 → 0 | 349. 8 | 187. 132 | 99. 9 | 1289. 772 |
| 2 | B | 0 → 28 → 18 → 16 → 37 → 9 → 2 → 40 → 0 | 496. 1 | 204. 838 | 99. 2 | |
| 3 | A | 0 → 34 → 25 → 29 → 0 | 287. 6 | 181. 683 | 82. 2 | |
| 4 | B | 0 → 13 → 14 → 31 → 32 → 6 → 21 → 26 → 15 → 7 → 33 → 30 → 1 → 3 → 0 | 475. 5 | 353. 880 | 95. 1 | |
| 5 | A | 0 → 23 → 12 → 10 → 17 → 20 → 22 → 27 → 39 → 35 → 0 | 334. 8 | 196. 281 | 95. 7 | |
| 6 | A | 0 → 4 → 8 → 0 | 141. 3 | 65. 957 | 40. 37 | |

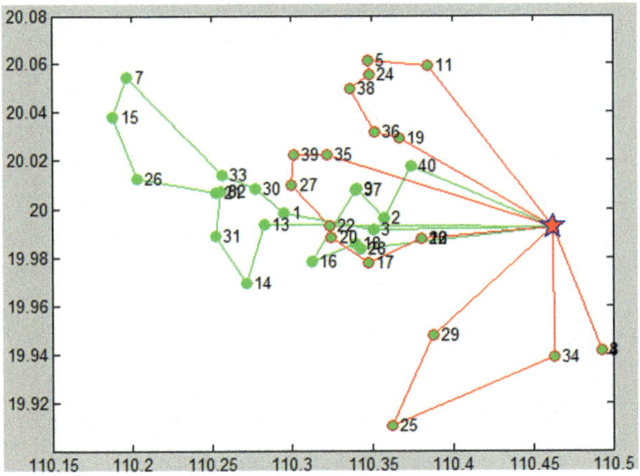

**Fig. 1.** Vehicle dispatch path

The optimal path result is showed as Fig. 1.

Next, the actual operation effect of the algorithm is tested by changing the genetic algebra. The algorithm is operated 10 times under different genetic algebras (Table 4).

**Table 4.** Algorithm effectiveness under different genetic algebras

| Genetic algebra | Max | Min | Average value |
|---|---|---|---|
| 600 | 1476.1112 | 1309.0416 | 1391.42721 |
| 800 | 1460.9066 | 1300.7312 | 1384.47232 |
| 1000 | 1454.6589 | 1289.7715 | 1370.16066 |

It can be seen from the above chart that the optimal individual fitness can always converge quickly and reach stability. The larger the genetic algebra, the more obvious this trend; after determining the population parameters, each randomly generated initial population can finally get the most problematic The optimal solutions are concentrated in a fixed range. This shows the effectiveness of the algorithm (Fig. 2).

### 4.4 Research Results

In this chapter, genetic algorithm is used to study the multi-type vehicle scheduling problem of urban distribution center. Firstly, based on the principle of minimum total distribution cost, a mathematical model of multi-model vehicle scheduling problem is established. The simulation calculation realizes the solution of the multi-model vehicle scheduling problem of 40 customers in a distribution center. The results show that the algorithm and genetic operation strategy designed in this chapter have good adaptability to the solution of the multi-vehicle scheduling problem.

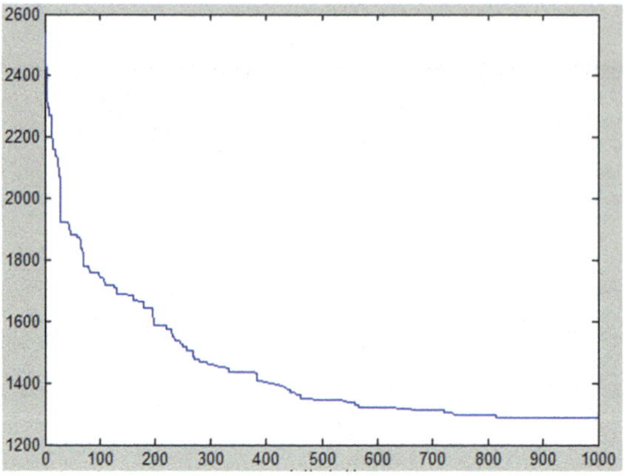


**Fig. 2.** Genetic evolution curve

# 5 Conclusion

The paper got mainly two conclusions.

1) *The model of urban common distribution platform*
This chapter studies the development model of the city's common distribution platform by comparing and analyzing the development status of various domestic cities and enterprises. First, by comparing the operating modes of 22 pilot cities for joint distribution in cities, four platform operating modes were summarized, and their development status and development obstacles were compared and analyzed. Secondly, by studying the modes of two same-city distribution companies with different target users, the feasibility of the commercialization of the city's common distribution platform is analyzed.
2) *Vehicle route optimization plan for urban common distribution platform*
In this paper, genetic algorithm is used to study the multi-type vehicle scheduling problem of urban distribution center. Firstly, based on the principle of minimum total distribution cost, a mathematical model of multi-model vehicle scheduling problem is established. The simulation calculation realizes the solution of the multi-model vehicle scheduling problem of 40 customers in a distribution center. The results show that the algorithm and genetic operation strategy designed in this chapter have good adaptability to the solution of the multi-vehicle scheduling problem.

# References

1. Kambalimath, M.G., Kakkasageri, M.S.: Cost optimization based resource allocation scheme for vehicular cloud networks. Int. J. Comput. Netw. Inf. Secur. **12**(2), 22–31 (2020)
2. Osypchuk, O.: The possibilities of using IT platforms for urban freight distribution in transport SMEs from the West Pomeranian region. Transp. Res. Procedia (39), 381–388 (2019)

3. Rosenzweig, E.: A contingent view of e-collaboration and performance in manufacturing. J. Oper. Manag. **27**(6), 462–478 (2009). https://doi.org/10.1016/j.jom.2009.03.001
4. Bai, S., Wang, Y.: Summary of research on urban joint distribution. Econ. Res. Guide (08), 116–117 (2016). (in Chinese)
5. Fahim, A.: Finding the number of clusters in data and better initial centers for K-means algorithm. Int. J. Intell. Syst. Appl. **12**(6), 1–20 (2020)
6. Deng, W.: Research on building urban distribution platform based on highway port mode. Logist. Eng. Manag. **38**(02), 11–13 (2016). (in Chinese)
7. Jia, H.: Study on the status quo, problems and countermeasures of Xiamen's urban distribution development. China Logist. Purchasing (17), 64–65 (2018). (in Chinese)
8. Wu, Y.: Four major challenges faced by urban common distribution. China Bus. News (A11), 79–81 (2019). (in Chinese)
9. Qayyum, R.: A roadmap towards big data opportunities, emerging issues and hadoop as a solution. Int. J. Educ. Manag. Eng. **10**(4), 8–17 (2020)
10. Zhang, R., Wang, Z.: Bi-level programming model and algorithm for the layout of urban distribution end nodes. J. Tongji Univ. (Nat. Sci. Ed.) (7), 1035–1040 (2012). (in Chinese)
11. Guyon, O., Absi, N., Feillet, D., Garaix, T.: A modeling approach for locating logistics platforms for fast parcels delivery in urban areas. Procedia-Soc. Behav. Sci. (39), 360–368 (2012)
12. Buhrkal, K., Larsen, A., Ropke, S.: The waste collection vehicle routing problem with time windows in a city logistics context. Procedia-Soc. Behav. Sci. (39), 241–254 (2012)

# Interdisciplinary Training Models for Shipping Talents in the Universities

Ying Zhang[1,2](✉), Haojie Zhu[1], Dexin Tao[3], and Zhiping Liu[1,2]

[1] School of Logistics Engineering, Wuhan University of Technology, Wuhan 430063, China
zhangying2003@whut.edu.cn

[2] Engineering Research Center of Port Logistics Technology and Equipment, Ministry of Education, Wuhan 430063, China

[3] Yangtze River Shipping Industry Research Center, Wuhan University of Technology, Wuhan 430063, China

**Abstract.** With the advent of the new shipping era, shipping industrial requires more talents on both quantity and quality. Although shipping training in Wuhan has formed a mature professional education system, its traditional methods can no longer meet the needs of shipping market in this new era, especially the need for high-level shipping talents. As a result, universities must innovate their cultivation models to achieve a comprehensive improvement of the quality and applicability of their shipping talents. Accordingly, this paper explores the interdisciplinary training models for shipping talents in the universities. It gives a literature review on the previous researches on shipping talent cultivation. Then, it analyzes the existing situation and points out the problems in shipping college academic education from the perspective of talents cultivation in universities of Wuhan. Based on that, it proposes two interdisciplinary shipping talents training models and their supporting systems for undergraduates and postgraduates respectively, which provide practical references for the shipping training in Wuhan.

**Keywords:** Shipping talents · Interdisciplinary training · Higher academic education

## 1 Introduction

Driven by modern information technologies, the shipping service industry is developing rapidly, which makes the contradiction between the supply and demand of shipping talents, especially high-level shipping talents, become increasingly prominent as the bottleneck that constrains the development of shipping centers. In Wuhan City, it has outstanding shipping professional education [1]. However, due to the limitation of companies' ability to train talents, the growth of talents in the enterprises is very slow. Besides, the training models are lagging behind the market demand. The shortage of high-level shipping talents and the brain drain are still the difficulties in the construction and development of the shipping center in Wuhan.

Actually, many scholars have done researches on the cultivation of shipping talents. For instance, based on the actual construction of Yangtze River Shipping Center, Zhang


Z. Hu et al. (Eds.): ICAILE 2021, LNDECT 82, pp. 228–237, 2021.
https://doi.org/10.1007/978-3-030-80475-6_23

Yu et al. put forward specific countermeasures for various levels of shipping talents from the training-education system and the talents service system [2]. Li Yuan summarizes the advantages and disadvantages in the training of shipping legal talents in Chinese universities, and puts forward his own suggestions on shipping-related service industry and its talents training [3]. Qu Qunzhen et al. summarize the existing research achievements of talent training models under interdisciplinary cooperation and establish a scientific and effective model for entrepreneurial shipping talents cultivation [4]. Lai Yongsheng analyzes the development status and existing problems of Chinese shipping talents, providing the Objective-System-Institution talent training model and corresponding countermeasures and suggestions [5]. Yu Jie et al. propose to explore methods in talents cultivation, such as the home-abroad joint training model [6]. Tao Dexin et al. from three aspects, including enterprises, universities and industries, propose the idea of strengthening the cultivation of the Yangtze River shipping interdisciplinary talents [7]. Ding Zihua analyzes the cultivation structure for international maritime talents under *One Belt and One Road*, providing ideas on the curriculum construction [8]. Zhang Yuanyuan innovates the professional instruction course and builds the teaching practice platform with school-enterprise mode to cultivate the application-oriented talents [9].

With regard to interdisciplinary education, Beemt et al. summarize the characteristics, instruction practice, and supporting factors of interdisciplinary education from the perspective of curricula and courses [10]. Self et al. point out the barriers in the design of interdisciplinary education based on a series of student interviews [11]. Klaassen makes case studies and illustrates that the interaction between different disciplines and constructive alignment are important to the training model design [12].

Since very few researches focus on the cultivation model for interdisciplinary talents, this paper analyzes the existing situation and the problems of shipping college academic education from the perspective of talents cultivation in universities of Wuhan, and proposes an interdisciplinary shipping talents training model [13].

## 2 Status and Challenges

Wuhan is one of the most concentrated shipping scientific research and education resources in China. The shipping education and training system in Wuhan has formed a complete system of different levels, such as undergraduate education, higher vocational college, secondary vocational school and on-the-job crew training, etc. [14]. Many universities in Wuhan offer shipping-related majors such as shipping, water transport and port trade. They also provide supporting majors, such as finance, law, and trade. Moreover, there are more than a dozen professional shipping universities. Apart from that, there are dozens of universities offering shipping and its related majors for junior college students, undergraduates and postgraduates, which cultivates shipping talents to meet the different needs in shipping industry. As one of the best shipping educators in China, Wuhan trains a large number of talents for the shipping industry every year.

### 2.1 Capabilities of Shipping Education and Training

Currently, there are 20 universities and professional organizations provide shipping education and crew training in Wuhan, including 9 universities and 11 organizations.

### 2.1.1 Universities

There are 9 universities offering shipping-related majors in Wuhan that cultivate a large number of shipping talents every year.

Among them, Wuhan University of Technology and Huazhong University of Science and Technology are national key universities under the Education Ministry of China. The former is supported by Project 211 and Project 985 of National Innovative Platform of Dominant Discipline, and the latter is supported by Project 985. Wuhan Marine College is the only maritime vocational college in Central China that has been ratified by China Maritime Administration. Wuhan Waterway School is the only secondary vocational school in China that trains talents for inland waterway construction.

### 2.1.2 Training Organizations

There are 11 organizations in Wuhan that are mainly engaged in mariner training. They provide 22 training programs including professional training, special training and others. These training are characterized by multiple levels, large scale and high quality.

Shipping colleges in Wuhan cultivate a large number of talents each year, which is in the forefront of China, shown in the Table 1, Table 2 and Table 3. As one city for the largest training scale in the world, Wuhan trains the greatest number of senior, intermediate and general mariner annually. Shipping companies around the world are optimistic about its market. As a result, they not only develop mariner training business with local organizations in Wuhan, but also cooperate with universities and colleges in the mode of order training. For example, it is reported that Maersk trains at least 60 mariners every year in Wuhan Shipping College.

**Table 1.** Data of mariner training in major domestic port cities in 2019

| City | Number of training organizations | Number of training programs | Training person |
|---|---|---|---|
| Wuhan | 11 | 22 | 3518 |
| Dalian | 9 | 26 | 3497 |
| Shanghai | 8 | 25 | 3420 |
| Tianjin | 6 | 14 | 1960 |
| Guangzhou | 7 | 20 | 2270 |

**Table 2.** Student enrollment of academic education in the major of mariner driving vessel in recent three years

| City | 2016 | 2017 | 2018 |
|---|---|---|---|
| Wuhan | 4569 | 3039 | 2704 |
| Chongqing | 609 | 405 | 361 |
| Others | 914 | 608 | 541 |
| Total | 6092 | 4052 | 3606 |

**Table 3.** Student enrollment of training organizations in inland river ship crew education in recent three years

| City | 2016 | 2017 | 2018 |
|---|---|---|---|
| Wuhan | 267 | 275 | 197 |
| Chongqing | 446 | 459 | 329 |
| Others | 179 | 185 | 132 |
| Total | 892 | 919 | 658 |

## 2.2 Challenges

In spite of outstanding shipping education, there are still some problems.

### 2.2.1 Shortage of Senior Talents

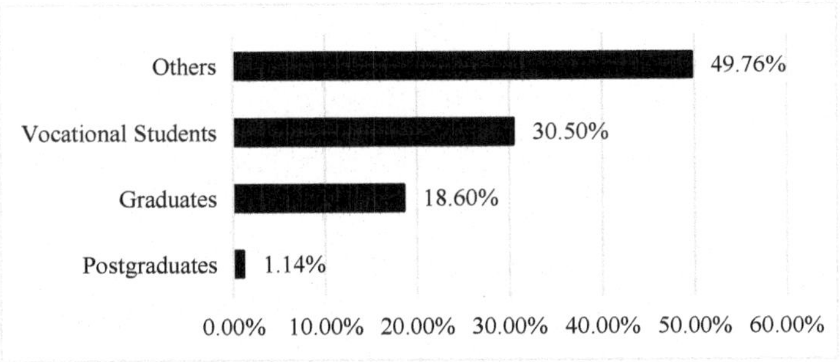


**Fig. 1.** Education background of staff in the shipping industry

The construction of shipping industry requires a large number of complex transportation senior professionals who are familiar with international trade and transportation. At present, most of the universities lack high-level and compound shipping talents training plans or methods. Besides, the knowledge structure they provide the students is unreasonable. They mainly focus on navigation technology knowledge, lacking the instruction of soft knowledge, such as corporate management, logistics management, shipping management, international maritime law, international economic law, international freight, etc. Moreover, they do not hold enough knowledge-expanding activities for students, such as lectures by enterprise experts.

The education background of staff in the shipping industry is showed as Fig. 1:

#### 2.2.2 Lagging Training for Senior Talents

Although Chinese shipping education has developed rapidly, it still cannot meet the needs of economy development. On the one hand, the quality education needs to be further strengthened. In terms of quality, the students still have problems such as inability to bear hardships or hard work, poor technical operation ability in practice, low foreign language proficiency and foreign-related ability, lack of knowledge on ship operation management, and weak safety awareness. On the other hand, our investment in shipping education is still insufficient. In the cultivation process, the cost of simulation facilities and practice training ships is very high. Therefore, the cultivation costs a lot and most shipping colleges are incapable to pay for the equipment, which directly affects the quality of talents training. Moreover, there is a lack of a teacher, and training content is limited.

Briefly, current training of senior shipping talents in China not only requires an increase in quantity, but also needs further improvement in quality to participate in the construction of global shipping industry.

## 3 Interdisciplinary Training Model in Shipping Universities

For shipping universities, it is crucial to improve shipping disciplines, innovate talents training models, and optimize the allocation of shipping education resources. In view of the current shortage of high-level and interdisciplinary shipping talents, it is suggested to establish a diversified shipping education system, explore an interdisciplinary and profession-expanding cultivation model for undergraduates and postgraduates, shown as Fig. 2.

### 3.1 Interdisciplinary Training

It mainly refers to the joint training talents among different disciplines. From the perspective of talents training, talents with comprehensive knowledge, ability and qualities can better adapt to the requirements of the times. In order to cultivate such talents, colleges are expected to become more comprehensive or multi-disciplinary, which is not only conducive to the cultivation of high-quality talents, but also beneficial to the improvement of academic capabilities.

Different from the traditional single-disciplinary professional training model, it refers to interdisciplinary training. It helps to reform education system, implement quality education, improve the quality of talents, realize resource sharing and achieve complementary advantages. Its advantages are mainly manifested in:

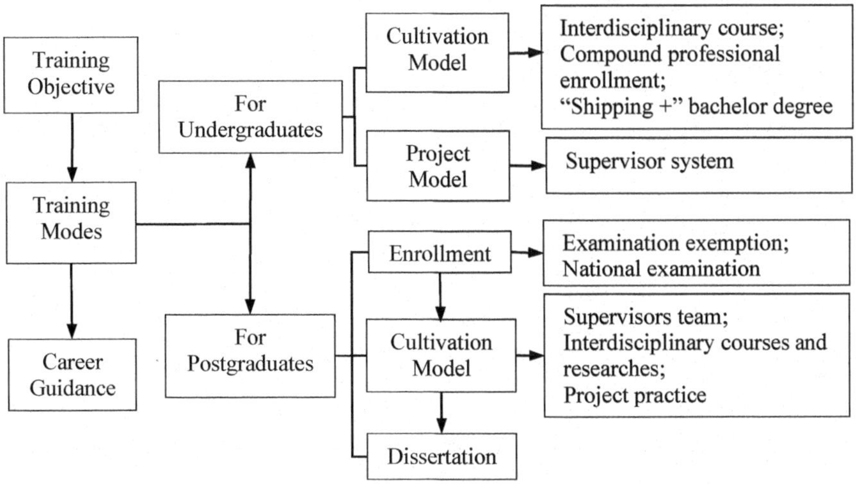


**Fig. 2.** Interdisciplinary training model in universities

1) Conducive to cultivating new talents. It can unify the specialization and integration of professional education by providing students with an opportunity to learn multi-disciplinary professional knowledge, build their professional qualities and skills, which improves their comprehensive adaptability and innovation capabilities.
2) Conducive to discipline reformation and construction. In the reformation of higher education and the construction of disciplines, it is necessary to broaden the scope, enrich the content, and enhance the adaptability of the profession. Fortunately, the interdisciplinary cultivation model is exactly an effective measure for this.

## 3.2 Profession-Expanding Cultivation for Undergraduates

### 3.2.1 Cultivation Models

The model expands the professional scope of the shipping by introducing courses in the supporting fields under the original professional pattern. It can adopt the following four programs:

1) Introduce courses of different disciplines to the compulsory courses and elective courses, which are also instructed by teachers in those fields.
2) Increase compound professional enrollment in shipping trade, shipping finance, shipping law, port and shipping engineering, port and shipping information, etc.
3) Cultivate double-bachelor degree based on "Shipping +". In this way, students are cultivated with an interdisciplinary plan and are guided to cross-training among shipping, traffic management, logistics management, finance, law, information and other disciplines. Afterwards, they are encouraged to apply double Bachelor degree in other colleges or universities.

### 3.2.2 Supervisor System

The supervisor system in the first and second grades for compound undergraduate education helps teachers to take full advantages of their leading role in students training, which enables them to participate more in the guidance of undergraduates and guide their students to study interdisciplinary courses and participate in academic research as soon as possible, thereby improving the quality of students training and the level of students' education. The specific operations are as follows:

1) Establish a group for the undergraduate supervisor system. The group's main work include selecting undergraduate supervisors, arranging students to choose their supervisors, and coordinating the supervisor system work, etc..
2) Build the competent departments for the system, which consists of the academic affairs office, the teaching office and the administration office. It is responsible for the post review and adjustment. Besides, it cooperates with the college to carry out the supervisor system work, report the school's assessment, conduct supervision and inspection, organize and select outstanding undergraduate supervisors and manage coordination.
3) Take full advantages of supervisors. They help students understand the teaching resources in the campus and guide students to make full use of them. Besides, they introduce the teaching content and development frontiers of the subjects and majors to students, so that students can understand their profession and its development direction. Moreover, they provide academic guidance for students in selecting courses, majors, second majors, learning methods, career design, etc.

## 3.3 Interdisciplinary Cultivation Model for Postgraduates

In higher education, especially postgraduate education, interdisciplinary graduates are still scarce. In fact, the interdisciplinary in postgraduate education is vital in cultivating high-level compound talents. Accordingly, it is possible to conduct pilot postgraduate cultivation plans, such as shipping-finance project or shipping-law project for shipping majors. It is expected to integrate the interdisciplinary construction and the supervisor team management, combine the enrollment system and the training model, and create an atmosphere that encourages innovation. Moreover, it is supposed to reform interdisciplinary award standards, innovate administrative management, eliminate the restrictive factors in talents training, and establish a campus-wide interdisciplinary management platform.

### 3.3.1 Enrollment

1) Examination exemption for recommended students. It can attract innovative and interested students to join in interdisciplinary research. Universities can combine this mode with the national summer camp to select suitable undergraduates. Its enrollment plan can be implemented particularly.
2) National postgraduate entrance examination. As a useful supplement to the exemption method, it can maximize the source of enrollment by attracting every student who is interested in interdisciplinary study.

Additionally, the specific enrollment quota of the two enrollment methods depends on the specific majors. For instance, for the bases or platforms that have made good achievements, they can increase the enrollment quota.

### 3.3.2 Cultivation Methods

The methods need to combine both the curriculum mode and the project mode.

1) Selection and formation of the supervisors team. The cultivation of interdisciplinary postgraduates generally requires jointly cultivated by dual supervisors or a supervisors team. Accordingly, the faculty must be selected with strict standards. First of all, an interdisciplinary supervisor must have certain conditions, such as interdisciplinary knowledge background, interdisciplinary research interest, interdisciplinary projects, etc. Secondly, the team should consist of a main supervisor and several assistant supervisors. Thirdly, the faculties must have evaluation criteria for labor division and team work performance, which defines the powers and responsibilities of each member, decides the reward distribution system, and keeps a certain proportion of mobile research personnel.
2) Interdisciplinary courses and researches. Based on students' interests, knowledge background and project needs, the supervisor group make a comprehensive cultivation plan, which includes the credit requirements for compulsory courses, optional courses and others. Besides, teachers can adjust the subject and professional catalog and formulate personalized training programs for students.
3) Project practice. In order to promote the development of interdisciplinary, the universities can establish research groups, research centers or other organizations to encourage teachers and students of different majors to participate. For the research group, it can recruit personnel from different disciplines to jointly tackle key problems through interdisciplinary projects. For the research center, it can be set up independently or jointly with enterprises and scientific research institute.

### 3.3.3 Dissertation System

The dissertations of interdisciplinary graduates must be novel and creative. They are expected to combine multiple fields and disciplines and provide their own insights. Different from traditional dissertations, they are the academic achievements in two or more disciplines. Accordingly, it is necessary to establish a complete evaluation system to ensure the quality of graduate cultivation. For universities, they can establish a committee who make a detailed plan to promote the interdisciplinary integration. Besides, they can provide suggestions for research bases, new majors, degree points, etc. Moreover, it can conduct the construction, inspection and evaluation of interdisciplinary. Apart from that, they can delegate the right of academic judgments to supervisors or the defense committee. At the same time, in order to enhance the quality of talents and drive innovation, they are expected to take students as the center, breaking the existing disciplines and institutional barriers, providing the most institutional support and convenience for talents training, and effectively promoting the development of interdisciplinary.

## 4 Career Guidance

### 4.1 Strengthening the Employment Guidance Role of Supervisors

Supervisors should help students to understand the development trend of the modern shipping industry, build their professional interest, make their career plan, and enhance their consciousness and confidence in interdisciplinary. Due to the differences between undergraduates and graduates, majors and majors, supervisors should employ appropriate cultivation methods for each student. In this process, compound supervisors in campus and off-campus can both be employed. In addition, as students' referrers in internships and working, supervisors should undertake the employment guidance work, strengthening the cooperation research with employers from shipping enterprises and institutions, and learning the requirements and talent needs of enterprises to help students to adapt to the working environment as soon as possible.

### 4.2 Building a Career Guidance Platform

Firstly, it can collect and study employment information in the port and shipping industry and subdivided business areas, providing employment information and guidance for graduates. Secondly, it is supposed to cooperate with enterprises and public institutions, holding employment activities and providing excellent graduates for them. Thirdly, it can establish an information tracking mechanism to collect the feedback from the employers. Based on that, it helps universities to continuously improve and optimize their cultivation model and professional curriculum settings. Finally, for those who have difficulties in employment, it reports the problems to the university and provides useful guidance for them.

### 4.3 Building a School-enterprise Training Platform

Universities can broaden employment channels with their abundant alumni resources. Besides, they can strengthen exchanges and cooperation with port and shipping companies to facilitate graduates to work in mainstream industries. Moreover, they can take the initiative to contact employers to recommend their compound majors.

In the cultivation process, they can improve students' practical ability by combining production, education and research, expanding the cooperation between schools and enterprises, universities and universities, and universities and scientific research units. At the same time, they can implement enterprise-oriented training through school-enterprise research.

## 5 Conclusion

With the advent of the new shipping era, traditional methods can no longer meet the needs of shipping market. As a result, universities must innovate their shipping talent cultivation models, change training concepts, and comprehensively improve the quality and applicability of talents, providing a steady stream of high-level shipping compound talent resources with solid professional foundation, broad professional knowledge, strong practical ability, and high comprehensive quality for the shipping industry.

## References


1. Xu, Y.: A new education mode of engineering specialty graduate students—co-founding by college and enterprises, uniting of production, learning and researching. Int. J. Educ. Manag. Eng. **2**(12), 590–593 (2012)
2. Zhang, Y., Han, X., Zhang, B., et al.: Research on the training and service system of shipping talents in the shipping center construction for the upper reaches of Yangtze River. Maritime Educ. Res. (03), 20–27 (2017). (in Chinese)
3. Li, Y.: The Enlightenment of Singapore shipping law personnel training mode to the construction of Shanghai international shipping center. China Water Transp. **11**(11), 33–35 (2012). (in Chinese)
4. Zhai, Q., Wei, Y.: Study on the training mode of entrepreneurial shipping talents under interdisciplinary cooperation. Logist. Eng. Manag. **12**(12), 179–186 (2017). (in Chinese)
5. Lai, Y.: Implementation guarantee and evaluation of interdisciplinary cooperation in training innovative shipping talents. Market Weekly (10), 136–137 (2017). (in Chinese)
6. Yu, J., Li, J., Dong, H.: Survey on applications of wide area measurement system in power system analysis. World Shipping **11**(233), 58–59 (2014). (in Chinese)
7. Tao, D., Zhou, H., Zhang, Y., et al.: Research on the training mode of shipping compound talents. China Water Transp. (10), 18–20 (2018). (in Chinese)
8. Ding, Z.: A research on the construction of international training course for maritime talents based on the initiative of "One Belt and One Road". J. Shipping Ocean Eng. (08), 69–75 (2018)
9. Zhang, Y.: Research on the teaching mode of innovation and entrepreneurship course for students of maritime vocational colleges from the perspective of applied talents training. Front. Educ. Res. **02**(10), 48–53 (2019)
10. Beemt, A.V.D., Macleod, M., Veen, J.V.D., et al.: Interdisciplinary engineering education: a review of vision, teaching, and support. J. Eng. Educ. (01), 508–555 (2020)
11. Self, J.A., Mark, E., Thomas, J., et al.: Interdisciplinary: challenges and opportunities for design education. Int. J. Technol. Des. Educ. **29**(04), 843–876 (2019)
12. Klaassen, R.G.: Interdisciplinary education: a case study. Eur. J. Eng. Educ. **43**(06), 842–859 (2018)
13. Yu, L.: Research on the "Cooperative Education" model cultivating in higher vocational education. Int. J. Educ. Manag. Eng. **2**(1), 521–524 (2012)
14. Zhang, L., Wei, D., Huang, L.: The research on the program of educational technology ability building for vocational education teachers. Int. J. Educ. Manag. Eng. **2**(11), 46–51 (2012)

# The Application of Data-Driven Technologies to Enhance Supply Chain Resilience in the Context of COVID-19

Zhong Zheng[1,2](✉), Yun Lin[3], Liwei Li[1], Liuqing Lu[1], and Yanfang Pan[1]

[1] Nanning College, Nanning 530200, China
[2] King Mongkut's Institute of Technology Ladkrabang, Bangkok 10520, Thailand
[3] Chongqing University, Chong Qing 400030, China

**Abstract.** The sudden outbreak of COVID-19 has wreaked havoc on global supply chains. Fortunately, digital drive technology can improve supply chain resilience. This paper is based on literature review to research the impact of COVID-19 to supply chain, to research the interplay of supply chain resilience and data-driven technology, focuses on the operation mechanism of data-driven technology (digital twin, blockchain, big data) and integration them together to enhance the resilience of supply chain. This paper try to presents an effective perspective of supply chain resilience by data-driven technology, and it can provide some reference and guidance for improving the resilience of supply chain during and after the COVID-19.

**Keywords:** Supply chain resilience · Data-driven technologies · COVID-19 · Big data · Digital twins · Blockchain

## 1 Introduce

### 1.1 Background

There are now confirmed individuals with COVID-19, which is caused by SARS-CoV-2, in nearly every country. Worldometer website real-time statistics show that as of November 12, 2020, the COVID-19 has caused about 1. 6 million deaths, about 71. 41 million people worldwide infected with the disease. The COVID-19 has not only taken the health and even the life of people but also produce a serious and incalculable impact on productive activities and the economy.

COVID-19 is wreaking havoc on global supply chains, which typically have a low probability of disruption due to an emergency, but once delivered, the impact can be profound [1]. Different from general operational risks, including routine machine failures, inadequate supply or fluctuating demand, the disruption caused by COVID-19 will have a chain reaction, the consequences of which will permeate the whole supply chain, and even impact global business perform and all mankind. [2–8]. Due to lockdown, most of the logistics activities has been suspend. Global supply chain due to node structural shock and the interruption of trans-regional channels form a robust rupture, resulting in


Z. Hu et al. (Eds.): ICAILE 2021, LNDECT 82, pp. 238–253, 2021.
https://doi.org/10.1007/978-3-030-80475-6_24

production stagnation, sales decline, slow logistics, liquidity difficulties and other problems [9] According to an Accenture report (17 March 2020), 94% of the supply chains of Fortune 1000 companies have been impacted by this epidemic, 75% of their businesses have been hit or severely hit, and 55% have or plan to cut their growth forecasts.

Because of these serious consequences caused by COVID-19, it is rather important and urgent to research operation of supply chain in the context of the epidemic. This paper mainly studies the following three questions:

RQ1: To research how the COVID-19 influences on supply chain.
RQ2: To find out how data-driven technology to enhance supply chain resilience.
RQ3: To integrate several data-driven technologies to enhance resilience and reduce the damage caused by the epidemic.

## 1.2 The Impact of COVID-19 on Supply Chain

Uncertain and turbulent market environment bring about issue of supply chain vulnerability, The epidemic has caused serious damage to the global supply chain. COVID-19 has seriously affected the auto industry, tourism, aviation, oil, construction, telecommunications, food and health care [8]. COVID-19 has caused serious impacts on the supply chain in different fields. Many researchers have studied the impact on supply chains in different areas, as shown in Table 1.

**Table 1.** A summary of the impact of COVID-19 on supply chains in different areas [2, 8, 10–14].

| The authors | Areas | Findings |
|---|---|---|
| Christophe Béné (2020); Rizou, M., Galanakis, I. (2020); Hobbs, J. E. (2020) | Food supply chain | Greater than the risk of virus (infection, disease or death) are the consequences of reduced income and impaired purchasing power resulting from blockade controls and business closures, which have put the food security of millions of people at risk. So for food suppliers, more flexibility is needed to ensure safety |
| Abhijit Majumdar (2020), Simone Preuss (2020) | Clothing supply chain | Many of the enterprise in clothing supply chain suffered heavy losses, and most of them must take at least six to nine months to recover. Most of the clothing SC is vulnerability, lack of resilience |

(*continued*)

**Table 1.** (*continued*)

| The authors | Areas | Findings |
|---|---|---|
| Cai, M., & Luo, J. (2020) | Manufa-cturing supply chain | During the outbreak, the supply of spare parts for raw materials is interrupted logistics setbacks gradually intensified, SC is lack of resilience. Market demand is decreasing and the operation cost is increasing, and longtime shutdown will lead SMEs' operation difficulties |
| Gary Gereffi(2020); Whalen, J. (2020); Lapook, J. (2020); Govindan, K. (2020) | Medical supply chain | The serious lack of key drugs and medical equipment, poor supply capacity and long replenishment time cause serious problems to medical treatment. COVID-19 has clearly exposed the deficiencies in current national response policies and global supply chains in dealing with major public health and economic crises caused by pandemics |

It can be concluded from the table that the supply chain of different industries suffers from many problems, among which the outstanding problem is that the supply chain is relatively fragile and lacks resilience [15]. Therefore, it is very necessary and urgent to explore how to increase the resilience of supply chain.

At the same time, some experts use simulation experiments to research the influence of this epidemic on supply chain resilience and get some useful insights [16]. Scholars through experiments show that compared with the pandemic itself to the direct impact of supply chain, the facilities caused by the outbreak of closed and open time affect decision more epidemic on supply chain performance, other important factors affecting the performance of supply chain also includes the mould replacement, speed of the spread of the epidemic and upstream of the supply chain interruption time [17]. And many researchers believe that the epidemic like COVID-19 is a special risk for supply chain, this risk has three distinct features: (I) There is a risk of long-term disruptions of an unpredictable scale, (ii) the risk of disruptions spreading around in supply chain and the spread of epidemics among populations is uncertain and random, and (iii)the risk of interruption occurs simultaneously throughout the process from the supplier to the end user. Unlike other disruptive risks, outbreaks start small, but are large and spread over many geographical areas, and their risks grow amid uncertainty, such as SARS, Ebola, and most recently COVID-19 [18].

But most businesses were obviously not adequately prepared for this outbreak, Carsten Knauer, head of logistics at BME, said: "Only 22 percent of the companies surveyed had considered a pandemic scenario in their risk management activities and only half had an action plan in place." That means 89% of companies were unprepared for the crisis. Pettit et al. consider that supply chain resilience is an important means to deal with disasters, mainly through reducing vulnerability and improving processing capacity to deal with emergencies. [19].

Therefore, during this epidemic and in the post-epidemic era, it is of great necessity and urgency to study how to enhance the supply chain resilience. Due to the continuous and sustainable development of digital technology, especially the popularization of data-driven technology, many methods and means are provided for improving supply chain resilience [3, 9, 11, 20].

## 2 Review of Supply Chain Resilience

### 2.1 Definition of Supply Chain Resilience

Supply chain resilience (SCR) is a relatively new concept. The study of SCR is based on the study of resilience. Taking supply chain as a system or a whole, then to study the resilience of this system. In 2012 the National Academy of Sciences (NAS) gave a comprehensive and systematic definition of resilience: resilience means having the ability to prepare advance, effectively absorb, and quickly recover from adverse events and thus to adapt better. [2, 21]. Since Martin Christopher and Helen Peck published seminal research on SCR, the concept of SCR has attracted more and more attention and research in the early 21st century. [22]. And the definition of SCR has become one of the topic of concern among researchers in the field of supply chain. Typical SCR definitions in literature are summarized as the Table 2 below [22–26].

From these definitions by different authors we can conclude that supply chain resilience need to possess the ability which include: preparation for disruption, absorb disruption, respond to disruption, maintain controlling and regain its original configuration and so on. And all of these abilities are important to handle the supply chain risk like COVID-19.

### 2.2 The Approaches for Supply Chain Resilience

With the continuous advancement of research, the analysis of SCR has changed from traditional qualitative analysis to quantitative analysis, at the same time, many different effective models have been produced for the analysis and evaluation of supply chain resilience. These methods provide a good perspective for the in-depth understanding of SCR. A typical approach is shown in Table 3 below [27–33].

From these researches we can learn lots of approach to performing quantitative analysis for SCR, and the key of driven of SCR is quite clear. And it can be find that although many researches focus on the supply chain resilience, most of current researches focus on static analysis, and focus on the resilience of individual enterprises, rarely combining the across enterprises. At present, the mathematical model on SCR is more

**Table 2.** Typical SCR definitions

| Term Using | Definition | Reference |
|---|---|---|
| Flexibility and agility | SCR means the ability to quickly recover to its initial condition or to new and adaptive state after the supply chain was disturbed, resilience to some extent refers to flexibility and agility | Christopher, M., & Peck, H. (2004) |
| Absorb disruption and recovery | SCR is "the ability of enterprises to absorb interruptions (natural disasters or other emergencies), and the SC network is better restored to its original state, thus having a good impact on enterprise implementing" | Sheffi and Rice (2005) |
| Respond to disruption and recovery | SCR refers to respond quickly to interruptions and operational vulnerabilities and quickly return to equilibrium, thus ensuring excellent performance and high efficiency of the supply chain | Longo and Oren (2008) |
| Preparation, response, and restore | SCR refers to the overall preparation for emergencies, rapid response to interruptions, and through various preparatory programs to maintain operational continuity, through a strong control of the structure and function of the supply chain to restore | Ponomarow and Holcomb(2009) |
| Maintaining control and restore | SCR refers to the ability to resist risk proliferation by maintaining control, and to respond quickly and accurately through timely and efficient prepared measures, so as to overcome risk interference and recover the supply chain to a stable operating condition | Kamalahmadi and Mellat-Parast(2016) |
| Regain its original configuration | SCR refers to the supply chain network has a risk-resistant characteristics, so that it can quickly regain its original configuration and structure after being damaged by various natural disasters and emergencies | Sheffi Y, Rice J B (2003) |

**Table 3.** Example of approaches for SCR

| Author | Modeling | Key drivers of SCR |
|---|---|---|
| Chowdhury and Quaddus (2017) | Dynamic capability theory | Response capacity, Financial solvency, Flexibility |
| Day, J. M. (2014) | Complexity system theory | Network layout, Flexibility, warning |
| Hosseini S, Al Khaled A (2016) | Hybrid ensemble and AHP | Geographic segregation, Raw material replacement, Budget resource, Surplus inventory, Reliability of supplier |
| Li H, Pedrielli G (2017) | The hree-echelon dynamic models | Information sharing, Cooperation relationship |
| Ivanov et al. (2014) | Dynamic control | Additional inventory, Backup Sources, Rerouting |
| Liu X, Zhang K. (2018) | Structural equations modeling | Agility, Capacity, Integration, Technology |
| Ivanov (2018) | Dynamic control | Backup supplier, Surplus inventory, Capacity buffer, Multiple sourcing |

about static analysis, less consideration of the dynamic nature of resilience, and few papers have paid attention to the time constraints and economic costs of resilience recovery. However, in the process of enterprise practice, the limitation of time and budget needs to be considered, but the current model does not focus on it. The optimization model can consider further widening the time window and considering the time period before and after the supply chain interruption. And most of the literatures fails to consider the effect of technology on SCR in current model.

## 2.3 The Countermeasures to Supply Chain Resilience

Researchers have generated some methods to increase supply chain resilience especially in the context of the epidemic. Lohmer, J (2020) establish an event-based supply chain early warning system at the enterprise level to improve the resilience [23]. Haobin Li (2017) researched through a three-echelon dynamic models and find that improving information sharing helps to enhance the resilience of supply chain. [30]. LIU Da-cheng (2020) suggest to apply 5G technology to construct digital -oriented smart logistics supply chain system to respond for the epidemic crisis [34]. Rizou, M. (2020) consider that digital risk management will become increasingly important to provide a transparent overview of the entire supply chain [10]. Many scholars believe that data-driven technology is conducive to supply chain resilience, so as to solve the problems during the COVID-19. Typical technologies to increase the resilience of supply chain include digital twin technology, big data, blockchain, etc. This paper will take these technologies as the research focus to analyze how data-driven technology improves supply chain resilience [6, 20, 35–37].

# 3 Data-Driven Technology for Supply Chain Resilience

During the epidemic period, the overall supply chain monitoring and response plan lacks lower level of transparency, supply chains are fragile and lack of resilience, which is due to the low degree of digitalization of the supply chain. The application of digital technology is not widespread in the whole supply chain, which brings difficulties to timely response and scheduling in emergency logistics. After the epidemic, the Data-driven technology will be required to be used by more and more enterprises, and the whole society will require the digital technology to quickly understand the changes, so as to achieve timely, transparent and controllable logistics status. The application of digital technology to enhance supply chain resilience will be very important and urgent.

## 3.1 Digital Twin for Supply Chain Resilience

The technology of a digital twin has been around since 2002, due to the continuous development of the Internet of Things, digital twin become so essential to business today, which was listed as one of the top ten strategic technology trends of Goldner in 2017. Digital twin technology is considered as a combination of all kinds of advanced digital technology including Cyber Physical System, Big Data, IoT, AI, Wearables, Additive Manufacturing and etc. [38]. Digital twins combine the virtual and physical worlds to create a virtual model of a process, and through data analysis and system monitoring to prevent problems before they occur, prevent accidents, model analysis and develop new opportunities, use simulations to plan for the future, and provide high system adaptability and flexibility [39].

Digital twin technology is very important to increase SCR. At present, the supply chain is faced with more and more complex environment, extensive network, and uncertain events happen from time to time. Data twin technology provides a good technique for supply chain resilience management. Its main function is to provide a visual representation of where your supplies are coming from, where they are going, what the ideal state is, and any obstacles that prevent them from getting there. Digital twin technology can present the network state at any given moment in real time and dynamically, and realize the end-to-end visualization of the whole process, so as to increase the resilience and emergency capability. [40–42]. The digitalization of supply chain by digital twin technology enables the whole supply chain to be controlled and optimized in real time, which will lead the supply chain into a virtuous cycle of continuous improvement and adjustment [19, 43], which can not only optimizes the current supply chain, but also use visual models to plan the future, especially for businesses with global suppliers and complex supply chains, and after being affected by COVID-19, digital twin is helpful to plan these potential vendor switches and understand how they affect your provisioning path and what types of events might affect these new partners. So as to improve the resilience of the supply chain to deal with risks.

The operation mechanism of digital twin technology is developing and maturing constantly, which is more and more widely used in supply chain resilience. Application of digital twin in SCR system model-based decision support can simulate the dynamic scenario of supply chain in case of interruption. In addition, the potential impact on SC performance can be assessed before the interruption occurs, and then the recovery strategy can be optimized. The data analysis is based on the historical interruption risk data and other data in each stage of SC design to build a real interruption scenario (such as sudden COVID-19 outbreaks, earthquakes, wars, etc.). In the response phase, data analysis is used to identify interrupts timely using all kinds of process data, such as from sensors, RFID, ERP, and WMS. The purpose of data analysis is to input real-time risk information into the model to realize recovery and optimization, which is useful to build a data-driven learning system, contributing to create enough interruption scenarios for supply chain resilience which is useful for designing and planning, as well as providing enough countermeasures and feasible solutions [6, 18]. The operation mechanism of digital twin in supply chain is shown in Fig. 1.

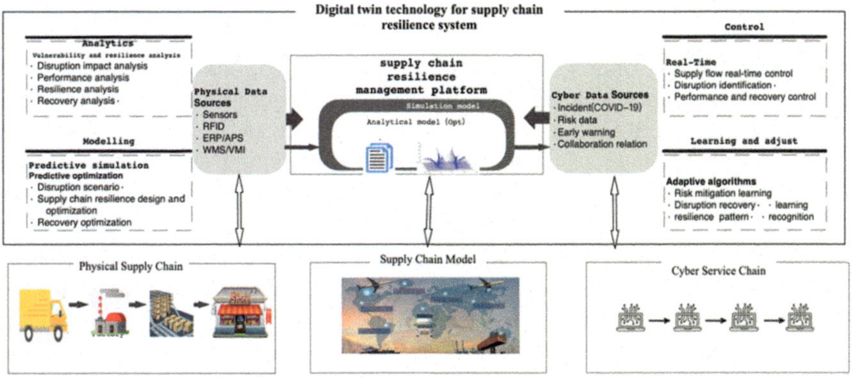


**Fig. 1.** The framework of digital twin technology for supply chain resilience system (adopted with changes from Dmitry Ivanov, 2020)

The core of digital twin technology applied in supply chain is to connect the physical world with the Cyber world. By collecting and modeling data in the real world, including data from RFID and various sensors, ERP /APS, WMS/WMI and other systems, then to perform vulnerability and resilience analysis (disruption impact analysis, performance analytics, resilience analysis, recovery analysis). In this case, the Cyber part can dynamically identify, monitor, and track transportation, inventory, demand, and capacity along the supply chain. On this basis, the simulation model will support the advanced identification and prediction of interruption caused by various risks, and the quantitative prediction of the possible influence. At the same time, this model will offer effective recovery strategy effectiveness tests to adapt plans according to different situations, such as choosing alternative network layout and backup routes of the supply chain during COVID-19. And the system use data twinning technology to construct learning system, by simulating all kinds of interruption scenarios which supply chain may encounter, providing basis for decision-making such as raw material purchasing, logistics activities and supply chain backup, digital twinning system builds perfect framework of elastic evaluation, recovery decision and real-time control development, combines information world with physical world, better combines simulation with model analysis, and more reliable identification of potential interrupts and risks, and improves the resilience of supply chain, thus minimizing losses in dealing with risks and effectively eliminating negative effects. [18].

### 3.2 Big Data for Supply Chain Resilience

With 5 key attributes: "volume", "velocity", "variety", "verification" and "value", big data is well suited for use in complex, large, systematic supply chains and its application effect is highly effective [44]. With accurate predictive insight, big data allows supply chain teams to set up pre-emptive response mechanisms where appropriate, improving supply chain resilience, helping to minimize damage and maintain value. The study of the interaction between big data and supply chain is in its infancy and has attracted the attention of many scholars. Cai, M., & Luo, J. (2020) suggest that enterprises should take full advantage of big data, the successful conduct of business activities depends on accurate decisions, and continuous calibration and optimization of supply chain planning through extensive data mining and analysis is key[5]. Big data technology has been widely used in all aspects of supply chain management, from procurement to end users, as well as traffic monitoring, active safety management and other professional fields [45], Hazen, B. T(2014) focus on digital quality and study how to make good use of big data for better prediction and strategic selection in supply chain [46].

Big data has a strong application scenario and role in improving the elasticity of supply chain. The key to supply chain resilience lies in visibility, so visibility needs to be further enhanced to assist more informed decisions, and big data analysis provides a good help to enhance visibility. Many decision makers have ignored the impact of resilience on supply chain performance and failed to make accurate decisions, resulting in frequent errors and inefficiency [20]. Big data can effectively help mitigate and recover from disasters. It can also help build social and natural capital to improve resilience for the future [37, 46–48]. Rameshwar Dubey (2018) research how to build big data analytics to enhance the resilience, and find that the application of big data can significantly improve the competitive advantage and performance of supply chain [49].

With the continuous maturity and wider application of big data, the operation mechanism of this technology in supply chain resilience is becoming more and more mature. According to the achievements of many scholars, this paper summarizes the technical framework of big data for resilience analysis, as shown in Fig. 2. The main processes of big data operation in supply chain resilience include: connect to all kinds of data sources, ingest the data (including structured and disorganized data), then to interconnect these data together, and perform big data analysis, finally to visualize the data to users [17, 44]. And the data sources generally fall into two broad categories: internal and external, internal data resources mainly including: different kinds of system like: WMS, TMS, CRM, SAP, ERP and web, more specifically, these data come from the company's various sensors like: RFID, GIS, ordering and so on [50]. The external data resources generally include data on economic society and nature, as well as data on various risk events [51]. Through big data analysis, hidden and valuable information

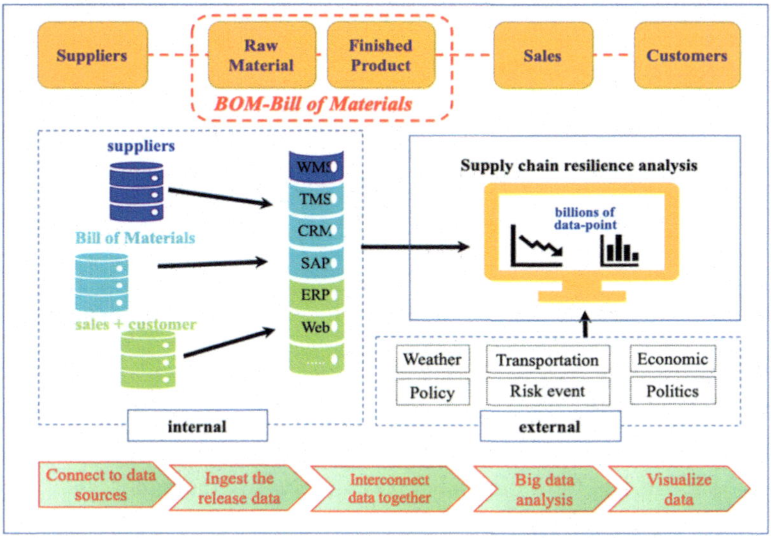


**Fig. 2.** The framework of supply chain analysis with big data

and conclusions can be found, so as to increase the efficiency and operation efficiency of the supply chain, and to enhance the ability and resilience of supply chain to deal with emergencies like COVID-19.

Although the research of big data is very hot, the understanding and application of it in the field of supply chain resilience is not deep enough and systematic [48]. And Chae (2015) noted that this technology has not been thoroughly examined in supply chain [52]. The current research and application are advancing vigorously. With the continuous maturity of big data technology, the application of big data in supply chain resilience will be more and more widespread and important.

### 3.3 Blockchain for Supply Chain Resilience

With the rapid raising value of Bitcoin and increasing attention to cryptocurrency, the blockchain which is revolutionized by Bitcoin has attracted great attention from every area. Blockchain refers a peer-to-peer (P2P) IT network that differs from traditional central databases in that IT USES a distributed ledger to keep records of transactions involving digital assets. It is this dispersion, openness, and cryptography that produce unprecedented benefits in secrecy [21, 36, 53]. Supply chains are usually complex, widely distributed, and with a wide range of global business partners have a multinational multi-regional and cross-industry cooperation. There are too many uncertainties, such as COVID-19 and trade wars, and these interactions add the vulnerability of supply chains and more easily lead to disruptions. To deal with such vulnerabilities, it is necessary to identify potential vulnerabilities and assess risk level, and blockchain provides a good tool [54, 55]. This kind of operation mechanism is very helpful for supply chain. We can see the operation mechanism of blockchain from Fig. 3.

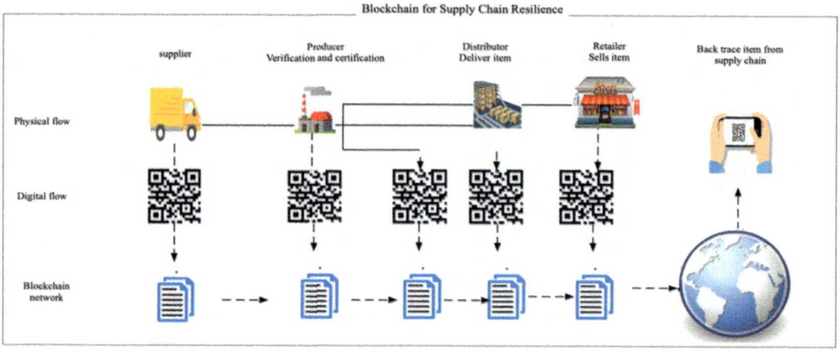


**Fig. 3.** The operation framework of blockchain for supply chain resilience (adopted with changes from dreamztechusa. com)

It can be found that from Fig. 3 that both of them are the same "chain", as a "trust machine" blockchain and supply chain can be said to be a match made in heaven. Take the goods as the starting point, in order to establish the identity information of the goods, constitute a security logistics network covering the whole process, open and transparent, traceable, can make the fake goods have nowhere to hide, the inferior goods can be found. At the same time, a block chain information summary platform is established based on the collected data, and each node on the chain is synchronized in real time, making it possible for large-scale and fast distributed collaboration. In this process, product information, capital flow and amount details will also be clearly recorded on the block chain, which cannot be tampered with, increasing transparency, reducing trust cost, and thus enhancing the resilience of supply chain [56].

Because of the excellent operation mechanism, during the COVID-19, it can solve the problem of security and trust which is very important, such as application block chain to food supply chain, can ensure that every link are health and safety, to avoid more people be contaminated food and infection, once there was a problem, the strong traceability ability of blockchain can quickly find the root cause of the problem, to efficiently solve the crisis, so the chain block application can greatly improve supply chain resilience to deal with the risk event.

Due to the lack of maturity and acceptance of blockchain technology, there are few research achievements on the connection between blockchain and supply chain at present, and even fewer papers on the interaction between blockchain and supply chain elasticity. However, scholars all agree that blockchain plays a great role in supply chain and will be more and more extensive [36]. Because blockchain can solve an important problem in supply chain—mutual trust [57]. Supply chain is very focused on cooperation and trust, and blockchain technology can promote faster, auditable interaction and immutable data exchange between supply chain partners, make the cooperation more transparent, safe, reliable, and can be further establish a stable supply chain upstream and downstream cooperation mechanism, build a more stable and resilient supply chain.

### 3.4 A Mix of Data-Driven Technologies for Supply Chain Resilience

Both of digital twin, big data and blockchain are effective in enhancing supply chain resilience, this paper intends to further study how these three technologies work together in order to play a better role in enhancing the supply chain resilience.

We have designed a system framework scheme to improve the resilience with a mix data-driven technology, as shown in Fig. 4.

The combination of these three technologies is expected to improve supply chain resilience more comprehensively and effectively, as shown in Fig. 5. The integration of the three technologies may play a better overall role and be more effective and secure than a single technology.

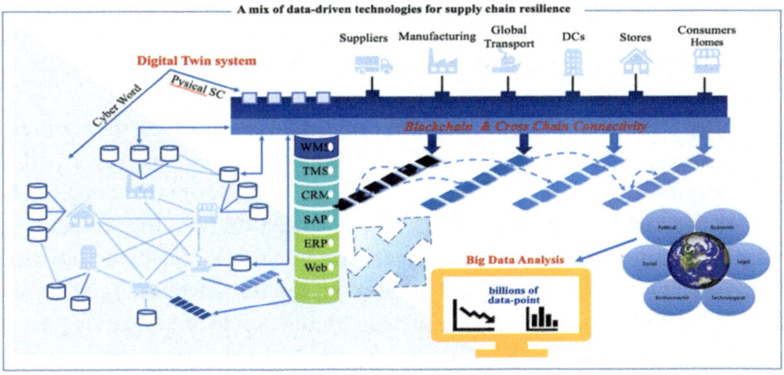


**Fig. 4.** A framework of mixed data-driven technologies for supply chain resilience

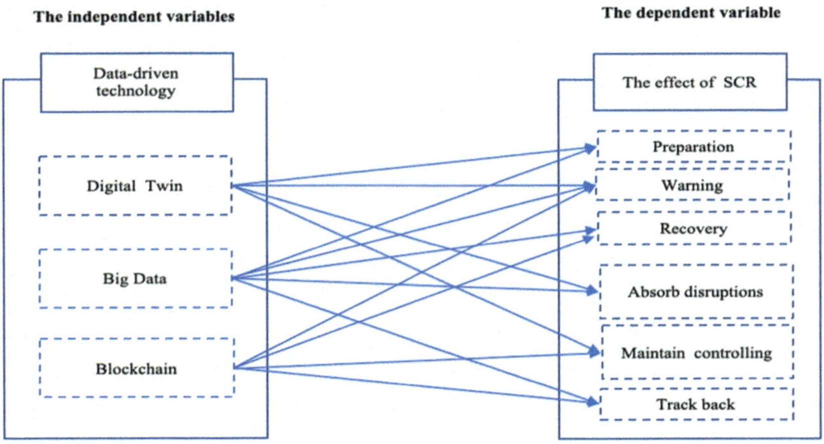


**Fig. 5.** The effect of data-driven technology on supply chain resilience

## 4 Conclusion

Through the analysis of the influence of COVID-19 of the supply chain, this paper draws forth the necessity and importance of data-driven technique used in supply chain resilience. The papers review the definition of supply chain flexibility, method, driving factors, further study of supply chain resilience and find that less attention has paid on the related between technical operation mechanism and supply chain research. The paper then studies the interaction of digital twin, big data and block chain on the resilience of supply chain, research how their operation mechanism improves the supply chain resilience, and finally integrates the three technologies to enhance the resilience of supply chain. This paper provides a certain innovative perspective for research and application, and provides certain reference and significance for increasing supply chain resilience.

## 5 Limitation and Future Research

This paper only studies the effects of three important data-driven technologies on the resilience of supply chain. In fact, many other data-driven technologies will affect the supply chain resilience to different degrees. Therefore, the research scope can be further expanded. At the same time, the paper is mainly based on literature review, qualitative analysis as the main method, lack of quantitative analysis and data statistics, can be further improved and in-depth research. Although the three technologies are integrated in this paper, the fusion and influence mechanism of each technology still needs further study.

**Acknowledgment.** This project is supported by Guangxi philosophy and social science planning research project(17FJY014) and Guangxi young and middle-aged teacher's basic ability enhancement project (2018KY0744).

## References

1. Qin, C., Shuyi, T.: Manufacturer's pricing strategy for supply chain with service level-dependent demand. Int. J. Eng. Manuf. **4**(4), 1–13 (2014)
2. Majumdar, A., Shaw, M., Sinha, S.K.: COVID-19 debunks the myth of socially sustainable supply chain: a case of the clothing industry in South Asian countries. Sustain. Prod. Consum. **24**, 150–155 (2020)
3. Anner, M.: Abandoned? The impact of covid-19 on workers and businesses at the bottom of global garment supply chains[EB/OL]. Center for Global Workers' Rights Research Report, 27 March 2020. https://www.workersrights.org/wp-content/uploads/2020/03/Abandoned-Penn-State-WRC-Report-March-27-2020.pdf. Accessed 10 May 2020
4. Béné, C.: Resilience of local food systems and links to food security–A review of some important concepts in the context of COVID-19 and other shocks. Food Secur. **12**, 1–18 (2020)
5. Cai, M., Luo, J.: Influence of COVID-19 on manufacturing industry and corresponding countermeasures from supply chain perspective. J. Shanghai Jiaotong Univ. (Sci.) **25**(4), 409–416 (2020)

6. Ivanov, D.: Predicting the impacts of epidemic outbreaks on global supply chains: a simulation-based analysis on the coronavirus outbreak (COVID-19/SARS-CoV-2) case. Transp. Res. Part E: Logist. Transp. Rev. **136**, 101922 (2020)
7. Linton, T., Vakil, B.: Coronavirus is proving that we need more resilient supply chains[EB/OL]. Harvard Business Review, March 2020. https://hbr.org/2020/03/coronavirus-is-proving-that-we-need-more-resilient-supply-chains
8. Gereffi, G.: What does the COVID-19 pandemic teach us about global value chains? The case of medical supplies. J. Int. Bus. Pol. **3**(3), 287–301 (2020). https://doi.org/10.1057/s42214-020-00062-w
9. Golan, M., Jernegan, L., Linkov, I.: Trends and applications of resilience analytics in supply chain modeling: systematic literature review in the context of the COVID-19 pandemic. Environ. Syst. Decis. **40**(2), 222–243 (2020). https://doi.org/10.1007/s10669-020-09777-w
10. Béné, C.: Resilience of local food systems and links to food security – A review of some important concepts in the context of COVID-19 and other shocks. Food Secur. **12**(4), 805–822 (2020). https://doi.org/10.1007/s12571-020-01076-1
11. Rizou, M., Galanakis, I.M., Aldawoud, T.M.S., et al.: Safety of foods, food supply chain and environment within the COVID-19 pandemic. Trends Food Sci. Technol. **102**, 293–299 (2020)
12. Hobbs, J.E.: Food supply chains during the COVID-19 pandemic. Canadian J. Agric. Econ./Revue canadienne d'agroeconomie **68**(2), 171–176 (2020)
13. Govindan, K., Mina, H., Alavi, B.: A decision support system for demand management in healthcare supply chains considering the epidemic outbreaks: a case study of coronavirus disease 2019 (COVID-19). Transp. Res. Part E: Logist. Transp. Rev. **138**, 1–13 (2020)
14. Lapook J.: Surgical gown recall leaves medical centers scrambling. CBS News, 31 January. https://www.cbsnews.com/news/surgical-gown-recall-leaves-medical-centers-scrambling-2020-01-31/
15. Li Feng, T., Yijian, G.X.: Who is the free rider in the drop-shipping supply chain? Int. J. Inf. Eng. Electron. Bus. **6**(17), 44–51 (2011)
16. Cheowsuwan, T., Arthan, S., Tongphet, S.: System design of supply chain management and Thai food export to global market via electronic marketing. Int. J. Math. Sci. Comput. **9**(8), 1–8 (2017)
17. Ivanov, D.: Simulation-based single vs. dual sourcing analysis in the supply chain with consideration of capacity disruptions, big data and demand patterns. Int. J. Integr. Supply Manag. **11**(1), 24 (2017)
18. Ivanov, D., Dolgui, A.: A digital supply chain twin for managing the disruption risks and resilience in the era of Industry 4.0. Prod. Plan. Control **32**, 775–788 (2020)
19. Pettit, T.J., Croxton, K.L., Fiksel, J.: The evolution of resilience in supply chain management: a retrospective on ensuring supply chain resilience. J. Bus. Logist. **40**(1), 56–65 (2019)
20. Jessop, K.: Big data analytics to improve supply chain resilience (2020). https://cerasis.com/supply-chain-resilience/
21. Fiksel, J.: The new resilience paradigm-essential strategies for a changing risk landscape i. IRGC. Resour. Guide Resil. **29**(07), 1–5 (2016)
22. Christopher, M., Peck, H.: Building the resilient supply chain. Int. J. Logist. Manag. **15**(2), 1–13 (2004)
23. Longo, F., Oren, T.: Supply chain vulnerability and resilience: a case study of the art overview. In: Giovani Campora, S. (ed.) Proceedings of European Modeling & Simulation Symposium, 17–19 September 2008
24. Kamalahmadi, M., Parast, M.M.: A review of the literature on the principles of enterprise and supply chain resilience: major findings and directions for future research. Int. J. Prod. Econ. **171**, 116–133 (2016)

25. Sheffi, Y., Rice, J.B., Fleck, J.M., et al.: Supply chain response to global terrorism: a situation scan. Center for Transportation and Logistics, pp. 1–6 (2003)
26. Kamalahmadi, M., Mellat-Parast, M.: Developing a resilient supply chain through supplier flexibility and reliability assessment. Int. J. Prod. Res. **54**(1), 302–321 (2016)
27. Chowdhury, M.M.H., Quaddus, M.: Supply chain resilience: conceptualization and scale development using dynamic capability theory. Int. J. Prod. Econ. **188**, 185–204 (2017)
28. Day, J.M.: Fostering emergent resilience: the complex adaptive supply network of disaster relief. Int. J. Prod. Res. **52**(7), 1970–1988 (2014)
29. Hosseini, S., Al Khaled, A., Sarder, M.D.: A general framework for assessing system resilience using Bayesian networks: a case study of sulfuric acid manufacturer. J. Manuf. Syst. **41**, 211–227 (2016)
30. Li, H., Pedrielli, G., Lee, L., Chew, E.: Enhancement of supply chain resilience through inter-echelon information sharing. Flex. Serv. Manuf. J. **29**(2), 260–285 (2017). https://doi.org/10.1007/s10696-016-9249-3
31. Ivanov, D., Sokolov, B., Dolgui, A.: The ripple effect in supply chains: trade-off "efficiency-flexibility-resilience" in disruption management. Int. J. Prod. Res. **52**(7), 2154–2172 (2014)
32. Xiaotong, L., Kai, Z., Bokui, C., et al.: Analysis of logistics service supply chain for the one belt and one road initiative of China. Transp. Res. Part E: Logist. Transp. Rev. **117**, 23–39 (2018)
33. Ivanov, D.: Revealing interfaces of supply chain resilience and sustainability: a simulation study. Int. J. Prod. Res. **56**(10), 3507–3523 (2018)
34. Shohin, A., Xun, X., Zhong, R.Y., et al.: Digital twin as a service (DTaaS) in Industry 4.0: an architecture reference model. Adv. Eng. Inform. (47), 1–15
35. Liwei, C., Hongyan, D., Chi, Z.: A resilience measure for supply chain systems considering the interruption with the cyber-physical systems. Reliab. Eng. Syst. Saf. **199**, 1–16 (2020)
36. Min, H.: Blockchain technology for enhancing supply chain resilience. Bus. Horiz. **62**(1), 35–45 (2019)
37. Papadopoulos, T., Gunasekaran, A., Dubey, R., et al.: The role of big data in explaining disaster resilience in supply chains for sustainability. J. Clean. Prod. **142**, 1108–1118 (2017)
38. Kong, X.T.R., Luo, H., Huang, G.Q., et al.: Industrial wearable system: the human-centric empowering technology in Industry 4.0. J. Intell. Manuf. **30**(8), 2853–2869 (2019)
39. Marr, B.: What Is digital twin technology-and why is it so important? (2020). https://www.forbes.com/sites/bernardmarr/2017/03/06/what-is-digital-twin-technology-and-why-is-it-so-important/#39e2b9f72e2a
40. Hosseini, S., Ivanov, D., Dolgui, A.: Review of quantitative methods for supply chain resilience analysis. Transp. Res. Part E: Logist. Transp. Rev. **125**, 285–307 (2019)
41. Marmolejo-Saucedo, J., Hurtado-Hernandez, M., Suarez-Valdes, R.: Digital twins in supply chain management: a brief literature review. In: Vasant, P., Zelinka, I., Weber, G.-W. (eds.) Intelligent Computing and Optimization: Proceedings of the 2nd International Conference on Intelligent Computing and Optimization 2019 (ICO 2019), pp. 653–661. Springer, Cham (2020). https://doi.org/10.1007/978-3-030-33585-4_63
42. Park, K.T., Son, Y.H., Noh, S.D.: The architectural framework of a cyber physical logistics system for digital-twin-based supply chain control. Int. J. Prod. Res. **25**, 1–22 (2020)
43. Scholten, K., Stevenson, M., Van Donk, D.P.: Dealing with the unpredictable: supply chain resilience. Int. J. Oper. Prod. Manag. **40**(1), 1–10 (2020)
44. Addo-Tenkorang, R., Helo, P.T.: Big data applications in operations/supply-chain management: a literature review. Comput. Ind. Eng. **101**, 528–543 (2016)
45. Schoenherr, T., Speier-Pero, C.: Data science, predictive analytics, and big data in supply chain management: current state and future potential. J. Bus. Logist. **36**(1), 120–132 (2015)

46. Hazen, B.T., Boone, C.A., Ezell, J.D., et al.: Data quality for data science, predictive analytics, and big data in supply chain management: an introduction to the problem and suggestions for research and applications. Int. J. Prod. Econ. **154**, 72–80 (2014)
47. Folke, C., Carpenter, S.R., Walker, B., et al.: Resilience thinking: integrating resilience, adaptability and transformability. Ecol. Soc. **15**(4), 1–7 (2010)
48. Mishra, D., Gunasekaran, A., Papadopoulos, T., et al.: Big data and supply chain management: a review and bibliometric analysis. Ann. Oper. Res. **270**(1–2), 313–336 (2018)
49. Dubey, R., Gunasekaran, A., Childe, S.J.: Big data analytics capability in supply chain agility. Manag. Decis. **57**(8), 2092–2112 (2019)
50. Ivanov, D., Dolgui, A., Das, A., Sokolov, B.: Digital supply chain twins: Managing the ripple effect, resilience, and disruption risks by data-driven optimization, simulation, and visibility. In: Ivanov, D., Dolgui, A., Sokolov, B. (eds.) Handbook of ripple effects in the supply chain. ISORMS, vol. 276, pp. 309–332. Springer, Cham (2019). https://doi.org/10.1007/978-3-030-14302-2_15
51. Shan, L., Yubin, C.: Review of researches on analysis and application of big data in supply chain. J. Bus. Econ. **9**, 1–8 (2018)
52. Chae, B.K.: Insights from hashtag# supplychain and twitter analytics: considering twitter and twitter data for supply chain practice and research. Int. J. Prod. Econ. **165**, 247–259 (2015)
53. Brandon-Jones, E., Squire, B., Autry, C.W., et al.: A contingent resource-based perspective of supply chain resilience and robustness. J. Supply Chain Manag. **50**(3), 55–73 (2014)
54. Shengqiao, G., Xinliang, L., Yanping, G.: Optimized model of dual-chain storage of food supply chain data based on blockchain. Food Mach. **3**, 1–7 (2020)
55. IBM: The benefits of blockchain to supply chain networks[EB/OL]. IBM Corporation, Somers (2017). https://www.ibm.com/blockchain/industries/supply-chain
56. Li, Y., Zobel, C.: Exploring supply chain network resilience in the presence of the ripple effect. Int. J. Prod. Econ. **228**, 107693 (2020)
57. Lohmer, J., Bugert, N., Lasch, R.: Analysis of resilience strategies and ripple effect in blockchain-coordinated supply chains: an agent-based simulation study. Int. J. Prod. Econ. **228**, 51–62 (2020)

# Design and Application of Tag Identification APP for Warehouse Sorting System

Hao Li(✉), Yanjia Li, Zanxing Yang, Haiyuan Wang, and Gang'ao Chen

School of Logistics Engineering, Wuhan University of Technology, Wuhan 430070, China

**Abstract.** In the development process of automated warehouse, the tags in the sorting system are miscellaneous and disorderly, resulting in the labor and time consumption of sorting operations. Therefore, this paper designs a label recognition APP for warehouse sorting system. Based on Zxing barcode recognition technology and ultra-lightweight OCR technology, this paper fully taps the ability of the latest text detection algorithm DB and the classical text recognition algorithm CRNN, and designs an Android APP that can identify various labels. Finally, select a 13. 56 MHz high-frequency RFID tag, and read the information by using the NFC technology on Android phones. With the help of the automatic identification and extraction technology and the frequency identification reading and writing technology, the label identification of the automatic sorting system is completed.

**Keywords:** Tag identification APP · OCR technology · RFID tags · NFC technology

## 1 Introduction

With the improvement of China's economic level in recent years, the logistics industry is developing rapidly. At the same time, commodities tend to be "short, small and light," and circulation tends to be small batch, multi-species, and punctuality. An electronic manufacturing enterprise has at least 20–30 or more suppliers. Each supplier has different material label rules, which brings problems such as waste of human resources, increase of management cost, high error rate of material label and low efficiency to warehouse management. The efficiency of warehouse sorting is always difficult to improve, which has become an important problem that restricts the rapid development of logistics industry.

In the investigation and analysis of several logistics enterprises such as China's foreign transportation, it is found that there are three main problems in the current warehouse sorting system: 1) the label is not uniform; 2) Low efficiency; 3) Need manual review. The core problem in these three problems, is the first one, the label is not uniform, but also the focus of this paper.

Therefore, this paper will design a label recognition APP to solve the problem of label recognition and information extraction in warehouse sorting system [1, 2]. The tag recognition APP introduced in this paper uses ultra-lightweight OCR technology


Z. Hu et al. (Eds.): ICAILE 2021, LNDECT 82, pp. 254–262, 2021.
https://doi.org/10.1007/978-3-030-80475-6_25

Optical Character Recognition) and bar code recognition technology to achieve high-precision information extraction of bar code, two-dimensional code, electronic tag, and handwritten tag. Then the information obtained from the different rules of the material label text is written into the RFID (Radio Frequency Identification) tag with the NFC (Near-field communication) of the mobile phone to complete the tag identification of the automatic sorting system. This APP can reduce the workload of warehouse personnel, reduce errors, improve efficiency, and reduce the loss of goods and information in the warehouse (Fig. 1).

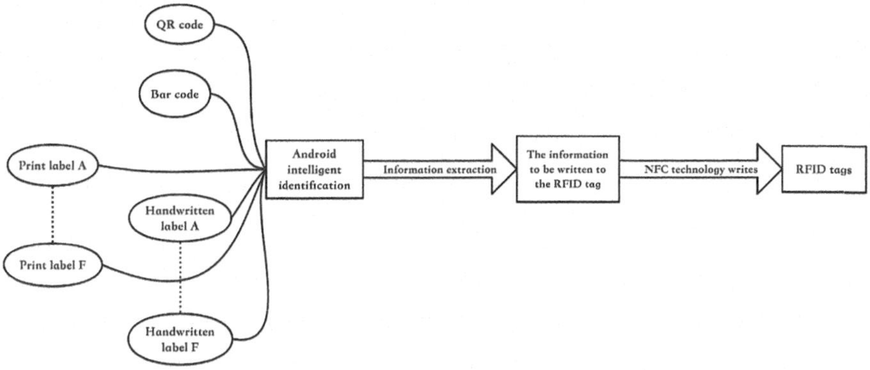


**Fig. 1.** Tag identification process analysis diagram

## 2 Multi-label Information Extraction Technology

The information extraction of one-dimensional barcode, two-dimensional code, electronic tags, handwritten tags, and other tags by tag recognition APP depends on Zxing image processing library and ultra-lightweight OCR technology [3]. The standardized record of extracted information depends on NFC technology and RFID tags.

### 2.1 Zxing

Zxing is a cross-platform Java-based image processing library for processing one-dimensional or two-dimensional barcodes [4], including ports linked to other languages. Zxing supports multiple formats, one-dimensional barcode supports UPC-A, UPC-E, EAN-8, Code 39, Code 93 and other formats, and two-dimensional barcode supports QR Code, Data Matrix, PDF 417, Maxi Code, and other formats. It can realize the scanning and decoding of barcode by the built-in camera of mobile phone [5].

### 2.2 Ultra-lightweight Text Recognition Algorithm PP-OCR

PP-OCR is a super lightweight OCR system, which is mainly composed of DB text detection, detection box correction and CRNN text recognition. Aiming at the OCR problem [6], this paper fully excavates the ability of the latest text detection algorithm Differentiability Binarization (DB) and the classical text recognition algorithm CRNN, and uses 19 effective strategies to optimize and slim the model of each module from eight aspects of backbone network selection and adjustment, prediction head design, data enhancement, learning rate transformation strategy, regularization parameter selection, pre-training model use and model automatic clipping and quantification.

1) *Text detection*
Differentiable Binarization is used in the text detection section of PP-OCR model, which is different from the previous segmentation-based text detection network. The final binary map is no longer obtained by a fixed threshold, but the adaptive binarization of each pixel. The binarization threshold is obtained by network learning, and the binarization step is added to the network to train together, which avoids the great influence on the performance caused by different thresholds, and makes the final output map very robust to the threshold.
At the same time, in the training process, the cosine learning rate is used to replace the fixed learning rate in the Adam optimizer, so that the effect is slightly improved, the comparison results are shown in Fig. 2.

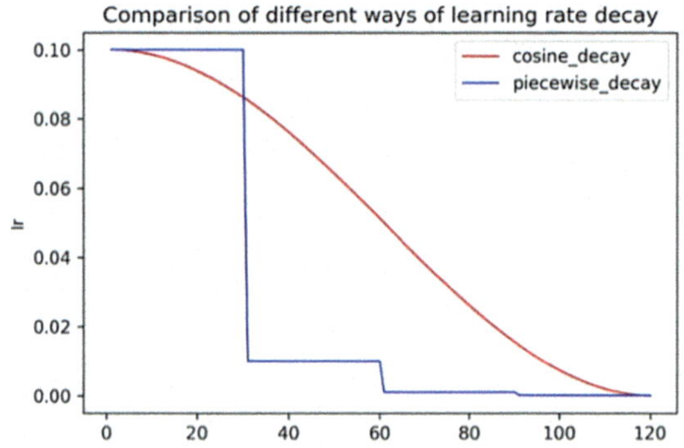


**Fig. 2.** Cosine learning rate and fixed learning rate comparison chart

2) *Text recognition*
In the text recognition module, a full connection layer is used to encode the sequence features, so that the dimension of the sequence features has a great influence on the size of the model. Therefore, the speed benefit is obvious when the dimension of the sequence features of the head is reduced. Similar to the text detection module, the cosine learning rate and the preheating learning rate also effectively improve the model effect. Similar to the text direction classifier module, PP-OCR also uses PACT quantization to reduce the size of the model and reduce the redundancy in the model, so as to improve the model effect.
3) *Detection box correction*
In the detection box correction module, namely text recognition module, it is available to use BDA + Ran Augment data enhancement method to enhance the data [7].

At the same time, the improved PACT quantization is used to improve the prediction efficiency. The original PACT activation function formula is as follows:

$$y = PACT(x) = 0.5(|x| - |x - a| + a) = \begin{cases} 0 \; x \in (-\infty, 0) \\ x \; x \in [0, a) \\ a \; x \in [a, +\infty) \end{cases} \tag{1}$$

The original PACT mainly considers the ReLU activation function, and all inputs are truncated when they are greater than a certain threshold. Therefore, the PACT quantization method used in PP-OCR is improved as follows to cut off in two directions so as to reduce the quantization error.

$$y = PACT(x) = \begin{cases} -a \; x \in (-\infty, -a) \\ x \quad x \in [-a, a) \\ a \quad x \in [a, +\infty) \end{cases} \tag{2}$$

Zxing image processing library and advanced ultra-lightweight OCR technology is introduced above. Based on the mature application of Zxing and the improvement and optimization of ultra-lightweight OCR technology, tag identification APP has been able to accurately identify and extract text information from tags of various styles and standards.

## 3 RFID Tag Information Entry Technology

Collect the above technology, design an Android APP that can identify various warehouse labels; The information extracted from APP identification is written into the RFID tags that need to record the information of items in the warehouse through NFC technology, that realize the tracking and monitoring of goods in warehouse [8].

### 3.1 High-Frequency RFID Tags

By discussion comprehensively, 13. 56 MHz high-frequency RFID tags are selected, which can penetrate most materials except metals, and the data storage is fast. At the same time, this frequency band has global anti-collision algorithms and communication protocols [9]. More importantly, 13.56 MHz high-frequency RFID tags are used to support data writing, which is the function required by this scheme.

The digital control unit for RFID tags is shown in Fig. 3. The functions of electronic tags must satisfy the following points:

1) Storage of data;
2) Encoding and decoding data;
3) Communication with reader by wireless frequency;
4) Conversion of RF signal to energy Data writing:

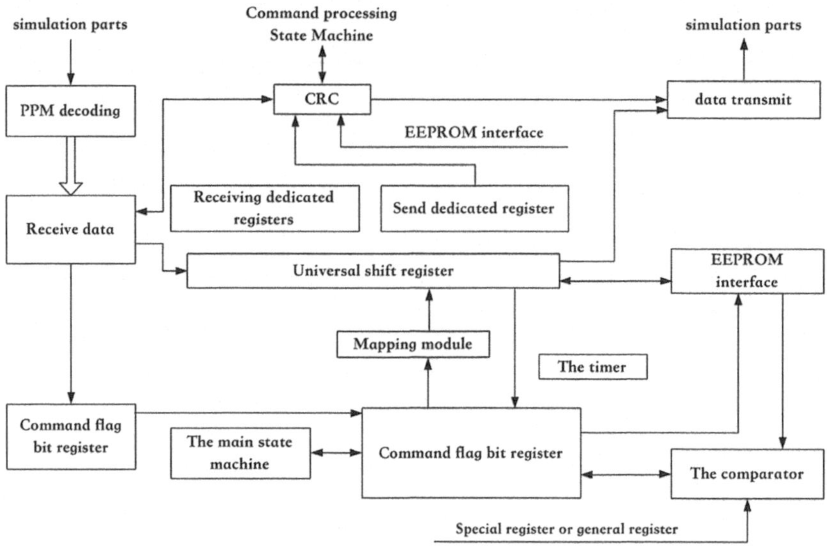


**Fig. 3.** Digital control unit for RFID tags

Data insertion: The electronic tag processes the received signal through the analog front-end, and then decodes the received information through PPM, and records the data to the register. Data reading: The data in the internal register of the electronic tag is encoded, transmitted out in the form of radio wave through the RF front end, and finally read by the reader or NFC mobile phone.

RFID tag complete system has four parts, including reader, electronic tag, antenna, and PC. The reader is the most critical part of data exchange with RFID [10–13], and it is also used by staff.

### 3.2 RFID Tag Read and Write Based on NFC

NFC, also known as short-range wireless communication, is a short-range high-frequency wireless communication technology, allowing non-contact point-to-point data transmission within 10 cm [14]. NFC reader mode can be used as a non-contact reader to read or write relevant information from the electronic tag.

Smart phones with NFC are equivalent to the combination of host computer and reader, which can realize the function of information processing and information exchange, as shown in Fig. 4.

After reading the tag information through the mobile phone APP, and then using the NFC function to write into the RFID tag in the warehouse, the standardization of storage and sorting information in the automated warehouse can be realized [15–17].

The research proposes the NFC function of high-frequency RFID tags and mobile phones, and puts forward the application idea of combining the two. The combination of RFID tag and NFC function can realize the writing and modification of information, and provide an effective solution for the standardization of goods information in automatic sorting warehouse.

**Fig. 4.** Mobile NFC function diagram

# 4 Application Testing

## 4.1 Label Identification

The text extraction effect of label recognition APP on material labels is shown in Fig. 5. The left figure is the extraction of printed material labels, and the right figure is the extraction of handwritten material labels. It can be seen that the PP-OCR model has a good recognition and extraction effect, regardless of the type and content of the label field, whether it is a handwritten label or a printed label.

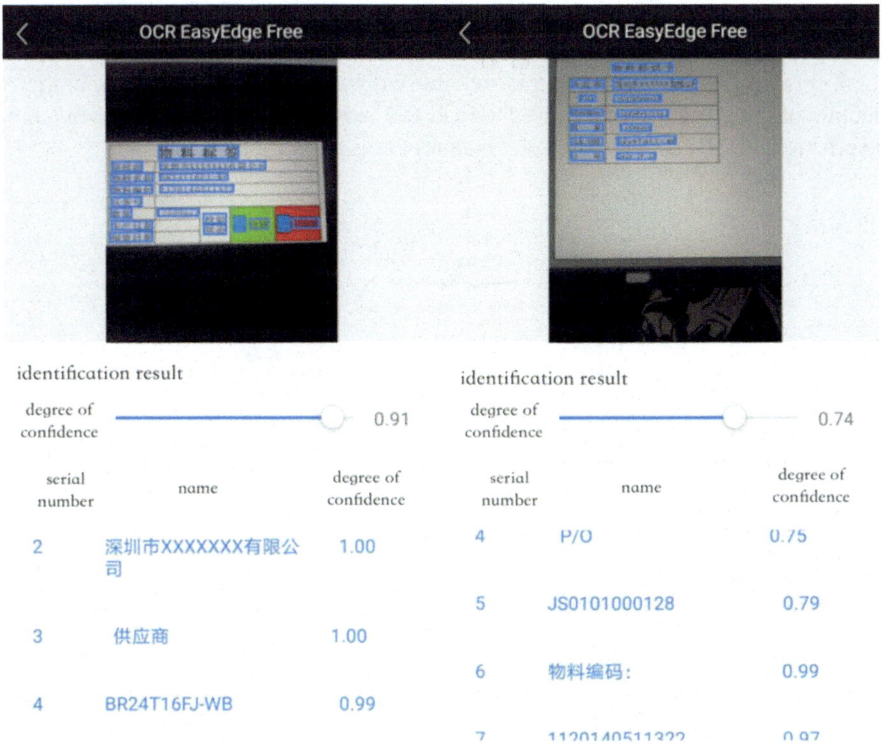


**Fig. 5.** Material label identification diagram

It can be seen from Fig. 5 that APP can not only identify the information in the label, but also determine the reliability of information read and give the confidence.

The left graph is a printed label, field standard, regular content, so the confidence of information recognition is high. The right map is a handwritten label, the font is more humanized, and the content layout is relatively miscellaneous, so the overall confidence is lower. But overall, this label recognition APP for various material labels recognition effect is better.

### 4.2 NFC Read and Write RFID Tags

Because the RFID tags written by NFC need to be designed separately, this paper takes Wuhan all-in-one card as an example to read and write with mobile phone NFC.

Open the NFC of the mobile phone, put the card or electronic tags to the NFC sensing area on the back of the mobile phone, can open the operation page for information reading and writing. After practical operation, the mobile phone NFC display can read the data in Wuhan Card, that is, to view the balance, use records, etc.; At the same time, it can also write data to Wuhan Card (a variety of ways of recharge services).

As shown in Fig. 6, the material labels on warehouse goods are identified by mobile phone APP, and the information is extracted and recorded in the mobile phone. Then, the information is written into the standard RFID tags in the warehouse in the form of radio waves, which helps the sorting work.

The main job of this part is to test the application of the designed tag identification APP. The results show that whether it is a handwritten label or a printed label, APP can accurately identify it and extract reliable information. At the same time, the NFC function of the mobile phone is used to read and write the Wuhan Card, which confirms the practical effect of NFC function on smart cards and RFID tags.

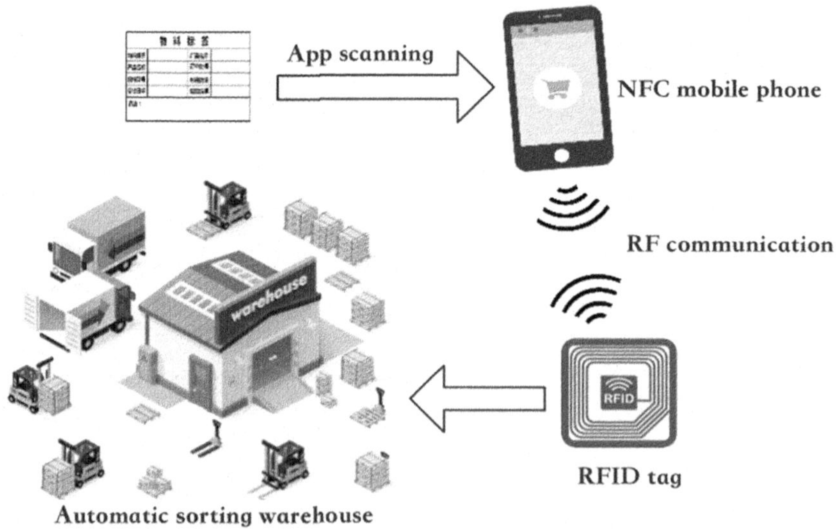


**Fig. 6.** NFC read-write label flow chart

## 5 Conclusion

Aiming at the problems of label recognition and information unification in the process of logistics warehouse sorting and disassembly, this paper designs a label recognition APP for warehouse sorting system. The ultra-lightweight OCR technology is used to extract high-precision information of barcodes, two-dimensional codes, electronic tags and handwritten tags with different specifications, and the recognition information is written into the corresponding RFID tags by NFC technology on Android phones. Finally, the function of tag information recognition and information input is realized by APP.

Using tag identification APP to complete the tag identification of automatic sorting system can reduce the amount of manual work and lay the foundation for automatic integrated warehouse. In the subsequent research, if we can expand the corresponding APP function by combining the sorting process of various warehouses to make it more suitable for the current sorting operation, the practicability and operability of APP can be improved.

## References


1. Feng, M., Cao, D., Li, J.: Design and research of pig face recognition APP based on ResNet50 model. Comput. Age, (06), 46–50 (2020). (in Chinese)
2. Lei, G.: Design and research of mobile logistics APP based on handheld terminal. Digit. Technol. Appl. (08), 177–178 (2015). (in Chinese)
3. Wang, X., Gao, T.: Translation optimization method based on OCR optical character recognition. Laser J. **41**(12), 156–160 (2020)
4. Xu, A., Zhu, J., Chen, Y.: Research on the recognition technology of QR two-dimensional code based on fuzzy invoice. Mod. Commer. Ind. **40**(30), 212–213 (2019). (in Chinese)
5. Kong, B.: The application of Zxing barcode scanning technology in mobile digital library. Comput. Knowl. Technol. **12**(27), 196–198 (2016). (in Chinese)
6. Su, C.: Research on the design and implementation of intelligent tally system based on OCR technology. Res. Ind. Innov. (20), 115–116 (2020). (in Chinese)
7. Hao, Y., Qiao, G., Tan, Y.: Research on automatic word error correction method for OCR text recognition. Comput. Simul. **37**(09), 333–337 (2020). (in Chinese)
8. Vivaldi, F., et al.: A temperature-sensitive RFID tag for the identification of cold chain failures. Sens. Actuators A Phys. **313**, 112182 (2020)
9. Huang, Y.: RFID technology and application, p. 230. People's Posts and Telecommunications Press, Beijing (2013). (in Chinese)
10. Yang, Z.: Research and label design of UHF RFID system. Hunan University, Changsha (2009). (in Chinese)
11. Gan, W., Zhang, T., Zhu, Y.: On RFID application in the information system of rail logistics center. Int. J. Educ. Manag. Eng. **3**(2), 52–58 (2013)
12. Ting, P.-H.: An efficient and guaranteed cold-chain logistics for temperature- sensitive foods: applications of RFID and sensor networks. Int. J. Inf. Eng. Electron. Bus. **5**(6), 1–5 (2013)
13. Gan, W., Zhu, Y., Zhang, T.: Research on RFID application in the pharmacy logistics system. Int. J. Educ. Manag. Eng. **2**(8), 13–19 (2012)
14. Wang, J.: Intelligent display technology research and system design based on the internet of things. University of Electronic Science and Technology, Chengdu (2020). (in Chinese)

15. The application of intelligent labels and interconnected packaging. Green Packag. (10), 82–84 (2019). (in Chinese)
16. Viegas, V., Postolache, O., Pereira, J.M.D.: Transducer electronic data sheets: anywhere, anytime, anyway. Electronics **8**(11), 19 (2019)
17. Anzules, J.R., Uchubanda, W.L., Gurumendi, A.V., Cruz, N.E.: NFC solutions for purchasing public transport tickets at babahoyo terrestrial terminal. J. Sci. Res. (3), 50–55 (2018)

# Modeling in Logistics Engineering

# Game Analysis of Military-Civil Integration in Emergency Logistics Distribution Under the Context of Pandemic

Bingbing Li, Lulu Wang(✉), and Zili Liu

School of Logistics, Wuhan Technology and Business University, Wuhan 430065, China

**Abstract.** Under the context of COVID-19, country needs the military to take part in the lives of the residents for distribution of related products during special periods. The civil also need to cooperate with the military for products distribution, while the civil also need to improve their sales for growth. Nowadays, prior studies focused on the distribution of products about the cooperation with the civil and the military. Based on the theory of game analysis, we set numerical simulation to test the stability of the equilibrium point. Then, we find that set (non-compulsory, positive distribution) is the evolutionarily stable strategy. The more distribution quantity that the military require the civil delivery, and the more proportion of reputation that the civil get, the less time for the civil to consider, the win-win will realize. In the end, we provide some implications for the civil and the military to consider, and represents limitations and further studies.

**Keywords:** Military-civil integration · Evolution game · Emergency logistics distribution · Numerical simulation

## 1 Introduction

In recent years, large-scale public health events have occurred frequently, such as SARS, Bird flu, Influenza A (H1N1) virus and Ebola virus, which have caused varying degrees of loss of life and property. The global public health outbreak at the end of 2020, COVID-19, was a powerful one, involving many countries around the world. Facing the nationwide closure of cities in China, major hospitals and residents are short of supplies. In the face of the donations of many enterprises and individuals, it is urgent for the army to deal with them. The timely support provided by the army to Hubei province has greatly helped people overcome the difficulties caused by the pandemic, and the delivery of military supplies is particularly important. How to guarantee the material demand of the army and the material supply of major hospitals, facing the severe test of national and even global public health events, this is an unprecedented challenge, but also a war without smoke.

Among previous studies, there exist three types of military-civil integrated emergency logistics distribution, for example locally-orientated and military-collaborative distribution, the military-orientated and locally-collaborative distribution, military-civil integrated distribution [1]. While in practice, the military-civil integrated distribution is


Z. Hu et al. (Eds.): ICAILE 2021, LNDECT 82, pp. 265–276, 2021.
https://doi.org/10.1007/978-3-030-80475-6_26

more efficient. The emergency logistics distribution may lead to low efficiency because of the asymmetric information [2]. The high efficiency of military-civil integrated emergency logistics distribution is vital for the development of the country, so the military and the civil should pay attention to the concept of the advanced military-civil integration system construction, it should be suitable for practical needs [3]. The high efficiency of the military service for the people deserves to consider, the social values and the economic values should both deserve to consider for the civil.

The military-civil integration is a characteristic model in China which is suitable for the development of Chinese economy and people's livelihood. The current research about COVID-19 showed a trend of significant growth in 2020, as to June 20, 2020, 581 piece of CSSCI literature has been published, study of military-civil integration also showed a trend of rise, published 557 papers, published 26 papers in high quality papers in 2020, this means that China and the global pay an important attention to the study of pandemic, military-civil integration has certain requirements. Prior studies are mostly on the construction of military-civilian integrated logistics system [3, 4], the integrated supply chain management [1], the agile mobilization theory system [5], construction of equipment intelligent manufacturing system [6], the military logistics construction and military-civil integration development model, etc., but many studies tent to questionnaire investigation, theory construction, such as static data, less research on dynamic game model [7, 8]. However, the logistics involved in major public health events tend to be emergency logistics. When an event occurs suddenly, time has a great impact about the logistics distribution decision. Therefore, it is necessary to explore the decision-making mechanism of military-civil integrated emergency logistics distribution based on the dynamic model. At present, the 14th five-year Plan of China is under way. There are relatively few studies on the decision making of military-civil integration emergency logistics distribution under the pandemic situation. So, based on the COVID-19 pandemic, this study carries out the decision making research of military-civil integration logistics distribution based on game evolution, in order to make some reference suggestions for the logistics support of military-civil integration.

The structure of this paper is as follows: In Sect. 2, the analysis is around the behavior of dynamic evolution between the civil and military. Section 3 analyses evolution game by numerical simulation. Section 4 shows the discussion about the results. Section 5 presents the implications, the findings would help the civil to responding to the government's call, it promotes the limitary to service for the people much better. Finally, Sect. 6 shows conclusion, it concludes summary, limitations and suggestions for future research.

## 2 Theoretical Analysis of Dynamic Evolution

It is necessary that the civil and the military could cooperate with each other. To achieve the flexibility of the military using, the civil should be encouraging to reach the standard of the military data distribution [9]. Faced with the needs of the military sector, the civilian sector is actively taking measures to cooperate. The military-civil collaborative innovation is the common way they cooperate. Under the limited rationality, their cooperation is related with the incentive policy of government, and the reasonable benefits distribution coefficient is lead to social stability, it help to create conditions to make the military and the civil to cooperation [10]. Dual-use technology has been a popular way for military-civil integration, this technology can realize the demand of the civil, and it also meets the needs of the military [11].

Then, assume that the military and the civil are limited rational men. In the process of resource delivering, due to information asymmetry, opportunistic behaviors often occur, the military should ensure the sufficient supply of materials and the demand for combat, for example facemask in COVID-19 pandemic. While the civil should make profits and ensure the efficient operation of enterprises. In the event of a major pandemic, military need to immediately go to the region to fight and need a large amount of relevant materials. Military production alone is often not enough, so products from private enterprises need to be commanded. The military is closely connected with the government, and the policy and strategy set is (compulsory, non-compulsory), while the behavior set that the civil can make according to their own will is (positive distribution, negative distribution), that is, how many resources can be distributed when responding to the needs of the military.

Based on the management experience of major public health incidents in China, the military will cooperate with the civil (private enterprises) to distribute materials to serve consumers in the market in the face of the pandemic. We assumed that the probability of military adopting compulsory strategy is $x(0 < x < 1)$ the probability of non-compulsory strategy is $(1-x)$, the probability of the civil adopting active distribution strategy is $y(0 < y < 1)$, and the probability of adopting negative distribution strategy is $(1 - y)$. The cost of military collect material to service residents is $c_1$, the number of distribution is M, the price is $p_1$. While the cost of civil distributes the material is $c_2$.When the civil takes positive strategy, the proportion of civil gets reputation is b, the price is $p_2$; If the civil takes negative strategy, the proportion of civil would decline, the price of civil cooperating with the military is $p_1$, with residents is $p_2$. The distribution inventory to military is $aM$, to residents is $(1 - a) \cdot M$. And $0 < c_1 < p_1 < p_2$, $0 < c_2 < p_1 < p_2$, $0 < a < 1$, $0 < b < 1, 0 < M$.We also assumed that negative distribution is $aM$, positive distribution is $M$ (Table 1).

**Table 1.** Payment matrix of game theory

| Role | | Civil | |
|---|---|---|---|
| | | *PD (y)* | *ND (1−y)* |
| Military | *CM (x)* | $\frac{M(p_1-c_1)}{t}$ | $\frac{aM(p_1-c_1)}{t}$ |
| | | $\frac{M(p_1-c_1)}{t}$ | $\frac{abM(p_1-c_2)+M(1-a)(p_2-c_2)}{t}$ |
| | *NCM (1−x)* | $\frac{M(2p_1-p_2-c_1)}{t}$ | $\frac{aM(2p_1-p_2-c_1)}{t}$ |
| | | $\frac{M(p_2-c_2)}{t}$ | $\frac{aM(p_2-c_2)+M(1-a)(p_2-c_2)}{t}$ |

Note: Compulsory = CM; Non-compulsory = NCM; Positive distribution = PD; Negative distribution = ND.

First, when military adopt compulsory strategy, his benefit is

$$V_{11} = \frac{yM(p_1 - c_1)}{t} + (1-y)[\frac{aM(p_1 - c_1)}{t}] \tag{1}$$

When military adopt non-compulsory strategy, his benefit is

$$V_{12} = \frac{yM(2p_1 - p_2 - c_1)}{t} + (1-y)[\frac{aM(2p_1 - p_2 - c_1)}{t}] \tag{2}$$

The average benefit of the military is

$$V_1 = xV_{11} + (1-x)V_{12} \tag{3}$$

The replication dynamic equation of the military is:

$$F(x) = x(V_{11} \_ V_1) = \frac{x(1-x)M[y(1+a) - a]}{t} \tag{4}$$

Second, when civil adopt positive distribution strategy, his benefit is

$$V_{21} = xM\frac{p_1 - c_2}{t} + (1-x)M\frac{p_2 - c_2}{t} \tag{5}$$

When civil adopt negative distribution strategy, his benefit is

$$V_{22} = xM\left[\frac{ab(p_1 - c_2) + (1-a)(p_2 - c_2)}{t}\right] + (1-x)M[\frac{ab(p_2 - c_2) + (1-a)(p_2 - c_2)}{t}] \tag{6}$$

So the average of civil's benefit is

$$V_2 = yV_{21} + (1-y)V_{22} \tag{7}$$

The replication dynamic equation of the civil is:

$$F(y) = y(V_{21} - V_2) = yM(1-Y)\frac{x(p_1 - p_2)(1-ab) + a(1-b)(p_2 - c_2)}{t} \tag{8}$$

So, we obtain Jacobi matrix

$$\varphi = \begin{vmatrix} \frac{M(1-2x)[y(1+a)-a]}{t} & \frac{yM(1-y)(p_1-p_2)(1-ab)}{t} \\ \frac{xM(1-x)(1+a)}{t} & \frac{M(1-2y)[a(p_1-p_2)1-ab)+a(1-b)(p_2-c_2)}{t} \end{vmatrix} \tag{9}$$

$$tr\varphi = \frac{M(1-2x)[y(1+a)-a]}{t} + M(1-2y)\frac{x(p_1 - p_2)(1-ab) + a(1-b)(p_2 - c_2)}{t} \tag{10}$$

Then we calculate to obtain five dynamic equilibrium A(0, 0), B(1, 0), C(0, 1), D(1, 1) and O($x^*$, $y^*$). $x^* = \frac{a(1-b)(p_2-c_2)}{(p_2-p_1)(1-ab)}, y^* = \frac{a}{1+a}$. According to the theory of evolution game, $det\varphi > 0$ and $tr\varphi < 0$, the equilibrium points is the gradual steady point at the local part. We make a table showed equilibrium, determinant and trace, as shown in Table 2.

From Table 2, we find that point D (1, 1) is the solo evolutionary stable strategy. The results were surprising. It means that point D is locally asymptotically stable, the evolutionary stability strategy of the system at this time is (compulsory, positive distribution).

**Table 2.** Equilibrium, determinant and trace

| EP | EQ | QAT | LS |
|---|---|---|---|
| A | $\frac{-a^2M^2(1-b)(p_2-c_2)}{t^2}$ <br> $\frac{-aM+aM(1-b)(p_2-c_2)}{t}$ | $det\varphi(-)$ <br> $tr\varphi(+,-)$ | SD |
| B | $\frac{-aM^2(1-ab)(p_1-p_2)+a(1-b)(p_2-c_2)}{t^2}$ <br> $\frac{-aM+(p_1-p_2)(1-ab)+aM(1-b)(p_2-c_2)}{t}$ | $det\varphi(+,-)$ <br> $tr\varphi(+,-)$ | US |
| C | $\frac{-aM^2(1-b)(p_2-c_2)}{t^2}$ <br> $\frac{-M-aM(1-b)(p_2-c_2)}{t}$ | $det\varphi(-)$ <br> $tr\varphi(+,-)$ | SD |
| D | $\frac{M^2(1-ab)(p_1-p_2)+a(1-b)(p_2-c_2)}{t^2}$ <br> $\frac{-M-M(p_1-p_2)(1-ab)+aM(1-b)(p_2-c_2)}{t}$ | $det\varphi(+)$ <br> $tr\varphi(-)$ | ESS |
| O | 0 <br> 0 | $det\varphi(0)$ <br> $tr\varphi(0)$ | SD |

Note: EP = Equilibrium point; EQ = Equation; DAT = Determinant and trace; LS = Local stability; SD = Saddle point; US = Unstable;

# 3 Evolution Stability of Numerical Simulation

In the third section, Simulation software is used for numerical simulation to verify the theoretical analysis results and the stability of the equilibrium point. Then we set the value of the constant component parameter. They are as follows: b = 0.6, p2 = 1.2, p1 = 1.0, c2 = 0.8, M = 100, t = 10. By adjust the value of parameter a, we finally obtain the replication dynamic equation of the civil which is shown in Fig. 1.

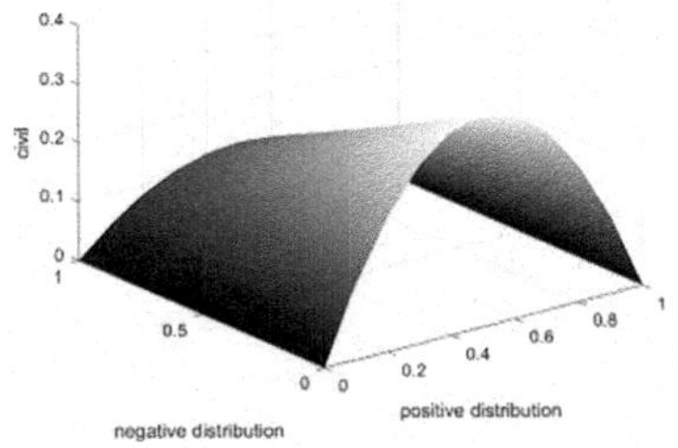


**Fig. 1.** Dynamic evolution analysis

Then, we get the evolution of the military. We find that the final result of evolution tends to D (1, 1), which is the equilibrium stable point. If we the military wants to get enough materials to service the residents and the civil wants to create much sales and reputation in the market, they should take positive cooperation, that is to say, the military should take compulsory strategy and the civil should take positive distribution strategy, not negative distribution strategy. From the Fig. 2, it tells us that D (1, 1) is the solo equilibrium, it reaches the evolution strategy stimulation. To test the validity of the model and the correctness of the theoretical analysis, we set four arrays [0.2, 0.1], [0.2, 0.5], [0.6, 0.1], [0.6, 0.8], and then set $a = 0.50$, 075 to test the optimal results. We observe the simulation results. Figure 2 and Fig. 3 shows the results of four arrays.

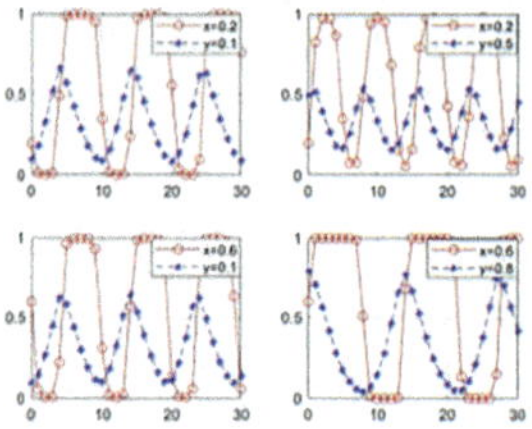

**Fig. 2.** The results of four arrays ($a = 0.50$).

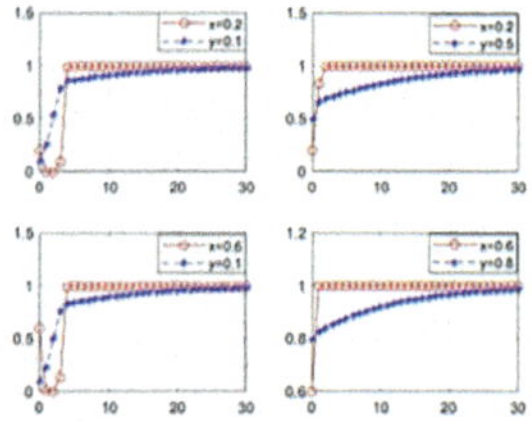

**Fig. 3.** The results of four arrays ($a = 0.75$).

When the military adopt compulsory strategy and the civil adopt positive distribution strategy, how many materials does the civil should distribute to the military, the civil could make money and also service the military. The value of "a" is vital. When $a = 0.75$, the strategy is to positive cooperation. Then we set $a = 0.70, 0.75, 0.80, 0.85, 0.90, 0.95$, shows in Fig. 4, Fig. 5. Only $a \geq 0.90$, it will taste five days to reach the cooperation. In five days, the military can get a*M pieces of materials, and the civil can get the sales. Only $a \geq 0.80$, it will taste ten days to reach the cooperation. In ten days, the military can get a*M pieces of materials, and the civil can get the sales.

To test how the strategy that the civil gets, we set $b = 0.25, 0.50, 0.75$, shows in Fig. 6 to, Fig. 7, Fig. 8. When $b < 0.75$, the trend of compulsory and positive distribution strategy. And then we set $b = 0.70$, and test it every 0.1 decreasing spacing, the results are shown in Fig. 9, Fig. 10 and Fig. 11. If $b \leq 0.70$, in 30 days the cooperation tends to the compulsory and positive distribution strategy. Also, we set $b = 0.60$, and test it every 0.1 decreasing spacing, the results are shown in Fig. 12 and Fig. 13. If $b \leq 0.59$, the civil tends to select positive distribution, he could also obtain high performance in the end in ten days. If $b \leq 0.40$, the civil tends to select positive distribution, he could also obtain enough performance in the end in five days.

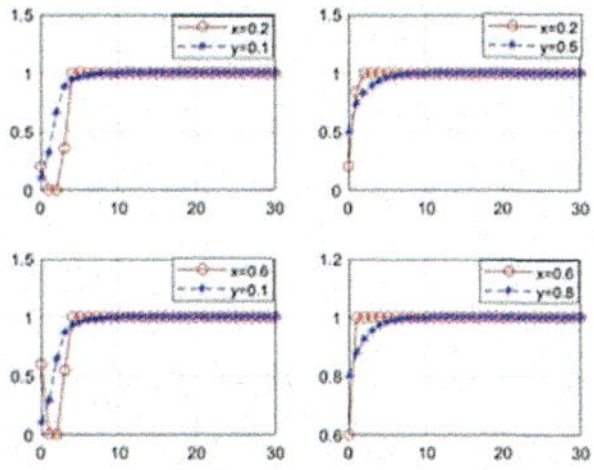


**Fig. 4.** The evolution results ($a = 0.90$)

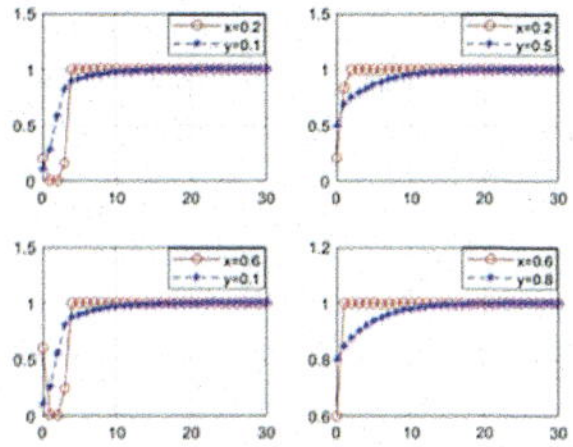


**Fig. 5.** The evolution results ($a = 0.80$)

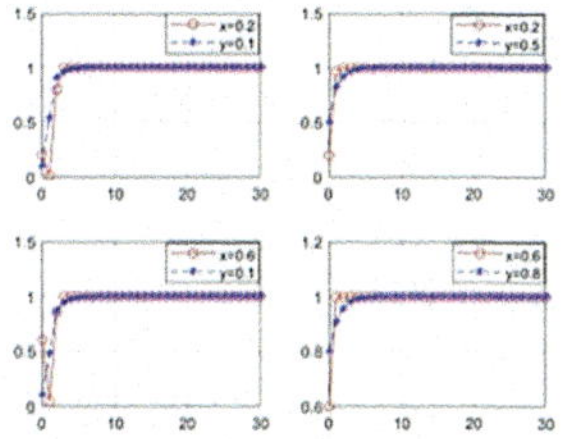


**Fig. 6.** The results of four arrays (b = 0.25).

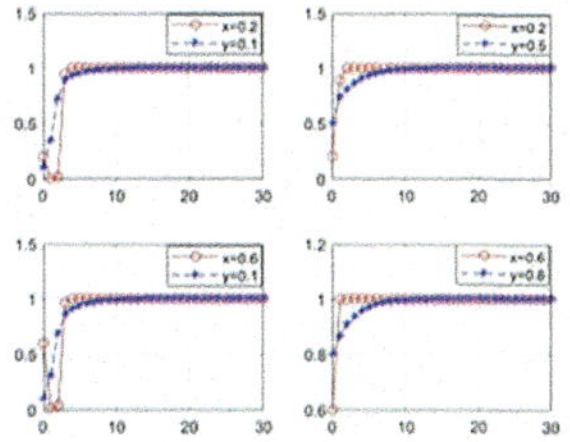


**Fig. 7.** The results of four arrays (b = 0.50).

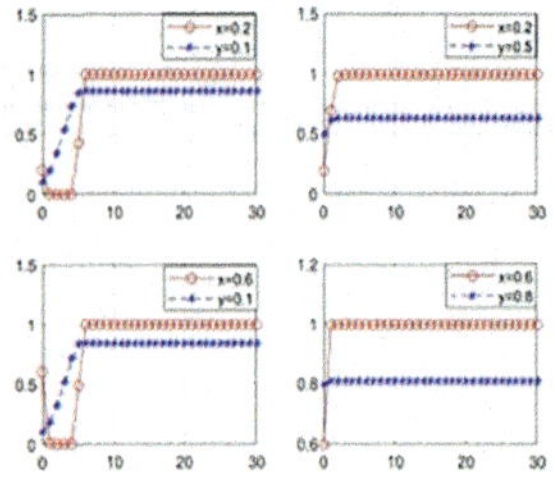


**Fig. 8.** The results of four arrays (b = 0.75).

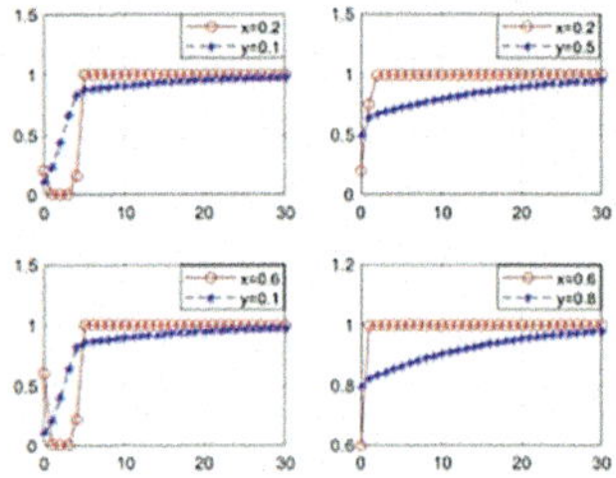


**Fig. 9.** The evolution results (b = 0.70).

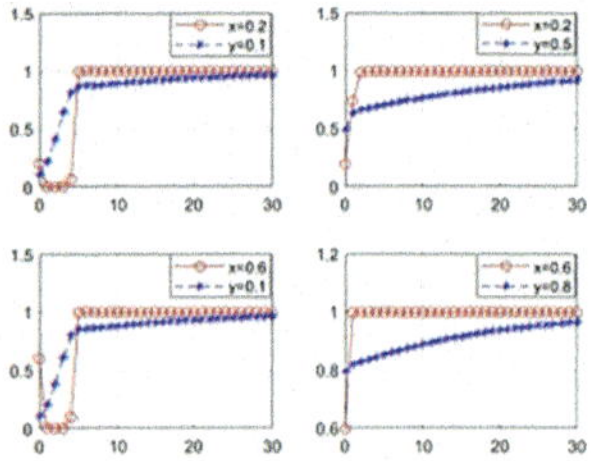


**Fig. 10.** The evolution results (b = 0.71).

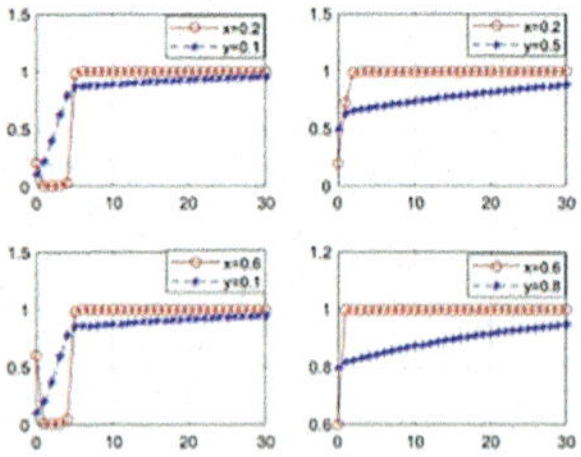


**Fig. 11.** The evolution results (b = 0.72).

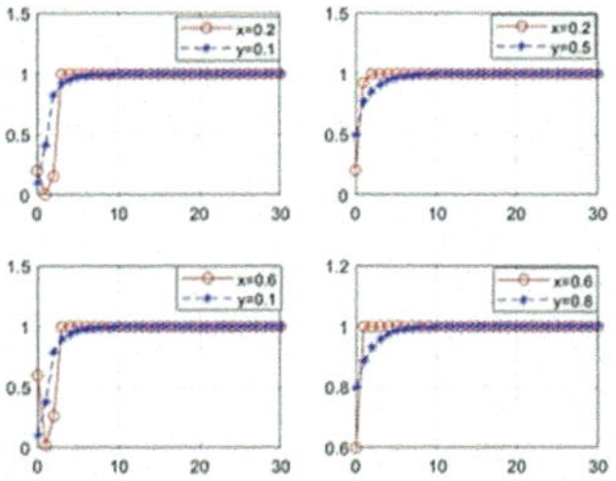


**Fig. 12.** The evolution results (b = 0.40).

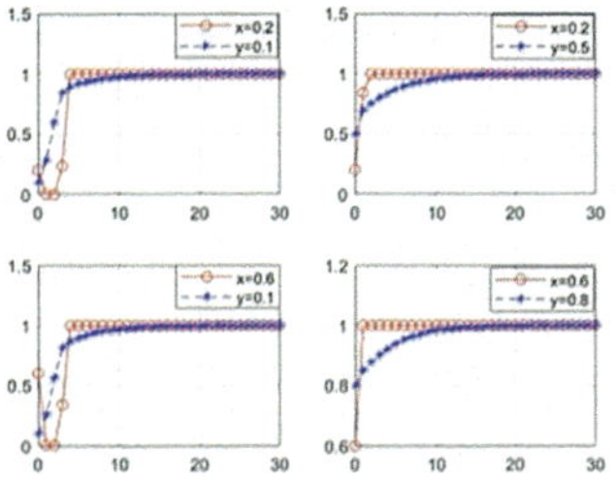


**Fig. 13.** The evolution results (b = 0.59).

## 4 Discussion

From the results we get above, we discuss three some interesting findings.

Firstly, the point (1,1) is the solo evolutionary stable strategy from the perspective of evolutionary game theory, which refers to the strategy of compulsory and positive distribution. It is appropriate for China's context, and it enhances the strength of government making compulsory strategy. This strategy is the best way for them to cooperate. But if the civil sends larger number of distribution to residents, that is to say they select negative distribution strategy, they will get enough benefits, because the market shares bring large performance to the military. Why the civil selects the positive strategy? From the mode that we construct, the opportunity cost that the service to the social bring much reputation to the civil, it will enhance the brand efficiency, which could help the civil create the long term of performance. It helps the civil survive longer and better. So it could offset the lost that sold the materials to the military at little price, not to the customers at larger price.

Secondly, the number of the distribution "a" that the military requires the civil to deliver to them becomes much smaller, the more possibility the civil could create more benefits when they reach the cooperation. And the time that the civil decides to select positive distribution strategy becomes much shorter, we can conclude that less time may lead the civil to make an irrational decision. On the opposite, the more time may make the civil to make a rational decision. It gives less time for the civil to consider which strategy to choose, because of the emergency logistics distribution's particularity, the more time the military could save, the more performance they will create. In fact, the military's compulsory strategy makes the civil has no time to consider enough. But the number of the distribution is big, when in the emergency logistics distribution, the policy about the military's need may be immediately promulgated and implemented as soon as possible, it may reach win-win condition.

Thirdly, the value of reputation that the civil may get from the cooperation tends to much smaller, the more possibility it intent to the compulsory-positive distribution strategy which could create much benefit. And when the value of reputation is stable, the more time may lead the compulsory-positive distribution strategy. It is just the opposite with the second findings. When the civil know the demand of the military, he would decide which strategy to choose, whether positive or negative distribution, the civil would consider the social values and economic values. From the results we can predict that in more time the civil may recognize more benefits they can get from cooperation with the military. So the value of cooperation the two sides get may take some time for it.

## 5 Implication

Based on China's national conditions, China is a country under the rule of law. When the important pandemic accident occurs, Chinese government would send the military to help people, many armies flied to Wuhan to help the local people to go through troubles. But the military need related materials, the civil need sales to make firm grow up and retain employees. So how does the cooperation need to solve. From the results that we get, if both of them want to achieve the demand, no matter how the external conditions change, they should take the (compulsory, positive distribution) strategy.

Some management implications have been got. First, the government should provide various preferential policies for firms to help facilitate the normal and efficient operation of firms. They could make market survey, take some measures to get know the actual production of firms, for example distribution. Second, we suggest that the government cooperate with civil to build the information system, use information automation technology to monitor the distribution schedule of firms. For example, firms could buy come radio frequency identification tools and then utilize into the internet of things, the government can subsidize some of the procurement costs to the firms. The firms share the data to the government, the government provides the policies for firms to rent the land warehouse and port warehouse. As for the civil, firms need to survive, it is necessary to sell products at an appropriate price in order to obtain profits. It is not a profit-making enterprise that blindly serves the society even though it is a social company. So, the firms need to make some measures to entry into the market, for example cooperating with the military, to enhance the firms' reputation, and then taking the meager profit but high turnover strategy. Not only it does create jobs, help many people to feed themselves, but also the firms could create high performance.

## 6 Conclusion

The paper focuses on the evolution game analysis between the military and the civil, following with the change of external environment, the military adopts the compulsory strategy and the civil adopts positive distribution strategy, both of them could realize their demands. And in ten days, the civil at least 0.80 times the amount of all supplies distributed to the military, he would benefit the maximum values. If the military wants five days earlier, the military should require the civil distributes at least 0.90 times the amount of all supplies. What's more, less proportion of reputation that the civil will get, the possibility of strategy tends to compulsory-positive distribution, but it will long time for the civil to select the positive distribution, more proportion of reputation that the civil will get, it tastes short time for the civil to get the values of reputation. These findings help the civil and the military to do more appropriate decisions.

However, this paper also exist some limitations. First, the evolution game model that we build didn't take the opportunity cost that the military has into consideration. If the military didn't realize their demands, it will lead much lost for the people and the civil. Second, the time that the civil distributes to the military is also an important factor for the performance of cooperation, the value of distribution should take factor time into consideration. The cooperation about the civil and the military still remains further study. In the further, we will take opportunity cost into the evolution game model, as well as the factor time into it.

**Acknowledgment.** This research is supported by project of Wuhan Technology and Business University. Project number (A2020004) (Research on the Decision Making of Military-civil Integrated Emergency Logistics Distribution based on Evolutionary Game under Epidemic Situation).

## References

1. Benotmane, Z., Belalem, G., Neki, A.: A cost measurement system of logistics process. Int. J. Inf. Eng. Electron. Bus. **10**(5), 23–29 (2018)
2. Huo, X.: Study on the optimization of emergency logistics distribution resources under emergencies. Rural Econ. Sci.-Technol. **31**(6), 92–93 (2020)
3. Gu, Z., Tan, Q.: Research on construction of military-civil integration logistics system. Sci. Technol. Manag. Res. **20**, 228–235 (2019). (in Chinese)
4. Gan, W., Zhang, T., Zhu, Y.: On RFID application in the information system of rail logistics center. Int. J. Educ. Manag. Eng. **3**(2), 52–58 (2013)
5. Zhang, J., Fan, W., Shi, Y.: The framework design of agile mobilization theory under the strategy of military-people integration. J. Beijing Inst. Technol. (Soc. Sci. Ed.) **1**, 121–127 (2019). (in Chinese)
6. Du, R.: The intelligent manufacturing system of military and civilian integrated equipment and its construction – based on the perspective of the whole manufacturing industry Chain. Mod. Econ. Res. **5**, 121–127 (2017). (in Chinese)
7. Ting, P.-H.: An efficient and guaranteed cold-chain logistics for temperature-sensitive foods: applications of rfid and sensor networks. Int. J. Inf. Eng. Electron. Bus. **5**(6), 1–5 (2013)
8. Ma, H., Dong, J., Zhou, N., et al.: Military-civilian cooperative emergency response to infectious disease prevention and control in China. Mil. Med. Res. **01**, 1–5 (2017)
9. Liu, X., Shao, F., Zhang, X.: Research on the civil-military integration logistics support system of warships. In: IOP Conference Series: Materials Science and Engineering, vol. 853, no. 1, pp. 1–5 (2020)

10. Fang, W., Wang, C.: The cooperative stability evolutionary game analysis of the military-civilian collaborative innovation for China's satellite industry. Math. Probl. Eng. **12**, 1–17 (2019)
11. Liu, X., Liang, H.: Research on civil-military integration support of Chinese warship equipment. J. Nav. Univ. Eng. **13**(1), 59–63 (2016)

# Pricing Game Model of Pallet Sharing Platform Under Bidding

Yu Li, Yong Gu(✉), and Yangyang Zheng

School of Logistics Engineering, Wuhan University of Technology, Wuhan 430063, China

**Abstract.** Pallet sharing market which dominated are selected by the pallet sharing platform under the competitive market environment, and apply game theory and multilateral market theory to analyze the pallet platform pricing strategy under the competitive bidding mode of the participants. The mechanism of platform pricing is analyzed from the perspectives of the complementarity of participants' needs, user's network externalities, government policy subsidies, the scale of the platform, the degree of differentiation of the platform's service quality, and the attributes of the participants. In a competitive environment, a multi-party game model of pallet sharing platform, transporter, and supply and demand side considering multiple factors was constructed. The impact of the above factors on the platform pricing strategy and the attribution of the participants was analyzed and analyzed based on mathematical derivation and numerical examples. The platform pricing analysis in the environment puts forward suggestions to promote the development of the pallet sharing market.

**Keywords:** Pallet sharing platform · Multilateral users · Multi-stage game

## 1 Instruction

In China's pallet sharing market, there are some serious problems such as the difficulty of matching the pallet supplier and the high empty rate in the process of empty pallet scheduling. In order to alleviate the severe problems and in response to the national "Internet+" and "sharing economy" policies, pallet sharing platform emerges as The Times require. However, the level of each platform is uneven, and there are many problems such as "difficult pricing". In order to promote the healthy development of sharing platform, give play to its expected utility value, improve the relationship between the supplier and demander, transportation and the managers, improve the efficiency of the pallet, reduce the attrition rate and the logistics cost of enterprise, it's urgent to study systematically and the improvement of the build pallet sharing platform pricing mechanism, promote the balanced development of pallet share market [1].

## 2 Literature Review

To study the pricing mechanism of pallet-shared platform, it is necessary to understand its platform model, pricing strategy and relevant theoretical model.


Z. Hu et al. (Eds.): ICAILE 2021, LNDECT 82, pp. 277–287, 2021.
https://doi.org/10.1007/978-3-030-80475-6_27

In terms of pallet sharing platform mode, based on literature [2–5], the pallet sharing system is mainly divided into open and closed. The former is jointly built by the pallet supplier, service outlets and management platform, and the user, so that the pallet circulates among enterprises. The latter shall set up the rental outlets within the scope of its services, purchase the pallets and set up the pallet pool, and the ownership of the pallets shall belong to the common platform.

In terms of pallet-shared platform pricing, Fang H and Xiaolei W formulated and solved the optimization design of taxi platform pricing, punishment and compensation strategies through penalty continuous linear programming algorithm based on the equilibrium model [6]. Zhu, Ling and Lin, Jie studied the impact of factors such as the proportion of revenue commission, the price of advertising products and the different brand influence on the pricing strategy of the platform [7].

In terms of theoretical model research, Hans Peters studied a variant of the Hotelling model and found that for two or more enterprises, the equilibrium position of these enterprises was different from that of the Hotelling model [8]. Loizou proposed a distributed robust Nash equilibrium model. The distributed robust games studied by the model are equivalent to the Nash equilibrium games without considering the hidden information [9]. Zahra Saberi classified online retailers in the Kei Stackelberg model and analyzed the pricing strategies of small retailers in different modes [10].

So, the pricing mechanism of pallet sharing platform is studied under the mode of participants' bidding optional and construct multistage game model based on multilateral market theory and game theory [11]. The model fully considers participants demand complementarity, the participants of network externalities, the policy of government subsidies, platform scale, the differentiation degree of service quality of the platform, the attribution of participants and other factors.

# 3 Pricing Mechanism of Pallets Sharing Platform

## 3.1 Analysis of Market Environment of Pallet Sharing Platform

At present, the construction of domestic pallet sharing system has just begun in China, and the overall coordination construction of the whole logistics industry supply chain is not perfect, and the process of pallet standardization lags behind [12]. Pallet sharing companies have just emerged in the market, and dozens of pallets sharing companies have appeared in the market in one or two years. At this time, there is no difference in the choice of pallet sharing companies by multilateral users, and the market characteristics show that the competition between pallet sharing platforms is not hindered, and it belongs to a perfectly competitive market.

Therefore, market environment is chosen to study and analyze the pricing problem of pallet sharing platform.

## 3.2 The Charging and Pricing Game Analysis of Pallet Sharing Platform

Pallet sharing is based on the cooperation of the pallet sharing platform, the pallet supplier, the pallet demander and the transporter to realize the circular flow of pallets in the whole logistics system. Pallet supplier, pallet demander and transporter are connected through the platform to trade. The demander meets the needs of their own free pallet through leasing supplier and produces rent. When supplier and demander reach an agreement, the transporter will transport the pallet from the supplier to the demander and produce freight. In order to simplify the research, the game is considered between the supplier and the transporter to determine the freight. The "open" pallet sharing platform includes three kinds of participants. It is a kind of research on the pricing of multilateral platform.

Under the competitive market environment, the pricing game of pallet sharing platform mainly includes the following four stages:

1. The game between pallet sharing platforms. By analyzing the scale of the platform, the differentiated services provided by the platform and other factors to determine the final number of users connected to the platform.
2. The game between the platform and users to determine the user registration fee.
3. The game between the pallet supply and demand to determine the rental fee.
4. The supplier and the carrier game to determine the freight.

# 4 Pricing Game Model and Propositional Analysis in Competitive Environment

## 4.1 Pricing Game Model

The pricing model of pallet sharing platform in competitive environment is multi ownership of pallet supplier, pallet demander and transporter. The attributes of participants of pallets sharing platform in the competitive environment are shown in Fig. 1:

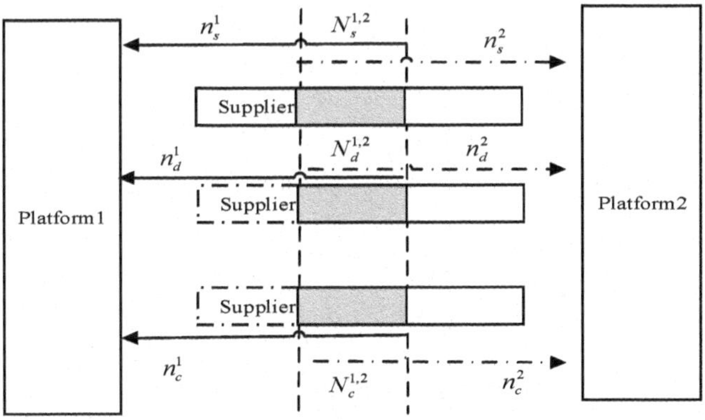


**Fig. 1.** Schematic diagram of multi-ownership of participants in multilateral market platform

Assumed that there are two platforms in the pallet sharing market, platform 1 and platform 2, two platforms in providing the services and functions of exist certain differences, so part suppliers $n_s^1$ select platform 1, part suppliers $n_s^2$ choose platform 2, and part suppliers $N_s^{1,2}$ select both platform 1 and 2. Part demand suppliers $n_d^1$ chooses platform 1, part demanders $n_d^2$ chooses platform 2, and part demanders $N_d^{1,2}$ accesses both platform 1 and 2.Some transporters $n_c^1$ choose platform 1, some transporters $n_c^2$ choose platform 2, and some transporters $N_d^{1,2}$ access both platform 1 and 2.

Model symbol description:

$U_s^1$, $U_s^2$, $U_d^1$, $U_d^2$, $U_c^1$, $U_c^2$: the utility of the supplier, the demander and the carrier in the pallet-shared platform 1 and 2;

$U_s^{1,2}$, $U_d^{1,2}$, $U_c^{1,2}$: the utility obtained by the multi-attribute supplier and demander on the pallet sharing platform;

$\pi_p^1$, $\pi_p^2$: the profit function of platform 1 and platform 2;
$n_s^1$, $n_s^2$: the amount of pallet suppliers that are finally added to platform 1 and 2;
$n_d^1$, $n_d^2$: the demand side of the pallet that is finally added to platform 1 and 2;
$n_c^1$, $n_c^2$: the amount of pallet carriers finally added to platform 1 and 2;
$N_s^{1,2}$, $N_d^{1,2}$, $N_c^{1,2}$: the quantity of multi-owned pallet suppliers, demanders and transporters;
$N_s^1$, $N_s^2$, $N_d^1$, $N_d^2$, $N_c^1$, $N_c^2$: the quantity of pallet supplier, demander and transporter;
$P_s^1$, $P_s^2$: the registration fee charged by Platform 1 and platform 2 to the supplier;
$P_d^1$, $P_d^2$: the registration fee charged by the demander on platform 1 and platform 2;
$P_c^1$, $P_c^2$: the registration fee charged by platform 1 and platform 2 to transporters;
$\delta_s^1$, $\delta_s^2$, $\delta_d^1$, $\delta_d^1$: the subsidy of platform 1 and 2 to the suppliers and the demanders;
$c^1$, $c^2$: the service cost of platform 1 and 2;
$t_s$, $t_d$, $t_c$: the unit conversion cost for supplier, demander and carrier to convert between pallet-shared platform 1 and 2.

The attributes of the pallet supplier, demander and transporter are manifested as partial multi-attribution, whose utility benefits are expressed as follows:

$$\begin{cases} U_s^1 = \left[A_s + P_r^i q(1-\alpha)\right] n_d^1 - P_t^i w n_c^1 - t_s\left(n_s^1 - N_s^{1,2}\right) - \left(P_s^1 - \delta_s^1\right) \\ U_s^2 = \left[A_s + P_r^i q(1-\alpha)\right] n_d^2 - P_t^i w n_c^2 - t_s\left(n_s^2 - N_s^{1,2}\right) - \left(P_s^2 - \delta_s^2\right) \\ U_s^{1,2} = A_s + \left[P_r^i q(1-\alpha) - P_t^i w\right] - t_s - \left(P_s^1 - \delta_s^1\right) - \left(P_s^2 - \delta_s^2\right) \end{cases} \tag{1}$$

$$\begin{cases} U_d^1 = \left(A_d - P_r^i q\right) n_s^1 - t_d\left(n_d^1 - N_d^{1,2}\right) - \left(P_d^1 - \delta_d^1\right) \\ U_d^2 = \left(A_d - P_r^i q\right) n_s^2 - t_d\left(n_d^2 - N_d^{1,2}\right) - \left(P_d^2 - \delta_d^2\right) \\ U_d^{1,2} = A_d - P_r^i q - t_d - \left(P_d^1 - \delta_d^1\right) - \left(P_d^2 - \delta_d^2\right) \end{cases} \tag{2}$$

$$\begin{cases} U_c^1 = n_s^1\left(A_{sc} + P_r^i w\right) - t_c\left(n_c^1 - N_c^{1,2}\right) - P_c^1 \\ U_c^2 = n_s^2\left(A_{sc} + P_r^i w\right) - t_c\left(n_c^2 - N_c^{1,2}\right) - P_c^2 \\ U_c^{1,2} = A_{sc} + P_r^i w - t_c - P_c^1 - P_c^2 \end{cases} \tag{3}$$

According to the benefit function of the three participants on platform 1 and 2, the final access quantity of the supplier, demander and transporter on the pallet-shared platform 1 and 2 can be solved as shown in Formula (4)–(6):

$$\begin{cases} n_s^1 = \dfrac{\left[A_s + P_r^i q(1-\alpha)\right]t_c P_d^2 - P_t^i w t_d P_c^1 - t_d t_c P_s^1 + \left[A_s + P_r^i q(1-\alpha)\right]t_c\left[t_d - \left(A_d - P_r^i q\right)\right]}{t_s t_d t_c - \left[A_s + P_r^i q(1-\alpha)\right]\left(A_d - P_r^i q\right)t_c + P_t^i w n_c^1\left(A_{sc} + P_t^i w\right)t_d} \\ \quad - \dfrac{P_t^i w t_d\left[t_c - \left(A_d - P_r^i q\right)\right]}{t_s t_d t_c - \left[A_s + P_r^i q(1-\alpha)\right]\left(A_d - P_r^i q\right)t_c + P_t^i w n_c^1\left(A_{sc} + P_t^i w\right)t_d} \\ n_s^2 = \dfrac{\left[A_s + P_r^i q(1-\alpha)\right]t_c P_d^1 - P_t^i w t_d P_c^2 - t_d t_c P_s^2 + \left[A_s + P_r^i q(1-\alpha)\right]t_c\left[t_d - \left(A_d - P_r^i q\right)\right]}{t_s t_d t_c - \left[A_s + P_r^i q(1-\alpha)\right]\left(A_d - P_r^i q\right)t_c + P_t^i w n_c^1\left(A_{sc} + P_t^i w\right)t_d} \\ \quad - \dfrac{P_t^i w t_d\left[t_c - \left(A_d - P_r^i q\right)\right]}{t_s t_d t_c - \left[A_s + P_r^i q(1-\alpha)\right]\left(A_d - P_r^i q\right)t_c + P_t^i w n_c^1\left(A_{sc} + P_t^i w\right)t_d} \end{cases} \tag{4}$$

$$\begin{cases} n_d^1 = \dfrac{(A_d - P_r^i q)t_c P_s^2 + [t_s t_d - P_t^i w(A_{sc} + P_t^i w)]P_d^1 + [A_s + P_r^i q(1-\alpha)]P_t^i w P_s^1}{t_s t_d t_c - [A_s + P_r^i q(1-\alpha)](A_d - P_r^i q)t_c + P_t^i w n_c^1(A_{sc} + P_t^i w)t_d} \\ \quad + \dfrac{+[A_s + P_r^i q(1-\alpha)]t_c\{t_s - [A_s + P_r^i q(1-\alpha)] + P_t^i w\}}{t_s t_d t_c - [A_s + P_r^i q(1-\alpha)](A_d - P_r^i q)t_c + P_t^i w n_c^1(A_{sc} + P_t^i w)t_d} \\ n_d^2 = \dfrac{(A_d - P_r^i q)t_c P_s^1 + [t_s t_d - P_t^i w(A_{sc} + P_t^i w)]P_d^2 + [A_s + P_r^i q(1-\alpha)]P_t^i w P_s^2}{t_s t_d t_c - [A_s + P_r^i q(1-\alpha)](A_d - P_r^i q)t_c + P_t^i w n_c^1(A_{sc} + P_t^i w)t_d} \\ \quad + \dfrac{[A_s + P_r^i q(1-\alpha)]t_c\{t_s - [A_s + P_r^i q(1-\alpha)] + P_t^i w\}}{t_s t_d t_c - [A_s + P_r^i q(1-\alpha)](A_d - P_r^i q)t_c + P_t^i w n_c^1(A_{sc} + P_t^i w)t_d} \end{cases} \tag{5}$$

$$\begin{cases} n_c^1 = \dfrac{\left(A_{sc} + P_t^i w\right)t_d P_s^2 + \left[A_s + P_r^i q(1-\alpha)\right]\left(A_{sc} + P_t^i w\right)P_d^1 + \left\{\left[A_s + P_r^i q(1-\alpha)\right]\left(A_d - P_r^i q\right) - t_s t_d\right\}P_c^1}{t_s t_d t_c - \left[A_s + P_r^i q(1-\alpha)\right]\left(A_d - P_r^i q\right)t_c + P_t^i w n_c^1\left(A_{sc} + P_t^i w\right)t_d} \\ \quad + \dfrac{\left(A_{sc} + P_t^i w\right)t_d\left\{t_s - \left[A_s + P_r^i q(1-\alpha)\right] + P_t^i w\right\}}{t_s t_d t_c - \left[A_s + P_r^i q(1-\alpha)\right]\left(A_d - P_r^i q\right)t_c + P_t^i w n_c^1\left(A_{sc} + P_t^i w\right)t_d} \\ n_c^2 = \dfrac{\left(A_{sc} + P_t^i w\right)t_d P_s^1 + \left[A_s + P_r^i q(1-\alpha)\right]\left(A_{sc} + P_t^i w\right)P_d^2 + \left\{\left[A_s + P_r^i q(1-\alpha)\right]\left(A_d - P_r^i q\right) - t_s t_d\right\}P_c^2}{t_s t_d t_c - \left[A_s + P_r^i q(1-\alpha)\right]\left(A_d - P_r^i q\right)t_c + P_t^i w n_c^1\left(A_{sc} + P_t^i w\right)t_d} \\ \quad + \dfrac{\left(A_{sc} + P_t^i w\right)t_d\left\{t_s - \left[A_s + P_r^i q(1-\alpha)\right] + P_t^i w\right\}}{t_s t_d t_c - \left[A_s + P_r^i q(1-\alpha)\right]\left(A_d - P_r^i q\right)t_c + P_t^i w n_c^1\left(A_{sc} + P_t^i w\right)t_d} \end{cases} \tag{6}$$

Taking $n_s^1 + n_s^2 = 1 + N_s^{1,2}$, $n_d^1 + n_d^2 = 1 + N_d^{1,2}$, $n_d^1 + n_d^2 = 1 + N_d^{1,2}$ into account, the number of multi-attribution participants can be calculated according to the number formula of the three access platforms 1 and 2, as shown in Formula (7):

$$
\begin{cases}
N_s^{1,2} = \dfrac{\left[A_s + P_r^i q(1-\alpha)\right]t_c\left(P_d^1 + P_d^2\right) - P_t^i w t_d\left(P_c^1 + P_c^2\right) - t_d t_c\left(P_s^1 + P_s^2\right)}{t_s t_d t_c - \left[A_s + P_r^i q(1-\alpha)\right]\left(A_d - P_r^i q\right)t_c + P_t^i w n_c^1\left(A_{sc} + P_t^i w\right)t_d} \\
\quad + \dfrac{2\left[A_s + P_r^i q(1-\alpha)\right]t_c\left[t_d - \left(A_d - P_r^i q\right)\right] - P_t^i w t_d\left[t_c - \left(A_d - P_r^i q\right)\right]}{t_s t_d t_c - \left[A_s + P_r^i q(1-\alpha)\right]\left(A_d - P_r^i q\right)t_c + P_t^i w n_c^1\left(A_{sc} + P_t^i w\right)t_d} - 1 \\
N_d^{1,2} = \dfrac{\left(A_d - P_r^i q\right)t_c\left(P_s^1 + P_s^2\right) + \left[t_s t_d - P_t^i w\left(A_{sc} + P_t^i w\right)\right]\left(P_d^1 + P_d^2\right) + \left[A_s + P_r^i q(1-\alpha)\right]P_t^i w\left(P_c^1 + P_c^2\right)}{t_s t_d t_c - \left[A_s + P_r^i q(1-\alpha)\right]\left(A_d - P_r^i q\right)t_c + P_t^i w n_c^1\left(A_{sc} + P_t^i w\right)t_d} \\
\quad + \dfrac{2\left[A_s + P_r^i q(1-\alpha)\right]t_c\left\{t_s - \left[A_s + P_r^i q(1-\alpha)\right] + P_t^i w\right\}}{t_s t_d t_c - \left[A_s + P_r^i q(1-\alpha)\right]\left(A_d - P_r^i q\right)t_c + P_t^i w n_c^1\left(A_{sc} + P_t^i w\right)t_d} - 1 \\
N_c^{1,2} = \dfrac{\left(A_{sc} + P_t^i w\right)\left\{t_d\left(P_s^1 + P_s^2\right) + \left[A_s + P_r^i q(1-\alpha)\right]\left(P_d^1 + P_d^2\right)\right\}}{t_s t_d t_c - \left[A_s + P_r^i q(1-\alpha)\right]\left(A_d - P_r^i q\right)t_c + P_t^i w n_c^1\left(A_{sc} + P_t^i w\right)t_d} \\
\quad + \dfrac{\left\{\left[A_s + P_r^i q(1-\alpha)\right]\left(A_d - P_r^i q\right) - t_s t_d\right\}\left(P_c^1 + P_c^2\right)}{t_s t_d t_c - \left[A_s + P_r^i q(1-\alpha)\right]\left(A_d - P_r^i q\right)t_c + P_t^i w n_c^1\left(A_{sc} + P_t^i w\right)t_d} \\
\quad + \dfrac{2\left(A_{sc} + P_t^i w\right)t_d\left\{t_s - \left[A_s + P_r^i q(1-\alpha)\right] + P_t^i w\right\}}{t_s t_d t_c - \left[A_s + P_r^i q(1-\alpha)\right]\left(A_d - P_r^i q\right)t_c + P_t^i w n_c^1\left(A_{sc} + P_t^i w\right)t_{dd}} - 1
\end{cases}
\tag{7}
$$

According to the assumptions and relevant definitions, the profit functions of pallet Shared platform 1 and 2 in the competitive environment are shown in Formula (8):

$$
\begin{cases}
\pi_P^1 = (P_s^1 - c^1 - \delta_s^1)n_s^1 + \alpha q P_r^i n_s^1 n_d^1 + (P_d^i - c^1 - \delta_d^1)n_d^1 + (P_c^i - c^1)n_c^1 + S_P \\
\pi_P^2 = (P_s^2 - c^2 - \delta_s^2)n_s^2 + \alpha q P_r^i n_s^2 n_d^2 + (P_d^i - c^2 - \delta_s^2)n_d^2 + (P_c^i - c^2)n_c^2 + S_P
\end{cases}
\tag{8}
$$

Considering the profit function of platform 1 and 2, the revenue source of the platform mainly includes user registration fee and government subsidy. The cost of expenditure includes the profit function of the pallets sharing platform given to drivers. Function $\pi_P^1$, $\pi_P^2$ is a multivariate function about $\{p_s^1, p_s^2, p_d^1, p_d^2, p_c^1, p_c^2\}$.When you solve for $\pi_{P_1}(p_s^1, p_d^1, p_c^1)$, think of $(p_s^2, p_h^2, p_c^2)$ as a constant, and vice versa. For the convenience of calculation, in the equilibrium solution, take the first derivative of $\pi_P^1$ with respect to $p_s^1, p_d^1$ and $p_c^1$, and set $\frac{\partial \pi_P^1}{\partial p_s^1}, \frac{\partial \pi_P^1}{\partial p_d^1}, \frac{\partial \pi_P^1}{\partial p_c^1}$ equal to 0 respectively, and the registration fee of pallet Shared platform 1 and platform 2 for the three participants in the equilibrium state can be solved, as shown in Formula (9):

$$
\begin{cases}
(P_s)^* = \left(P_s^1\right)^* = \left(P_s^2\right)^* = t_s - \dfrac{t_c\left\{\alpha P_r^i q\left(A_d - P_r^i q\right) + 2t_s t_d\right\}\left\{\left[A_s + P_r^i q(1-\alpha)\right] + P_r^i w - 2t_s\right\}}{P_t^i w\left(A_{sc} + P_t^i w\right)t_d - 4t_s t_d t_c - \left\{\alpha P_r^i q - \left[A_s + P_r^i q(1-\alpha)\right]\right\}\left(A_d - P_r^i q\right)t_c} \\
\qquad + c + \delta \\
(P_d)^* = \left(P_d^1\right)^* = \left(P_d^2\right)^* = \dfrac{\alpha P_r^i q t_d t_c\left\{\left[A_s + P_r^i q(1-\alpha)\right] + P_t^i w - 2t_s\right\}}{P_t^i w\left(A_{sc} + P_t^i w\right)t_d - 4t_s t_d t_c - \left\{\alpha P_r^i q - \left[A_s + P_r^i q(1-\alpha)\right]\right\}\left(A_d - P_r^i q\right)t_c} \\
\qquad + c + \delta \\
(P_c)^* = \left(P_c^1\right)^* = \left(P_c^2\right)^* = \dfrac{\left(A_d - P_r^i q\right)t_d t_c\left\{\left[A_s + P_r^i q(1-\alpha)\right] + P_t^i w - 2t_s\right\}}{P_t^i w\left(A_{sc} + P_t^i w\right)t_d - 4t_s t_d t_c - \left\{\alpha P_r^i q - \left[A_s + P_r^i q(1-\alpha)\right]\right\}\left(A_d - P_r^i q\right)t_c} + c
\end{cases}
\tag{9}
$$

According to the registration fees of the platform in the equilibrium state for the three participants, the scales of the three participants joining platform 1 or 2 in the equilibrium state are $n_s^*$, $n_d^*$ and $n_c^*$, the scales of single ownership are $N_s^*$, $N_d^*$ and $N_c^*$, and the scales of multiple ownership are $N_s^{1,2*}$, $N_d^{1,2*}$ and $N_c^{1,2*}$, as shown in formula (10)–(12):

$$\begin{cases} (N_s)^* = \left(N_s^2\right)^* = \left(N_s^1\right)^* = \dfrac{2t_d t_c\left\{\left[A_s + P_r^i q(1-\alpha)\right] + P_t^i w - 2t_s\right\}}{P_t^i w\left(A_{sc} + P_t^i w\right)t_d - 4t_s t_d t_c - \left\{\alpha P_r^i q - \left[A_s + P_r^i q(1-\alpha)\right]\right\}\left(A_d - P_r^i q\right)t_c} \\ \left(N_s^{1,2}\right)^* = 1 - \dfrac{4t_d t_c\left\{\left[A_s + P_r^i q(1-\alpha)\right] + P_t^i w - 2t_s\right\}}{P_t^i w\left(A_{sc} + P_t^i w\right)t_d - 4t_s t_d t_c - \left\{\alpha P_r^i q - \left[A_s + P_r^i q(1-\alpha)\right]\right\}\left(A_d - P_r^i q\right)t_c} \\ (n_s)^* = \left(n_s^2\right)^* = \left(n_s^1\right)^* = 1 - \dfrac{2t_d t_c\left\{\left[A_s + P_r^i q(1-\alpha)\right] + P_t^i w - 2t_s\right\}}{P_t^i w\left(A_{sc} + P_t^i w\right)t_d - 4t_s t_d t_c - \left\{\alpha P_r^i q - \left[A_s + P_r^i q(1-\alpha)\right]\right\}\left(A_d - P_r^i q\right)t_c} \end{cases} \tag{10}$$

$$\begin{cases} (N_d)^* = \left(N_d^2\right)^* = \left(N_d^1\right)^* = 1 - \dfrac{\left(A_d - P_r^i q\right)t_c\left\{\left[A_s + P_r^i q(1-\alpha)\right] + P_t^i w - 2t_s\right\}}{P_t^i w\left(A_{sc} + P_t^i w\right)t_d - 4t_s t_d t_c - \left\{\alpha P_r^i q - \left[A_s + P_r^i q(1-\alpha)\right]\right\}\left(A_d - P_r^i q\right)t_c} \\ \left(N_d^{1,2}\right)^* = \dfrac{2\left(A_d - P_r^i q\right)t_c\left\{\left[A_s + P_r^i q(1-\alpha)\right] + P_t^i w - 2t_s\right\}}{P_t^i w\left(A_{sc} + P_t^i w\right)t_d - 4t_s t_d t_c - \left\{\alpha P_r^i q - \left[A_s + P_r^i q(1-\alpha)\right]\right\}\left(A_d - P_r^i q\right)t_c} - 1 \\ (n_d)^* = \left(n_d^2\right)^* = \left(n_d^1\right)^* = \dfrac{\left(A_d - P_r^i q\right)t_c\left\{\left[A_s + P_r^i q(1-\alpha)\right] + P_t^i w - 2t_s\right\}}{P_t^i w\left(A_{sc} + P_t^i w\right)t_d - 4t_s t_d t_c - \left\{\alpha P_r^i q - \left[A_s + P_r^i q(1-\alpha)\right]\right\}\left(A_d - P_r^i q\right)t_c} \end{cases} \tag{11}$$

$$\begin{cases} (N_c)^* = \left(N_c^2\right)^* = \left(N_c^1\right)^* = 1 - \dfrac{\left(A_{sc} + P_r^i w\right)t_d\left\{\left[A_s + P_r^i q(1-\alpha)\right] + P_t^i w - 2t_s\right\}}{P_t^i w\left(A_{sc} + P_t^i w\right)t_d - 4t_s t_d t_c - \left\{\alpha P_r^i q - \left[A_s + P_r^i q(1-\alpha)\right]\right\}\left(A_d - P_r^i q\right)t_c} \\ \left(N_c^{1,2}\right)^* = \dfrac{2\left(A_{sc} + P_t^i w\right)t_d\left\{\left[A_s + P_r^i q(1-\alpha)\right] + P_t^i w - 2t_s\right\}}{P_t^i w\left(A_{sc} + P_t^i w\right)t_d - 4t_s t_d t_c - \left\{\alpha P_r^i q - \left[A_s + P_r^i q(1-\alpha)\right]\right\}\left(A_d - P_r^i q\right)t_c} - 1 \\ (n_c)^* = \left(n_c^2\right)^* = \left(n_c^1\right)^* = \dfrac{\left(A_{sc} + P_t^i w\right)t_d\left\{\left[A_s + P_r^i q(1-\alpha)\right] + P_t^i w - 2t_s\right\}}{P_t^i w\left(A_{sc} + P_t^i w\right)t_d - 4t_s t_d t_c - \left\{\alpha P_r^i q - \left[A_s + P_r^i q(1-\alpha)\right]\right\}\left(A_d - P_r^i q\right)t_c} \end{cases} \tag{12}$$

Rent is the result of the game between the suppliers and the transporters in each platform, and freight is the result of the game between the suppliers and the transporters in each platform. In the following analysis, the lease fee and freight involved in the formula are regarded as constant analysis and calculation.

## 4.2 Propositional Analysis

This section mainly analyzes the registration fee and ownership of each participant. Under the equilibrium state, the two platforms have the same strategy, so the analysis and study of platform 1 are carried out. According to the registration fee and attribution of each participant under the equilibrium state of 4.1, the relationship between the registration fee and each influencing factor can be obtained, as shown in Formula (13):

$$Y\{p, n\} = F\{A, \delta, t, c, \alpha, w, q,\} \tag{13}$$

Considering the influence of $A_s$, $A_d$ and $A_{sc}$ on the registration fee of the supplier, the demander and the transport party, in order to simplify the calculation, the appropriate values of $t_s$, $t_d$, $t_c$, $q$, and w can be set by $\partial p_s^* / \partial A_s < 0$, $\partial p_s^* / \partial A_d < 0$, $\partial p_s^* / \partial A_{sc} < 0$;$\partial p_d^* / \partial A_s > 0$, $\partial p_d^* / \partial A_d < 0$, $\partial p_d^* / \partial A_{sc} < 0$;$\partial p_c^* / \partial A_s > 0$, $\partial p_c^* / \partial A_d > 0$, $\partial p_c^* / \partial A_{sc} > 0$, proposition 1 can be obtained.

**Proposition 1:** With the increase of the network externality of the supplier, the registration fee of the demander will be reduced, and the registration fee of the transporter will also increase with the increase of the network externality. The subsidy amount of the platform to one party will also decrease with the network externality. Moreover, the registration fee of one participant and the network externality of the other party exist Negative correlation and show the asymmetry of pricing strategy.

Considering the effect of conversion cost $t_s$, $t_d$, $t_c$ on the registration fees of the supplier, the demander and the transporter, in order to simplify the calculation and make $A_s$, $A_d$, $A_{sc}$, $q_s$, $w$ an appropriate value, proposition 2 can be obtained from $\partial p_s^* / \partial t_s < 0$, $\partial p_d^* / \partial t_d < 0$, $\partial p_c^* / \partial t_c < 0$.

**Proposition 2:** The pallet sharing platform the service ability of the mass, the greater the ability to attract users are, the greater the viscosity for the customer, the greater the unit conversion cost is higher, the supply, demand and transportation party platform of the registration fee is lower, when participants conversion cost is big enough, as a result of the participants need to consider their own benefits will choose a fixed platform, then the market will not exist many belonging of participants. At the same time, the pallet-sharing platform also needs to cooperate with the participants to maintain the operation of the platform.

Considering the influence of $c$ and $\delta$ on the registration fees of the supplier, the demander and the transporter, proposition 3 can be obtained from $\partial p_s^* / \partial c > 0$, $\partial p_c^* / \partial c > 0$, $\partial p_c^* / \partial c > 0$, $\partial p_s^* / \partial \delta > 0$, $\partial p_c^* / \partial \delta > 0$.

**Proposition 3:** The lower the service cost of the pallet-shared platform, the lower the registration fee charged by the pallet-shared platform to the participants. The bigger the platform, the greater the market influence, the better the software and hardware technology, and the lower the service cost. Only a certain registration fee is needed to maintain the normal profit level. The registration fee charged by the platform is closely related to the subsidy given by the platform to the participants. The subsidy paid by the platform will increase the registration fee to the supplier and the demander, so as to maintain the profit of the platform. Therefore, the subsidy from the supplier and the demander will not directly affect the profit of the pallet sharing platform and the utility of the supplier and the demander.

Considering the influence of $w$ and $q$ on the registration fees of the supplier, the demand side and the transport side, in order to simplify the calculation and make $A_s$, $A_d$, $A_{sc}$, $\alpha$ an appropriate value, proposition 4 can be obtained from $\partial p_s^* / \partial w < 0$, $\partial p_s^* / \partial q < 0$, $\partial p_d^1 / \partial q < 0$, $\partial p_c^1 / \partial w < 0$.

**Proposition 4:** The registration fee charged by the platform decreases with the increase of the transaction probability between the supplier and the transporter, and between the suppliers. With the continuous expansion of the platform scale, the number of participants joining the platform will also increase. With the improvement of the platform service capacity, the transaction probability between the participants will be increased, and the registration fee of the participants will be reduced.

Considering the influence of $\alpha$ on the registration fees of the supplier, the demand side and the transport side, in order to simplify the calculation and make $A_s$, $A_d$, $A_{sc}$, $q$, $w$

an appropriate value, proposition 5 can be obtained from $\partial p_s^* / \partial \alpha < 0$, $\partial p_d^* / \partial \alpha < 0$, $\partial p_c^* / \partial \alpha < 0$.

**Proposition 5:** When the platform is in the stable stage, the scale of the participants is relatively stable, and the transaction fee is negatively correlated with the registration fee. The registration fee charged by the platform for the participants is low. When it is in the stable stage, the platform is in a dominant position. In order to maintain its own profits, the platform will take more transactions fee.

Considering the influence of $p_r$, $p_t$ on the registration fees of the supplier, the demander and the transporter, proposition 6 can be obtained from $\partial p_s^* / \partial p_r < 0$, $\partial p_s^* / \partial p_t < 0$, $\partial p_d^* / \partial p_r < 0$, $\partial p_c^* / \partial p_t < 0$.

**Proposition 6:** For the demander, the increase of the rental fee will make the benefit lower. As the sole expense of the platform, the registration fee will be reduced to maintain the stickiness of the demander to the platform. For the supplier, the increase of freight will reduce its revenue, so the platform will reduce its registration fee; Transporters are beneficiaries of the whole platform. The increase of freight will, on the one hand, increase the benefits in transportation and increase their stickiness to the platform; on the other hand, it will attract transporters who are not connected to the platform. The increase in the number of transporters will promote the expansion of the platform and thus reduce the registration fee of transporters.

From $\partial N_s^* / \partial A_s < 0$, $\partial N_d^* / \partial A_d < 0$, $\partial N_c^* / \partial A_c < 0$, $\partial N_s^{1,2} \Big/ \partial A_s > 0$, $\partial N_d^{1,2} \Big/ \partial A_d > 0$, $\partial N_c^{1,2} \Big/ \partial A_c > 0$ you get proposition 7.

**Proposition 7:** The number of participants with single belonging will reduce with the increasing of network externalities, with many participants belonging will increase with the increase of network externalities, which means the participants network externalities, the greater the participants with more belonging the more, when participants network externality is big enough, the platform for participants with low viscosity, pallet sharing market will transform into many market participants belonging.

From $\partial N_s^* / \partial t_s > 0$, $\partial N_d^* / \partial t_d > 0$, $\partial N_c^* / \partial t_c > 0$, $\partial N_s^{1,2} \Big/ \partial t_s < 0$, $\partial N_d^{1,2} \Big/ \partial t_d < 0$, $\partial N_c^{1,2} \Big/ \partial t_c < 0$ you get proposition 8.

**Proposition 8:** If the conversion cost of the participant is large enough, the participant will only join a pallet sharing platform to maximize the revenue. The unit conversion cost of participants is positively correlated with the number of multi-ownership participants. When the conversion cost is very low, the pallet sharing market will become a multi-ownership market.

From $\partial N_s^* / \partial w > 0$, $\partial N_s^* / \partial q > 0$, $\partial N_d^* / \partial w > 0$, $\partial N_d^* / \partial q > 0$, $\partial N_c^* / \partial w > 0$, $\partial N_c^* / \partial q > 0$ you get proposition 9.

**Proposition 9:** When the transaction proportion is relatively low, the platform is in a disadvantaged position, the participants will choose multi-ownership, and the pallet sharing market will become a multi-ownership market. When the service level of the platform is relatively high, the transaction probability is relatively high, and the participants are more inclined to belong to a certain platform alone.

## 5 Conclusion

The pallet sharing market under the participation of pallet sharing platform has the characteristics of typical multilateral market. This paper draws the following conclusions by analyzing the proposition analysis under the competitive environment:

1) Only if the parties demand complementarity, the participants of network externalities, participants expected return and participants' unit conversion cost, platform and platform scale of service quality to satisfy certain constraint conditions, the pallet on the market share of all the participants to reach equilibrium;
2) According to the principle of complementarity of participants' demands, the pallets sharing platform can retain users by setting high conversion cost, and attract more "shortage" users by setting appropriate subsidy strategy, so as to increase the profits of pallets sharing platform and make pallets sharing platform develop more steadily.
3) According to the market on the pallet rental fees and freight and platform of the actual situation of the number of participants to clinch a deal the rental fee and the freight is one of the main responsibility is to control the platform;
4) Pallet the size of the Shared platform and service ability is closely related with
5) the development of the platform, the stronger the platform service ability, the scale, the greater the user in the platform dependent increase, it will also attract more potential users;
6) The government subsidy has a positive incentive effect on the pallet-sharing platform, and does not affect the registration fee, transaction fee, rental fee and transportation fee of the pallets.

**Acknowledgment.** This work is supported by the National Key R&D Program of China through Grant No. 2018YFC1407405.

## References

1. Shunli, Z., Qingshuang, Y.: Research of network teaching platform based grid. Int. J. Educ. Manage. Eng. **2**(6), 41–46 (2012)
2. Yang, M., Mahmood, M., Zhou, X., et al.: Design and implementation of cloud platform for intelligent logistics in the trend of intellectualization. China Commun. **14**(10), 180–191 (2017). (in Chinese)
3. Daning, X., Qilan, Z., Zhigang, S.: A study of service model innovation for logistics information platform based on cloud ecosystem. J. Bus. Econ. **8**, 5–15 (2016). (in Chinese)
4. Nathan, G., Horvath, P.L.A., Brad, G.: Investigation of new and recovered wood shipping platforms in the United States. BioResources **15**(2), 2818–2838 (2020)

5. Tong, Z., Yixue, H., Xin, Z., et al.: Endogenous third-degree price discrimination in hotelling model with elastic demand. J. Econ. **127**(2), 125–145 (2019)
6. Li, Z., Li, M.: Research on the operation model of regional logistics public information platform based on the PPP model. In: 15th COTA International Conference of Transportation Professionals, pp. 745–755 (2015)
7. Ling, Z., Jie, L.: A pricing strategy of e-commerce advertising cooperation in the stackelberg game model with different market power structure. Algorithms **12**(1), 24–26 (2019)
8. Peters, H., Schröder, M., Vermeulen, D.: Hotelling's location model with negative network externalities. Internat. J. Game Theory **47**(3), 811–837 (2018)
9. Shaojian, Q., Dehua, M., Yongyi, Z., et al.: Distributionally robust games with an application to supply chain. J. Intell. Fuzzy Syst. **33**(5), 2749–2762 (2017)
10. Saberi, Z., Saberi, M., Hussain, O., Chang, E.: Stackelberg model based game theory approach for assortment and selling price planning for small scale online retailers. Futur. Gener. Comput. Syst. **100**, 1088–1102 (2019)
11. Cheowsuwan, T., Arthan, S., Tongphet, S.: System design of supply chain management and Thai food export to global market via electronic marketing. Int. J. Mod. Educ. Comput. Sci. **9**(8), 1–8 (2017)
12. Cheng, Q., Tangshu, Y.: Manufacturer's pricing strategy for supply chain with service level-dependent demand. Int. J. Eng. Manuf. **4**(4), 1–13 (2014)

# Oblique Projection Operators in Matrix Representations of Genetic Coding Systems

Sergey V. Petoukhov[1,2](✉)

[1] Mechanical Engineering Research Institute, Russian Academy of Sciences, M. Kharitonievsky pereulok, 4, Moscow, Russia
[2] Moscow State Tchaikovsky Conservatory, Bolshaya Nikitskaya, 13/6, Moscow, Russia

**Abstracts.** The article is devoted to connections of matrix representations of the DNA alphabets of 16 doublets and 64 triplets with sparse matrices of oblique projection operators. The article shows that sums of pairs of such genetic projection operators have useful features for modeling genetically inherited biological phenomena including, in particular, cyclic processes and parallelization of control processes related to different parameters of holistic multi-parameter biological systems. It is noted that many scientific fields use mainly orthogonal projectors but the described approach to model the genetic system uses non-orthogonal projectors in multi-dimensional configuration vector spaces. Different features of this model approach related to logistics of interrelations of different parts of inherited biological phenomena are discussed.

**Keywords:** DNA alphabets · Matrix representation · Genetic matrices · Oblique projectors · Algebras of hypercomplex numbers · Cyclic processes

## 1 Introduction

Our visual perception and a "projection method" in the drawing are based on projections of external objects on retina and a drawing plane (Fig. 1). In mathematics, such operations of projections are expressed by means of square matrices, which are called "projection operators" (or "projectors").

**Fig. 1.** Examples of projection. Left: the path of light rays entering the eyes. Middle: Ray projection in the old manner of painting. Right: oblique and orthogonal projection.

The matrix P (Fig. 2) is an example of a projector, which makes a projection of vector [x, y, z] on the plane [x, y, 0]:


Z. Hu et al. (Eds.): ICAILE 2021, LNDECT 82, pp. 288–297, 2021.
https://doi.org/10.1007/978-3-030-80475-6_28

$$P = \begin{bmatrix} 1 & 0 & 0 \\ 0 & 1 & 0 \\ 0 & 0 & 0 \end{bmatrix}. \qquad P\begin{pmatrix} x \\ y \\ z \end{pmatrix} = \begin{pmatrix} x \\ y \\ 0 \end{pmatrix}.$$

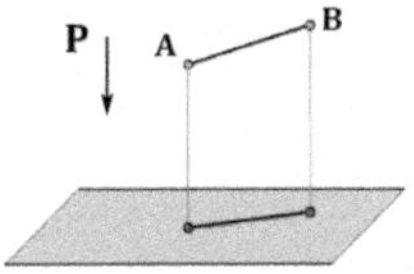


**Fig. 2.** Example of a matrix operator P, which provides a projection operation.

Projectors are expressed by means of square matrices (http://mathworld.wolfram.com/ProjectionMatrix.html). A necessary and sufficient condition that a matrix P is a projection operator is the fulfillment of the following condition: $P^2 = P$. A set of projectors is separated into two sub-sets:

- Orthogonal projectors, which are expressed by symmetric matrices. Their theory is well developed and has a lot of applications;
- Oblique projectors, which are expressed by non-symmetric matrices. Their theory and its applications are developed much weaker. Namely oblique pro-jectors will be the main objects of attention in this article.

Science has led to a new understanding of life itself: "*Life is a partnership between genes and mathematics*" [1]. But what kind of mathematics can be a partner for the genetic coding system? This article shows some pieces of evidence that algebra of projectors can be one of main parts of such mathematics. The notion of projection operators (or briefly, projectors) is one of the important notions in many non-biological sciences: physics, including quantum mechanics; mathematics; computer science and informatics, including theory of digital codes; chemistry; mathematical logic, etc. This article notes that projectors can be also used in studying and modeling properties of the genetic code system and many inherited biological ensembles including ensembles of cyclic processes, phyllotaxis patterns, etc.

In accordance with Mendel's law of independent inheritance of traits, information from the micro-world of genetic molecules dictates constructions in the macro-world of living organisms under strong noise and interference. This dictation is realized by means of unknown algorithms of multi-channel noise-immunity coding. For example, in human organism, his skin color, eye color and hair color are inherited genetically independently of each other. It is possible if appropriate kinds of information are conducted via independent informational channels and if a general "configuration space" (or "phase space") of living organism contains sub-spaces with a possibility of a selective control or a selective coding of processes in them. So, any living organism is an algorithmic machine of multi-channel noise-immunity coding with ability to a selective control and coding of different sub-spaces of its phase space (a model approach to configuration spaces with a selective control of their sub-spaces is presented in this article). This machine works in conditions of ontogenetic development of the organism when a multi-dimensionality of its configuration space is increased step by step.

To understand such genetic machine, it is appropriate to use the theory of noise-immunity coding and transmission of digital information, taking into account the discrete nature of the genetic code. In this theory, mathematical matrices have the basic importance. The use of matrix representations and analysis in the study of phenomenological features of molecular-genetic ensembles has led to the development of a special scientific direction under a name "Matrix Genetics" [2, 3]. Namely researches of the "matrix genetics" gave results that are represented in this article.

Concerning the theme of projectors in inherited biological phenomena, one can note that our genetically inherited visual system works on the principle of projection of external objects at the retina. This projection is modeled using projection operators. The author believes that the value of projectors for bioinformatics is not limited to this single fact of biological significance of projection operators, but that the whole system of genetic and sensory informatics is based on their active use in high degree. This ubiquitous use of projection operators reflects and ensures (in some degree) the unity of any organism and interrelations of its parts.

The aim of this article is to demonstrate that the set of oblique projection operators, which are associated with the matrix representation of the genetic code, has interesting algebraic properties to new opportunities for modeling genetically inherited biological phenomena. The article contains a few sections, conclusions, acknowledgments, and references.

## 2 Structured DNA Alphabets and Their Matrix Representations

As it is known, DNA-texts use different kinds of alphabets of $n$-plets: the alphabet of 4 nucleotides (adenine A, thymine T, cytosine C, and guanine G); the alphabet of 16 doublets (such as AA, AT, AC, ...); the alphabet of 64 triplets (such as AAA, AAC, ATT, ...). In the author's previous works [2, 3], it was shown that all such interconnected alphabets can be represented using a single tensor family of matrices, for example, such as $[C, T; A, G]^{(n)}$ where $(n)$ denotes a tensor power (Fig. 3).

$R_4$ =

| | | | |
|---|---|---|---|
| CC | CT | TC | TT |
| CA | CG | TA | TG |
| AC | AT | GC | GT |
| AA | AG | GA | GG |

$R_8$ =

| | | | | | | | |
|---|---|---|---|---|---|---|---|
| CCC | CCT | CTC | CTT | TCC | TCT | TTC | TTT |
| CCA | CCG | CTA | CTG | TCA | CG | TTA | TTG |
| CAC | CAT | CGC | CGT | TAC | TAT | TGC | TGT |
| CAA | CAG | CGA | CGG | TAA | TAG | TGA | TGG |
| ACC | ACT | ATC | ATT | GCC | GGT | GTC | GTT |
| ACA | ACG | ATA | ATG | GCA | GCG | GTA | GTG |
| AAC | AAT | AGC | AGT | GAC | GAT | GGC | GGT |
| AAA | AAG | AGA | AGG | GAA | GAG | GGA | GGG |

**Fig. 3.** Two representatives of the tensor family of DNA-alphabetic matrices $[C\ T; A\ G]^{(n)}$. Black color marks 8 strong duplets in the matrix $[C\ T; A\ G]^{(2)}$ (at left) and 32 triplets with strong roots in the matrix $[C\ T; A\ G]^{(3)}$ (at right) as it is explained in the text.

A combination of letters on the two first positions of each triplet is ususally termed as a "root" of this triplet [4–6]. Modern science recognizes many variants (or dialects) of the genetic code, data about which are shown on the NCBI's website http://www.

ncbi.nlm.nih.gov/Taxonomy/Utils/wprintgc.cgi. Most of these dialects (including the so called Standard Code and the Vertebrate Mitochondrial Code) have the symmetric general scheme of these correspondences, where 32 "black" triplets with "strong roots" and 32 "white" triplets with "weak" roots exist. The code value of 32 triplets with strong roots (CC, CT, CG, AC, TC, GC, GT, GG) is entirely determined by their root and does not depend on the letter in their third position (unlike 32 triplets with weak roots, in which the letter in the third position is essential for determining the type of the encoded amino acid). In Fig. 3, cells with 32 triplets with strong roots and with appropriate 8 "strong" doublets are marked by black color.

From the point of view of its black-and-white mosaic, each of columns of genetic matrices [C T; A G]$^{(2)}$ and [C T; A G]$^{(3)}$ has a meander-like character and coincides with one of Rademacher functions [https://mathworld.wolfram.com/RademacherFunction.html] that form orthogonal systems and well known in discrete signals processing. These functions contain only elements "+1" and "−1". Due to this fact, one can construct Rademacher representations of the symbolic genomatrices [C T; A G]$^{(2)}$ and [C T; A G]$^{(3)}$ by means of the following operation: each of black duplets and black triplets is replaced by number " +1" and each of white duplets and white triplets is replaced by number "−1". This operation leads immediately to the matrices $R_4$ and $R_8$ (Fig. 4), which are conditionally termed as Rademacher matrices and which are the Rademacher representations of the genetic symbolic matrices [C T; A G]$^{(2)}$ and [C T; A G]$^{(3)}$ from Fig. 3.

$$R_4 = \begin{bmatrix} 1 & 1 & 1 & -1 \\ -1 & 1 & -1 & -1 \\ 1 & -1 & 1 & 1 \\ -1 & -1 & -1 & 1 \end{bmatrix}; \quad R_8 = \begin{bmatrix} 1 & 1 & 1 & 1 & 1 & 1 & -1 & -1 \\ 1 & 1 & 1 & 1 & 1 & 1 & -1 & -1 \\ -1 & -1 & 1 & 1 & -1 & -1 & -1 & -1 \\ -1 & -1 & 1 & 1 & -1 & -1 & -1 & -1 \\ 1 & 1 & -1 & -1 & 1 & 1 & 1 & 1 \\ 1 & 1 & -1 & -1 & 1 & 1 & 1 & 1 \\ -1 & -1 & -1 & -1 & -1 & -1 & 1 & 1 \\ -1 & -1 & -1 & -1 & -1 & -1 & 1 & 1 \end{bmatrix}$$

**Fig. 4.** Rademacher matrices $R_4$ and $R_8$, which are Rademacher representations of the genetic mosaic matrices [C T; A G]$^{(2)}$ and [C T; A G]$^{(3)}$ from Fig. 3.

Each of these matrices can be decomposed into sum of sparse matrices, each of which contains only one non-zero column. Such decomposition can be conditionally termed a «column decomposition». The next section is devoted to the use of these column decompositions of genetic matrices for algebraic modeling of the features of genetically inherited biological structures.

## 3 Genetic Rademacher Matrices as Sums of Oblique Projectors

Let us begin with the column decomposition $R_4 = c_0 + c_1 + c_2 + c_3$ of the Rademacher (4*4)-matrix $R_4$ (Fig. 5).

$$R_4 = c_0+c_1+c_2+c_3 = \begin{vmatrix} 1&0&0&0 \\ -1&0&0&0 \\ 1&0&0&0 \\ -1&0&0&0 \end{vmatrix} + \begin{vmatrix} 0&1&0&0 \\ 0&1&0&0 \\ 0&-1&0&0 \\ 0&-1&0&0 \end{vmatrix} + \begin{vmatrix} 0&0&1&0 \\ 0&0&-1&0 \\ 0&0&1&0 \\ 0&0&-1&0 \end{vmatrix} + \begin{vmatrix} 0&0&0&-1 \\ 0&0&0&-1 \\ 0&0&0&1 \\ 0&0&0&1 \end{vmatrix}$$

**Fig. 5.** The column decomposition of the Rademacher matrix $R_4$ from Fig. 3.

Each of these sparse matrices $c_0, c_1, c_2, c_3$ is a projection operator because it satisfies the criterion of projectors $P^2 = P$ (for example, $c_0{}^2 = c_0$, etc.). Every of these projectors is an oblique (non-ortogonal) projector because it is expressed by means of a non-symmetrical matrix. The set of sparse matrices $c_0, c_1, c_2, c_3$ consists of non-commutative projectors.

Let us examine all possible variants of sums of pairs of the different oblique projectors $c_0$, $c_1$, $c_2$, and $c_3$: $(c_0 + c_1)$, $(c_0 + c_2)$, $(c_0 + c_3)$, $(c_1 + c_2)$, $(c_1 + c_3)$, $(c_2 + c_3)$. The result of this examination is the following: matrices $(c_0 + c_1)$ and $(c_2 + c_3)$ with a weight coefficient $2^{-0.5}$ lead to cyclic groups with their period 8 in cases of their exponentiation: $(2^{-0.5}*(c_0 + c_1))^n = (2^{-0.5}*(c_0 + c_1))^{n+8}$, $(2^{-0.5}*(c_2 + c_3))^n = (2^{-0.5}*(c_2 + c_3))^{n+8}$, where $n = 1, 2, 3, \ldots$. These two sums of column projectors $(c_0 + c_1)$ and $(c_2 + c_3)$ are marked by green colour in Fig. 6, where each of cells represents a sum of those projectors, which correspond to symbols of their column and row.

| | $c_0$ | $c_1$ | $c_2$ | $c_3$ |
|---|---|---|---|---|
| $c_0$ | - | $c_0+c_1$ | $c_0+c_2$ | $c_0+c_3$ |
| $c_1$ | $c_1+c_0$ | - | $c_1+c_2$ | $c_1+c_3$ |
| $c_2$ | $c_2+c_0$ | $c_2+c_1$ | - | $c_2+c_3$ |
| $c_3$ | $c_3+c_0$ | $c_3+c_1$ | $c_3+c_2$ | - |

**Fig. 6.** Tables of some features of sums of pairs of different projectors $c_0, c_1, c_2, c_3$ (from Fig. 5) in relation to their exponentiation. Explanations in the text.

Two other examined sums of column projectors $(c_0 + c_2)$ and $(c_1 + c_3)$ is doubled when squaring: $(c_0 + c_2)^n = 2^{n-1}*(c_0 + c_2)$, $(c_1 + c_3)^n = 2^{n-1}*(c_1 + c_3)$, where $n = 1, 2, 3, \ldots$ (the cells with these sums are marked by red colours in Fig. 6). This doubling reminds dichotomic dividing of biological cells in a result of mitosis when doubling of genetic information occurs.

The last examined sums of the column projectors $(c_0 + c_3)$ and $(c_1 + c_2)$ possess the following feature. Matrices of their second power is quadrupled in a result of exponentiation in integer powers: $((c_0 + c_3)^2)^n = 4^{n-1}*(c_0 + c_3)^2$, $((c_1 + c_2)^2)^n = 4^{n-1}*(c_1 + c_2)^2$, where $n = 1, 2, 3\ldots$ (this feature can be used to simulate a genetic phenomenon of tetra-reproduction of gametes and genetic information in a course of meiosis). The cells with these sums $(c_0 + c_3)$ and $(c_1 + c_2)$ are marked by yellow colour in Fig. 6.

Matrices of sums of pairs of projectors $c_0$, $c_1$, $c_2$, $c_3$ have additional interesting features illustrated by Fig. 7.

| | | | | | | |
|---|---|---|---|---|---|---|
| 1 1 0 0<br>-1 1 0 0<br>1 -1 0 0<br>-1 -1 0 0 | = $e_0$+$e_1$ = | 1 0 0 0<br>0 1 0 0<br>0 -1 0 0<br>-1 0 0 0 | + | 0 1 0 0<br>-1 0 0 0<br>1 0 0 0<br>0 -1 0 0 | ; |   $e_0$ $e_1$<br>$e_0$: $e_0$ $e_1$<br>$e_1$: $e_1$ $-e_0$ |
| 0 0 1 -1<br>0 0 -1 -1<br>0 0 1 1<br>0 0 -1 1 | = $e_0$+$e_1$ = | 0 0 0 -1<br>0 0 -1 0<br>0 0 1 0<br>0 0 0 1 | + | 0 0 1 0<br>0 0 0 -1<br>0 0 0 1<br>0 0 -1 0 | ; |   $e_0$ $e_1$<br>$e_0$: $e_0$ $e_1$<br>$e_1$: $e_1$ $-e_0$ |
| 1 0 1 0<br>-1 0 -1 0<br>1 0 1 0<br>-1 0 -1 0 | = $e_0$+$e_1$ = | 1 0 0 0<br>-1 0 0 0<br>0 0 1 0<br>0 0 -1 0 | + | 0 0 1 0<br>0 0 -1 0<br>1 0 0 0<br>-1 0 0 0 | ; |   $e_0$ $e_1$<br>$e_0$: $e_0$ $e_1$<br>$e_1$: $e_1$ $e_0$ |
| 0 1 0 -1<br>0 1 0 -1<br>0 -1 0 1<br>0 -1 0 1 | = $e_0$+$e_1$ = | 0 1 0 0<br>0 1 0 0<br>0 0 0 1<br>0 0 0 1 | + | 0 0 0 -1<br>0 0 0 -1<br>0 -1 0 0<br>0 -1 0 0 | ; |   $e_0$ $e_1$<br>$e_0$: $e_0$ $e_1$<br>$e_1$: $e_1$ $e_0$ |
| 1 0 0 -1<br>0 0 0 0<br>0 0 0 0<br>-1 0 0 1 | = $e_0$+$e_1$ = | 1 0 0 0<br>0 0 0 0<br>0 0 0 0<br>0 0 0 1 | + | 0 0 0 -1<br>0 0 0 0<br>0 0 0 0<br>-1 0 0 0 | ; |   $e_0$ $e_1$<br>$e_0$: $e_0$ $e_1$<br>$e_1$: $e_1$ $e_0$ |
| 0 0 0 0<br>0 1 -1 0<br>0 -1 1 0<br>0 0 0 0 | = $e_0$+$e_1$ = | 0 0 0 0<br>0 1 0 0<br>0 0 1 0<br>0 0 0 0 | + | 0 0 0 0<br>0 0 -1 0<br>0 -1 0 0<br>0 0 0 0 | ; |   $e_0$ $e_1$<br>$e_0$: $e_0$ $e_1$<br>$e_1$: $e_1$ $e_0$ |

**Fig. 7.** Two upper rows: decompositions of (4*4)-matrices ($c_0 + c_1$) and ($c_2 + c_3$) into sum of two sparse matrices $e_0$ and $e_1$ whose multiplication tables coinside with the multiplication tables of algebra of 2-dimensional complex numbers (at right). Four bottom rows: decompositions of (4*4)-matrices ($c_0 + c_2$), ($c_1 + c_3$), $0.5*(c_0 + c_3)^2$ and $0.5*(c_1 + c_2)^2$ into sum of two matrices $e_0$ and $e_1$. The table shows direct relations of these matrices with matrix representations of 2-dimensional hyperbolic numbers (at right). Here $c_0$, $c_1$, $c_2$ and $c_3$ are column projectors from Fig. 5.

It should be noted that cyclic features of (4*4)-matrices $2^{-0.5}*(c_0 + c_1)$ and $2^{-0.5}*(c_2 + c_3)$ in cases of their exponentiation exist due to their connections with matrix representations of 2-parametric complex numbers in 4-dimensional space; these connections are shown by means of a new decomposition of each of these matrices into sum of new sparse matrices $e_0$ and $e_1$ (Fig. 7, two upper levels, where the multiplication table for $e_0$ and $e_1$ in the right column is identical to the multiplication table of algebra of complex numbers). The dichotomic features of (4*4)-matrices ($c_0 + c_2$) and ($c_1 + c_3$), and tetra-reproduction features of (4*4)-matrices $((c_0 + c_3)^2)^n = 4^{n-1}*(c_0 + c_3)^2$ and $((c_1 + c_2)^2)^n = 4^{n-1}*(c_1 + c_2)^2$ exist due to their connections with matrix representations of 2-parametric hyperbolic numbers in 4-dimensional space (Fig. 7, four bottom levels, where the multiplication table for $e_0$ and $e_1$ in the right column is identical to the multiplication table of algebra of hyperbolic numbers). Synonyms of hyperbolic numbers are

Lorentz numbers, split-complex numbers, double numbers, perplex numbers (http://en.wikipedia.org/wiki/Split-complex_number).

Now let us turn to the Rademacher (8*8)-matrix $R_8$ (Fig. 4) to analyze its column decomposition $R_8 = s_0 + s_1 + s_2 + s_3 + s_4 + s_5 + s_6 + s_7$ (Fig. 8).

$$R_8 = s_0+s_1+s_2+s_3+s_4+s_5+s_6+s_7 = \begin{vmatrix} 1&0&0&0&0&0&0&0\\ 1&0&0&0&0&0&0&0\\ -1&0&0&0&0&0&0&0\\ -1&0&0&0&0&0&0&0\\ 1&0&0&0&0&0&0&0\\ 1&0&0&0&0&0&0&0\\ -1&0&0&0&0&0&0&0\\ -1&0&0&0&0&0&0&0 \end{vmatrix} + \begin{vmatrix} 0&1&0&0&0&0&0&0\\ 0&1&0&0&0&0&0&0\\ 0&-1&0&0&0&0&0&0\\ 0&-1&0&0&0&0&0&0\\ 0&1&0&0&0&0&0&0\\ 0&1&0&0&0&0&0&0\\ 0&-1&0&0&0&0&0&0\\ 0&-1&0&0&0&0&0&0 \end{vmatrix} +$$

$$\begin{vmatrix} 0&0&1&0&0&0&0&0\\ 0&0&1&0&0&0&0&0\\ 0&0&1&0&0&0&0&0\\ 0&0&1&0&0&0&0&0\\ 0&0&-1&0&0&0&0&0\\ 0&0&-1&0&0&0&0&0\\ 0&0&-1&0&0&0&0&0\\ 0&0&-1&0&0&0&0&0 \end{vmatrix} + \begin{vmatrix} 0&0&0&1&0&0&0&0\\ 0&0&0&1&0&0&0&0\\ 0&0&0&1&0&0&0&0\\ 0&0&0&1&0&0&0&0\\ 0&0&0&-1&0&0&0&0\\ 0&0&0&-1&0&0&0&0\\ 0&0&0&-1&0&0&0&0\\ 0&0&0&-1&0&0&0&0 \end{vmatrix} + \begin{vmatrix} 0&0&0&0&1&0&0&0\\ 0&0&0&0&1&0&0&0\\ 0&0&0&0&-1&0&0&0\\ 0&0&0&0&-1&0&0&0\\ 0&0&0&0&1&0&0&0\\ 0&0&0&0&1&0&0&0\\ 0&0&0&0&-1&0&0&0\\ 0&0&0&0&-1&0&0&0 \end{vmatrix} +$$

$$\begin{vmatrix} 0&0&0&0&0&1&0&0\\ 0&0&0&0&0&1&0&0\\ 0&0&0&0&0&-1&0&0\\ 0&0&0&0&0&-1&0&0\\ 0&0&0&0&0&1&0&0\\ 0&0&0&0&0&1&0&0\\ 0&0&0&0&0&-1&0&0\\ 0&0&0&0&0&-1&0&0 \end{vmatrix} + \begin{vmatrix} 0&0&0&0&0&0&-1&0\\ 0&0&0&0&0&0&-1&0\\ 0&0&0&0&0&0&-1&0\\ 0&0&0&0&0&0&-1&0\\ 0&0&0&0&0&0&1&0\\ 0&0&0&0&0&0&1&0\\ 0&0&0&0&0&0&1&0\\ 0&0&0&0&0&0&1&0 \end{vmatrix} + \begin{vmatrix} 0&0&0&0&0&0&0&-1\\ 0&0&0&0&0&0&0&-1\\ 0&0&0&0&0&0&0&-1\\ 0&0&0&0&0&0&0&-1\\ 0&0&0&0&0&0&0&1\\ 0&0&0&0&0&0&0&1\\ 0&0&0&0&0&0&0&1\\ 0&0&0&0&0&0&0&1 \end{vmatrix}$$

**Fig. 8.** The column decomposition $R_8 = s_0 + s_1 + s_2 + s_3 + s_4 + s_5 + s_6 + s_7$ of the Rademacher (8*8)-matrix $R_8$ (Fig. 4) where each of matrices $s_0, s_1, s_2, s_3, s_4, s_5, s_6, s_7$ is a projection operator.

By analogy with the described case of the projection (4*4)-operators $c_0, c_1, c_2, c_3$ (Figs. 6, 7), one can analyse features of sums of pairs of the projection (8*8)-operators $s_0, s_1, s_2, s_3, s_4, s_5, s_6, s_7$ (Fig. 8) regarding their exponentiation. In other words, one can analyze features of matrices $(s_0 + s_1)^n$, $(s_0 + s_3)^n$,.... where $n = 1, 2, 3$,.... Such analysis leads to the resulting table in Fig. 9.

Each of cells in the table in Fig. 9 represents a sum of those projectors, which denote its column and row by analogy with Fig. 6. Again we have three types of such sums, which are marked by green, red and yellow and which possess the similar properties in comparison with the cases in Fig. 6.

In the table in Fig. 9, green cells correspond to those matrices whose exponentiations generate 8 cyclic groups. The table contains 16 green cells that correspond to the following cyclic groups with period 8: $(2^{-0.5}*(s_0 + s_2))^n$, $(2^{-0.5}*(s_0 + s_3))^n$, $(2^{-0.5}*(s_1 + s_2))^n$, $(2^{-0.5}*(s_1 + s_3))^n$, $(2^{-0.5}*(s_4 + s_6))^n$, $(2^{-0.5}*(s_4 + s_7))^n$, $(2^{-0.5}*(s_5 + s_6))^n$, $(2^{-0.5}*(s_5 + s_7))^n$.

Red cells in the table in Fig. 9 contain those matrices, which possess a doubling property regarding their exponentiation. The table contains 24 red cells, whose matrices

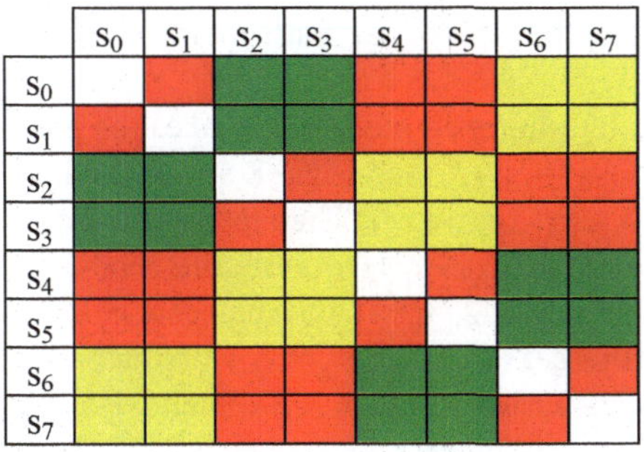


**Fig. 9.** Tables of some features of sums of pairs of different column projectors $s_0, s_1, \ldots, s_7$ (from Fig. 8) regarding their exponentiation. Explanations in the text.

satisfy the following feature: $(s_0 + s_1)^n = 2^{n-1}*(s_0 + s_1)$, $(s_0 + s_4)^n = 2^{n-1}*(s_0 + s_4)$, etc.

16 yellow cells in the table in Fig. 9 contain those matrices, which have «quadruplet property»: $((s_0 + s_6)^2)^n = 4^{n-1}*(s_0 + s_6)^2$, etc. The cells on the main diagonal correspond to sum of two projectors themselves: $(s_i + s_i)^n = 2^n*s_i$ where $i = 0, 1, \ldots, 7$.

## 4 On the Use of the Genetic Oblique Projectors in Modeling Inherited Biological Structures

The described sums of oblique projectors, which correspond to color cells in Figs. 6 and 9, can be used for algebraic modeling some well-known features of inherited biological phenomena, for example, the following ones: dichotomic dividing of biological cells under mitosis when doubling of genetic information occurs; tetra-reproducting of gametes and genetic information under meiosis; cyclic processes; parallelization of information and control processes related to different parameters of holistic multi-parameter biological systems. In this short article, let us illustrate the mentioned modeling opportunities by modeling cyclic processes.

Any living organism is an object with a huge ensemble of inherited cyclic processes. Even each of the proteins is involved in a cycle of the "birth and death," because after a certain time it breaks down into its constituent amino acids and they are then collected into a new protein. All inherited physiological subsystems of the body should be agreed with the structural organization of genetic coding for their coding and transmission to descendants; in other words, they bear the stamp of its features. The discovery of the described cyclic groups (on basis of the genetic projectors), which are connected with phenomenological properties of molecular-genetic systems in their matrix forms of representation, gives a mathematical approach to simulate ensembles of cyclic processes. In this approach, an idea of multi-dimensional configuration vector space is used to simulate cooperative ensembles of cyclic processes. Multidimensional vectors of this bioinformation space can be changed under influence of the described operators on the basis of sums of genetic oblique projectors. Due to special properties of these operators a useful possibility exists to provide a selective control (or a selective coding) of cyclic changes (and some other changes) of separate coordinates of multidimensional vectors in this space.

Let us explain this by one example. Let us take, for instance, the cyclic group of operators $Y^n = (2^{-0.5}*(s_0 + s_2))^n$ (see Figs. 8 and 9) and an arbitrary 8-dimensional vector $X = [x_0, x_1, x_2, x_3, x_4, x_5, x_6, x_7]$. Then let us analyze an expression $X*Y^n = [x_0, x_1, x_2, x_3, x_4, x_5, x_6, x_7]*(2^{-0.5}*(s_0 + s_2))^n = Z_n$ that leads to a new vector $Z_n = [z_0, z_1, z_2, z_3, z_4, z_5, z_6, z_7]$ (here $n = 1, 2, 3, \ldots$). One can check that only coordinates $z_0$ and $z_2$ have cyclic changes in this set of new vectors $Z_n$, all other coordinates are equal to zero. In other words, all cycles are realized on a 2-dimensional plane ($z_0$, $z_2$) inside the 8-dimensional space. If one uses another cyclic group of operators, for example, $(2^{-0.5}*(s_1 + s_3))^n$ ($s_1$ and $s_3$ are from Fig. 8) then the same initial vector $X = [x_0, x_1, x_2, x_3, x_4, x_5, x_6, x_7]$ will be transformed into a cyclic set of vectors in another 2-dimensional plane ($z_1$, $z_3$) of the same 8-dimensional space. One should conclude that, in this model approach, the same initial information in a form of a multidimensional vector X could generate a few cyclic processes in different planes of appropriate multidimensional space by means of using cyclic operators of the described type.

Our additional researches shown that similar approaches with sums of oblique projectors can be also used for modeling biological phenomena in cases of $2^k$-dimensional configuration spaces of multiparametric systems ($k = 4, 5, 6, \ldots$).

One can recall that method of projections is one of the basic mathematical tools in studying vision perception and creating of computer vision, recognition of images, realistic computer graphics, and systems of artificial intelligence; a great number of publications is devoted to these topics, including, for example [7–12]. One can note that such works traditionally use mainly orthogonal projections. Now we are drawing attention to another type of projection operators – non-orthogonal projectors -, which are useful to modeling of the genetic system and inherited biological phenomena.

## 5 Conclusions

1) The described representations of DNA-alphabets of 16 doublets and 64 triplets in the form of the mosaic square matrices reveal their connection with a set of sparse matrices of oblique projection operators;
2) The sums of pairs of such genetic projection operators have useful features for modeling genetically inherited biological phenomena like as cyclic processes, dichotomic dividing of biological cells under mitosis when doubling of genetic information occurs; tetra-reproducting of gametes and genetic information under meiosis; parallelization of information and control processes related to different parameters of holistic multi-parameter biological systems.
3) Similar approaches with sums of oblique projectors can be also used for modeling biological phenomena in cases of $2^k$-dimensional configuration spaces of multiparametric systems ($k = 4, 5, 6, \ldots$).
4) The use of the projection method and the corresponding projection operators is useful not only for studying visual perception, but also for modeling the organization of molecular genetic informatics and logistics of interrelations among different parts of the genetic system.

**Acknowledgments.** The authors are grateful to their colleagues M. He, Z. Hu, I. Stepanyan, V. Svirin, and G. Tolokonnikov for research assistance.

## References

1. Stewart, I.: Life's Other Secret: The New Mathematics of the Living World. Penguin, New-York (1999)
2. Petoukhov, S.V.: Matrix genetics, algebras of the genetic code, noise immunity. M., RCD, 316 p. (2008). (in Russian)
3. Petoukhov, S.V., He, M.: Symmetrical Analysis Techniques for Genetic Systems and Bioinformatics: Advanced Patterns and Applications. IGI Global, Hershey, USA (2009)
4. Konopelchenko, B.G., Rumer, Yu.B.: Classification of the codons in the genetic code. Doklady Akademii Nauk SSSR **223**(2), 145–153 (1975). in Russian
5. Rumer, Yu.B.: Systematization of the codons of the genetic code. sDoklady Akademii Nauk SSSR **183**(1), 225–226 (1968). in Russian
6. Fimmel E., Strüngmann L.: Yury Borisovich Rumer and his 'biological papers' on the genetic code. Phil. Trans. R. Soc. A **374**, 20150228 (2016). https://doi.org/10.1098/rsta.2015.0228
7. Khan, R., Debnath, R.: Human distraction detection from video stream using artificial emotional intelligence. IJIGSP **12**(2), 19–29 (2020)
8. Erwin, Ningsih, D.R.: Improving retinal image quality using the contrast stretching, histogram equalization, and CLAHE methods with median filters. IJIGSP **12**(2), 30–41, April 2020
9. Mostakim, M.N., Mahmud, S., Jewel, M.K.H., Rahman, M.K., Ali, M.S.: Design and development of an intelligent home with automated environmental control. IJIGSP **12**(4), 1–14 (2020)
10. Arora, N., Ashok, A., Tiwari, S.: Efficient image retrieval through hybrid feature set and neural network. IJIGSP **11**(1), 44–53 (2019)
11. Anami, B.S., Naveen, N.M., Surendra, P.: Automated paddy variety recognition from color-related plant agro-morphological characteristics. IJIGSP **11**(1), 12–22 (2019)
12. Ahmed, M., Akhand, M.A.H. Rahman, M.H.: Recognizing Bangla handwritten numeral utilizing deep long short term memory. IJIGSP **11**(1), 23–32 (2019)

# Simulation of AGV Dynamic Grouping Scheduling Considering Real-Time Operation of Quay Crane

Meng Yu(✉), Ruiyu Wang, and Zhiling He

School of Logistics Engineering, Wuhan University of Technology, Wuhan 430063, China

**Abstract.** In the automated container terminals, Automated Guided Vehicles (AGVs) are important equipment in the handling operation. Every handling operation of quay crane, all AGVs participate in the process, which causes the waste of equipment resources. In this paper, the AGV dynamic grouping scheduling strategy considering the real-time operation of quay crane is proposed, that is to dynamically configure AGV of corresponding scale according to the number of working quay cranes. This paper establishes a wharf handling operation model based on FLEXTERM simulation software. Data Envelopment Analysis (DEA) is used to evaluate the simulation data of different quantities of AGVs under the fixed quay crane quantity, and the optimal ratio of quay cranes and AGVs is obtained. According to the optimal ratio, a corresponding number of AGVs are dynamically configured for quay cranes.

**Keywords:** AGV configuration · Data Envelopment Analysis · FLEXTERM simulation

## 1 Instruction

The continuous growth of container transportation volume promotes the development of automated container terminals, and also puts forward higher requirements for the handling efficiency of terminals. AGV is an important equipment for horizontal transportation in automated container terminal, which transfers containers between the quay crane and the yard crane. Danijela Pjevcevic pointed out that the size of AGVs affects the efficiency of container handling in port container terminals [1]. To a certain extent, increasing the scale of AGVs can improve the handling efficiency of wharf. However, too many AGVs will cause traffic jams, thus reducing throughput of the terminal [2]. Therefore, it is an important guarantee for wharf efficiency to allocate an appropriate number of AGVs [3].

For the configuration of AGVs, Ma Yuehui explored the relationship between the end time of the last task and the number of AGVs. The result shows that the number of AGVs is not positively correlated with the overall operation efficiency of the system [4]. Huang Zeqian deconstructed AGV scheduling problem into AGV assignment problem and AGV task sequence problem, and studied the relationship between the number of


Z. Hu et al. (Eds.): ICAILE 2021, LNDECT 82, pp. 298–306, 2021.
https://doi.org/10.1007/978-3-030-80475-6_29

AGVs and total running time [5]. Liang Chengji set up a model aiming at minimizing the shore waiting time and the number of AGVs. He determined the minimum number of AGVs under the operation of 6 quay cranes in an example: 10 [6]. In order to minimize the end time of the last task of vessel unloading operation, Ma Sunyu analyzed the influence of different ratio of quay crane and AGV on the completion time under different task numbers, and determined the optimal configuration of quay crane and AGV ratio of 1: 3 [7]. Fan houming built a two-stage model to minimize the energy consumption of quay crane and AGV transportation process, and verified that minimizing the energy consumption can significantly reduce the number of AGVs, and the optimal allocation ratio of quay crane and AGV is about 1: 2.33 [8]. Han Xiaolong aimed at the scheduling strategy and quantity allocation of AGV in automated container terminals, and analyzed the influence of different AGV scheduling strategies and AGV allocation quantity on the efficiency of container port handling system. He finally obtained the optimal ratio of quayside: AGV = 1: 7 [9]. In order to minimize the cost, Yu Shaozheng proposed a heuristic optimization method for mechanical configuration scheme. The original mechanical configuration scheme is optimized as quay crane: gantry crane: AGV = 3: 5: 10 [10].

Most of the studies based on AGV configuration problems focus on the ratio number of quay cranes and AGVs, that is, the total number of AGVs in wharf is determined according to the number of quay cranes and the optimal ratio. However, these studies ignore the influence of the change of the number of quay cranes for handling operations on the configuration of AGVs, which may lead to the situation that large-scale AGV serves a small number of quay cranes, resulting in waste of resources.

Therefore, this paper considers the AGV configuration problem under the real-time operation of quay crane, and evaluates the simulation data of different quantities of AGV under the fixed quay crane quantity with DEA as the evaluation index, and obtains the optimal ratio of quay crane and AGV. According to the optimal ratio, the number of AGVs is dynamically allocated, and a dynamic grouping scheduling strategy is proposed to dynamically allocate AGVs of corresponding scale according to the number of working quay cranes.

## 2 Problem Description

After the vessel reaches the shore, the terminal configures a reasonable number of quay cranes according to the vessel's schedule, handling tasks, arrival time et al. The discharged containers are horizontally transported to the yard by AGV, and the yard crane will stack the containers to the corresponding container area until all the imported containers of the vessels are stored to complete the unloading of containers.

In the process of coordinated operation between quay crane and AGV, the number of AGVs allocated affects the handling efficiency of wharf. If the number of AGVs is too small, the quay crane will wait for AGV, which will reduce the operation efficiency of quay crane. While if the number of AGVs is too large, it will increase the waiting time of AGV and reduce the utilization rate of AGV. Unreasonable AGV configuration scheme will lead to waste of equipment resources, reduce the efficiency of handling equipment, and then reduce the handling efficiency of the wharf.

In the existing automatic terminal global scheduling mode, almost all AGVs start to participate in handling operation every time the quay crane works, but the quay crane in working state in each task changes according to the ship information. Mobilizing all AGVs to serve each arrival ship task can reduce the phenomenon of quay crane waiting for AGV operation, but for the period with few working quay cranes, large-scale AGVs serving a small number of quay cranes will increase the waiting time of AGV, reduce the AGV utilization rate, and increase the unnecessary equipment depreciation cost and power consumption cost of the wharf.

To solve this problem, this paper proposes a dynamic grouping scheduling model based on the common global scheduling model. Based on the dynamic configuration of AGV, the dynamic grouping scheduling mode realizes the dynamic change of the number of operating AGVs with the number of working quay cranes, which enables the automatic container terminals to configure AGVs of appropriate scale according to the real-time operation conditions of quay cranes.

Considering the dynamic and discrete characteristics of the automatic terminal operation system, this paper uses the discrete event virtual simulation software Flexterm to simulate the production process of the container terminal. The quay crane allocation logic is designed to allocate the quay cranes for arriving vessels, thus realizing the simulation of real-time operation changes of quay cranes [11]. The DEA method is used to evaluate the simulation experimental data, and the weight decision model is built to get the best ratio of quay crane and AGV. On this basis, the AGV dynamic grouping scheduling mode is proposed, that is, according to the number of working quay cranes, AGVs with corresponding scales are dynamically configured to serve them.

## 3 Quay Crane Allocation Algorithm

When the vessel arrives at the berth, the quay crane allocation algorithm is used to configure the quay crane for the arriving vessels to simulate the real-time operation changes of the quay crane [12]. The algorithm configures quay cranes for vessels based on the priority of vessels and the number of quay cranes to be allocated. When the vessel arrives or leaves, or the quay crane completes the handling task, the algorithm will be recalculated.

The algorithm follows the following steps:

Step1: Calculate the minimum and maximum number of quay cranes serving each vessel. The calculation method of the minimum number of quay cranes is based on the current task quantity and quay cranes operation efficiency, and ensures the number of quay cranes under the condition that vessels do not delay departure. The maximum number of quay cranes is set as the smaller of the number of ship cabins with handling operations minus 2 (to ensure the safe distance between quay cranes) and the number of quay cranes.
Step2: Determine the number of existing quay cranes and the maximum number and size of quay cranes allocated for vessels in port. When the number of existing quay cranes is less than the number of allocated quay cranes, there are not enough quay cranes to meet the requirements of all vessels in the berth, turn Step3. Otherwise, the allocation of quay cranes ends.

Step3: Sort in ascending order according to ship priority, and the vessel with the lowest priority is the first vessel.
Step4: According to the sorted list, remove one quay crane from each vessel at a time until the number of quay cranes available is reached. In particular, when a vessel reaches the minimum number of quay cranes allocated to it, it will not remove the quay crane serving the vessel, and directly judge whether to remove a quay crane serving the upper-level vessel.

## 4 Evaluation Method of Wharf Efficiency

Data Envelopment Analysis (DEA) method has advantages in evaluating problems with multiple input or output indexes, and does not need dimensionless processing of input and output data. Therefore, this paper uses DEA method to evaluate the simulation experimental data.

CCR model and BBC model are the most representative in DEA theory system. The results of CCR model represent the comprehensive technical efficiency under the assumption of constant returns to scale. The results of BBC model reflect pure technical efficiency and scale efficiency considering the change of returns to scale. In this paper, CCR model and BCC model are selected as evaluation models, and comprehensive technical efficiency, scale efficiency and pure technical efficiency are comprehensively used to analyze the operation efficiency of wharf under different simulation schemes.

It is assumed that there are t similar parts to be evaluated, which are called decision-making units (DMU). In this paper, DMU refers to the simulation scheme with different AGV configuration, each of which has m inputs and N outputs.

Each DMU has its corresponding efficiency evaluation index. The efficiency evaluation of the kth DMU is conducted with the weight coefficient v and u as independent variables, the efficiency index $h_k$ of the kth DMU as the target, and the efficiency index of all DMUs as the constraint. The CCR model is shown in model (1).

$$Maxh_k = \frac{\sum_{r=1}^{n} u_r y_{rk}}{\sum_{i=1}^{m} v_i x_{ik}} = \frac{u^T y_k}{v^T x_k} \tag{1}$$

$$s.t.\begin{cases} h_j = \dfrac{\sum_{r=1}^{n} u_r y_{rj}}{\sum_{i=1}^{m} v_i x_{ij}} \le 1, j = 1, 2 \ldots t \\ u_r \ge 0, v_i \ge 0, r = 1, 2 \ldots n; i = 1, 2 \ldots m \end{cases}$$

In the model (1), $x_{ik}$ represents the input amount of the kth DMU to the ith input, $x_{ik} > 0$. $y_{rk}$ represents the output amount of the rth outputs in the kth DMU. $u_i$ and $v_r$ respectively represent the weight of the ith input and the weight of the rth output.

If the comprehensive technical efficiency $h_k$ obtained by solving is less than 1, it indicates that decision-making unit k is not DEA efficient, which means that there is a hypothetical better decision-making unit. It can get more output with the same input as evaluation unit or get the same output with less input than evaluation unit.

Because the model (1) is a fractional model, set:

$$z = \frac{1}{v^T x_0}, \omega = zv, \tau = zu \tag{2}$$

When $t = \frac{1}{v^t x_0}$, simply getting the formula: $\omega^t x_0 = 1$.Therefore, the model (1) can be transformed into the model (3).

$$Maxh_k = \tau^T y_k$$

$$s.t.\begin{cases} \omega^T x_i - \tau^T y_i \geq 0, i = 1, 2...t \\ \omega^T x_k = 1 \\ \omega \geq 0, \tau \geq 0 \end{cases} \tag{3}$$

The dual programming model of model (3) is shown in model (4).

$$Min\theta$$

$$s.t.\begin{cases} \sum_{i=1}^{t} \omega_i x_i \leq \theta x_k \\ \sum_{i=1}^{t} \omega_i y_i \geq y_k \\ \omega_i \geq 0, i = 1, 2 \cdots t \end{cases} \tag{4}$$

BBC model can be used to evaluate the relative efficiency of each decision-making unit under different scale returns. The BBC model is showed in model (5).

In model (5), $\theta$ is the technical efficiency value of decision-making unit k. When $\theta = 1$, decision-making unit k is technically effective. Otherwise, decision-making unit k is technically ineffective.

$$Min\theta$$

$$s.t.\begin{cases} \sum_{i=1}^{t} \omega_i x_i \leq \theta x_k \\ \sum_{i=1}^{t} \omega_i y_i \geq y_k \\ \sum_{i=1}^{t} \omega_i = 1 \\ \omega_i \geq 0, i = 1, 2 \cdots t \end{cases} \tag{5}$$

According to the selection principle of evaluation index and referring to the research of Danijela Pjevcevic, there is a linear relationship between terminal operation cost and handling time [1], so handling time is selected as the input index. Combined with the actual operation characteristics of the wharf, the shorter the handling time and the higher the equipment utilization ratio are more beneficial to the port. So AGV utilization rate and quay crane utilization rate are selected as the output indicators.

## 5 AGV Configuration Research Based on Dynamic Operation of Quay Crane

Because the input cost of quay crane is the highest, this paper chooses quay crane as independent variable machine and AGV as dependent variable machine, and configures the corresponding quantity of AGVs according to the input of quay crane. The simulation process analyzes the simulation scenarios from 4 to 18 AGVs (the initial test results show that when there are less than 4 AGVs, the handling efficiency of quay crane cannot be guaranteed).

The simulation data obtained during the operation of a quay crane are showed in Table 1.

According to the data collected by simulation, in which different numbers of AGVs represent different DMUs, DEA method is used to evaluate and deap2.1 is used to analyze different simulation schemes. The results are shown in Table 2.

In Table 2, firm indicates the order of DMUs, where 1 indicates the first DMU, that is, the simulation situation when four AGVs are configured, and so on. Efficiency value equal to 1 indicates that DEA is effective, in which DMU 3(6 AGVs) and DMU 15(18 AGVs) are DEA effective units. The two units are technically and scale effective at the same time, and the difference lies in the different weights of handling time, AGV utilization rate and quay crane utilization rate.

**Table 1.** Operation simulation data of a quay crane

| AGV quantity | Handling time/h | AGV utilization rate | Quay crane utilization rate |
|---|---|---|---|
| 4 | 17.33 | 78.4 | 71.3 |
| 5 | 16.37 | 75.3 | 75.4 |
| 6 | 15.42 | 73.7 | 79.7 |
| 7 | 15.39 | 67.5 | 80.5 |
| 8 | 15.33 | 62.3 | 81.7 |
| 9 | 15.29 | 54.3 | 82.3 |
| 10 | 15.21 | 51.7 | 83.1 |
| 11 | 15.17 | 47.6 | 83.4 |
| 12 | 15.13 | 43.7 | 83.7 |
| 13 | 15.07 | 42.5 | 84.3 |
| 14 | 15.13 | 37.0 | 84.5 |
| 15 | 14.07 | 35.2 | 85.1 |
| 16 | 14.02 | 33.7 | 85.2 |
| 17 | 14.98 | 31.2 | 85.3 |
| 18 | 14.72 | 29.8 | 85.5 |

**Table 2.** Efficiency value of decision-making unit in operation of one quay crane

| Firm | Comprehensive technical efficiency | Pure technical efficiency | Scale efficiency |
|---|---|---|---|
| 1 | 0.947 | 1.000 | 0.947 |
| 2 | 0.962 | 0.982 | 0.980 |
| 3 | 1.000 | 1.000 | 1.000 |
| 4 | 0.995 | 0996 | 1.000 |
| 5 | 0.999 | 1.000 | 0.999 |
| 6 | 0.989 | 0.989 | 1.000 |
| 7 | 0.996 | 1.000 | 0.996 |
| 8 | 0.992 | 0.992 | 1.000 |
| 9 | 0.988 | 0.988 | 1.000 |
| 10 | 0.995 | 1.000 | 0.995 |
| 11 | 0.987 | 0.988 | 0.999 |
| 12 | 0.992 | 1.000 | 0.992 |
| 13 | 0.999 | 1.000 | 0.999 |
| 14 | 0.997 | 0.997 | 1.000 |
| 15 | 1.000 | 1.000 | 1.000 |

Formula (6) shows the weight decision model.

$$
\begin{aligned}
& Max\ \text{Max}\ (73.7x_1 + 79.7x_2)/\ 15.42x_3) \\
& (78.4x_1 + 71.3x_2)/\ 17.33x_3 \leq 1 \\
& (75.3x_1 + 75.4x_2)/\ 16.37x_3 \leq 1 \\
& (73.7x_1 + 79.7x_2)/\ 15.42x_3 \leq 1 \\
& (68.2x_1 + 80.9x_2)/\ 15.36x_3 \leq 1 \\
& (62.3x_1 + 82.2x_2)/\ 15.31x_3 \leq 1 \\
& (54.3x_1 + 82.7x_2)/\ 15.29x_3 \leq 1 \\
& (52.1x_1 + 83.1x_2)/\ 14.25x_3 \leq 1 \\
& (47.6x_1 + 83.4x_2)/\ 14.17x_3 \leq 1 \\
& (43.7x_1 + 83.7x_2)/\ 14.15x_3 \leq 1 \\
& (42.5x_1 + 84.3x_2)/\ 14.12x_3 \leq 1 \\
& (37.0x_1 + 84.5x_2)/\ 14.07x_3 \leq 1 \\
& (35.2x_1 + 85.1x_2)/\ 14.08x_3 \leq 1 \\
& (33.7x_1 + 85.3x_2)/\ 14.02x_3 \leq 1 \\
& (31.2x_1 + 85.3x_2)/\ 14.05x_3 \leq 1 \\
& (29.8x_1 + 85.5x_2)/\ 14.02x_3 \leq 1
\end{aligned}
\tag{6}
$$

The model (6) is solved by MATLAB, $x_1$, $x_2$, $x_3$ separately represent the weight of AGV utilization rate, quay crane utilization rate and handling time. As the primary goal of the wharf is to ensure the handling efficiency, on this basis, the operation efficiency of quay crane should be guaranteed. Secondly, in order to reduce the cost and protect the environment, the operation efficiency of AGV should also be considered. That is to say, the handling time weight > quay crane utilization weight > AGV utilization weight should be guaranteed.

1 quay crane, 6AGV: $x_1 = 0.0024$, $x_2 = 0.0103$, $x_3 = 0.0649$.

1 quay crane, 18 AGV: $x_1 = 0$, $x_2 = 0.0117$, $x_3 = 0.0679$.

When 18 AGVs are configured, $x_1 = 0$, which basically does not consider the importance of AGV and does not meet the requirements of the wharf. However, when six AGVs are configured, the weight constraints are satisfied. It is concluded that when there is only one quay bridge in the wharf, it is more appropriate to configure six AGVs.

When the number of quayside cranes of container terminal is set as two or three, the simulation data processing form is similar to the above, and the optimal configuration number of automatic quay crane and AGV is obtained, as shown in Table 3.

It can be seen from Table 3 that the comprehensive efficiency of handling operation is the best when the quay crane: AGV = 1: 6, 2: 10, 3: 14. When the number of quay cranes in working state changes, the corresponding number of AGVs are dynamically allocated to serve quay cranes according to the optimal ratio by writing codes, and a dynamic grouping scheduling mode is proposed, which enables the automated container terminals to schedule AGVs of appropriate scale according to the working conditions of quay cranes, thus achieving the purpose of improving the handling efficiency of the wharf.

**Table 3.** Quay crane -AGV configuration

| Quay crane quantity | AGV quantity |
|---|---|
| 1 | 6 |
| 2 | 10 |
| 3 | 14 |

## 6 Conclusion

In this paper, the AGV dynamic grouping scheduling considering the real-time operation of quay crane is proposed.

The innovations of this paper are as follows:

1) Considering the influence of the change of quay crane operation on AGV configuration, design the quay crane allocation logic to configure the number of quay cranes for arriving vessels. The allocation logic is used to simulate the real-time operation changes of quay cranes.

2) The DEA method is used to evaluate the simulation data of different quantities of AGV under the fixed quay crane quantity. This paper select CCR model and BCC model as evaluation models. And handling time is selected as the input index, AGV utilization rate and quay crane utilization rate are selected as the output indicators. The optimal ratio of quay crane: AGV = 1: 6, 2: 10, 3: 14 is obtained.
3) According to the optimal ratio, the number of AGVs is dynamically configured, and a dynamic grouping scheduling mode is proposed, that is, AGVs with corresponding scales are dynamically configured to participate in handling operations according to the number of working quay cranes.

**Acknowledgment.** This project is supported by National Natural Science Foundation of China (71672137) and National Key Research and Development Program of China (Grant No. 2019YFB1600404).

## References

1. Pjevcevic, D., Nikolic, M., Vidic, N., Vukadinovic, K.: Data envelopment analysis of AGV fleet sizing at a port container terminal. Int. J. Prod. Res. **55**(14), 4021–4034 (2016)
2. Debjit, R., Akash, G., Koster, D.: A non-linear traffic flow-based queuing model to estimate container terminal throughput with AGVs. Int. J. Prod. Res. **54**(2), 472–493 (2016)
3. Chengduan, W.: Research and implementation on the flexible feasibility of AGV routing. Int. J. Educ. Manage. Eng. **2**(9), 22–28 (2012)
4. Ma Yuehui, H., Zhihua. : The dispatching and scheduling problems of AGVs at automated container terminals under uncertainty conditions. J. Guangxi Univ. (Nat. Sci.) **41**(02), 589–597 (2016). (in Chinese)
5. Zeqian, H., Tao, D., Ya, W.: Research on AGV dispatching of automated container terminals. J. Wuhan Univ. Technol. **43**(06), 1165–1168 (2019). (in Chinese)
6. Chengji, L., Dengchuan, C.: Research on coupling problem of AGV configuration and scheduling in automated container terminal. Comput. Eng. Appl. **56**(14), 216–225 (2020). (in Chinese)
7. Sunyu, M., Yongsheng, Y., Chengji, L.: Collaborative scheduling of double-trolley quay cranes and AGV based on PSO at automated container terminal. Comput. Appl. Softw. **35**(10), 17–22 (2018). (in Chinese)
8. Fan, H., Guo, Z., Yue, L., Ma, M.: Joint configuration and scheduling optimization of dual-trolley quay crane and AGV for container terminal considering energy saving. Acta Automatica Sinica 1–16 (2019). (in Chinese)
9. Xiaolong, H., Jiawei, F.: Analysis of AGV dispatching and configuration simulation of automated container terminals. J. Chongqing Jiaotong Univ. (Nat. Sci.) **35**(5), 151–154 (2016). (in Chinese)
10. Shaozheng, Y., Dongshi, S., Jing, C., Yongchao, L.: Research on mechanical configuration of automatic container terminals based on simulation optimization method. J. Syst. Simul. **31**(12), 2750–2757 (2019). (in Chinese)
11. Kambalimath, M.G., Kakkasageri, M.S.: Cost optimization based resource allocation scheme for vehicular cloud networks. Int. J. Comput. Netw. Inf. Secur. **12**(2), 22–31 (2020)
12. Fahim, A.: Finding the number of clusters in data and better initial centers for K-means algorithm. Int. J. Intell. Syst. Appl. **12**(6), 1–20 (2020)

# Literature Review of Cold Chain Logistics Research in China Based on CiteSpace

Di Kang(✉), MengMeng Guo, and Gang Chen

School of Logistics, Beijing Wuzi University, Beijing 101149, China
kangdi@bwu.edu.cn

**Abstract.** In order to promote the development of cold chain logistics industry and help cold chain logistics researchers find new research points. This paper takes 681 cold chain logistics research literatures collected by core journals and CSSCI journals in CNKI database from 2000 to 2020 as samples, uses CiteSpace software to carry out descriptive statistics and visual analysis of the literatures, and comprehensively presents the current situation, hot spots and evolution trend of cold chain logistics research. The research shows that the research literature on cold chain logistics is large in number and wide in dimension, but the quality of literature needs to be improved. The research subjects show the characteristics of small concentration and large dispersion. Research focuses include the degree of marketization, infrastructure, demand, information construction and professional talents. The research trend is toward the combination of "smart cold chain logistics" and "green sustainable development".

**Keywords:** Chain logistics research · Knowledge map · CiteSpace · Bibliometric

## 1 Instruction

Cold chain logistics is a kind of supply chain system that keeps the perishable food in a low temperature environment through the whole process of processing, storage, transportation, distribution and sales from purchase to consumption [1]. In recent years, consumers pay more attentions to food quality, which lays a good market foundation for the development of cold chain logistics. Government has issued a system of favorable policies for cold chain logistics, which provides a clear policy guidance for the development of cold chain logistics. Driven by the good market foundation and national policies, the popularity of cold chain logistics is increasing rapidly, and the research achievements are constantly emerging. This paper uses CiteSpace software to analyze the existing research achievements, comprehensively combs the development context of cold chain logistics research in China, and presents the research status, hot spots and development trend in this field, which can provide new research direction and topic reference for scholars.


Z. Hu et al. (Eds.): ICAILE 2021, LNDECT 82, pp. 307–316, 2021.
https://doi.org/10.1007/978-3-030-80475-6_30

## 2 Research Design

The literature review of domestic cold chain logistics research based on CiteSpace software is divided into three parts: determining the theme, data preparation and data analysis. Firstly, based on the comprehensive analysis of the research progress in the field of logistics, "cold chain logistics" is determined as the research topic; then, the data selection method is determined according to the research topic, the data is filtered and denoised to obtain the sample data; finally, CiteSpace software is used to draw keywords co-occurrence, emergence, clustering map, quantitative statistics and qualitative analysis of cold chain logistics research literature are carried out, and the research conclusion is drawn.

### 2.1 Research Method

This paper mainly adopts bibliometric method and visual analysis method. With the help of bibliometric analysis, this paper analyzes the quantity change of literature and the distribution of journals, and draws the knowledge map of keywords co-occurrence, emergence and clustering in the field of cold chain logistics. with the help of visual analysis method, showing the cooperation network, frontier hot spots and evolution trend in the field. The CiteSpace software is a visualization tool based on CO citation analysis theory and path finding network algorithm [2]. The knowledge map obtained by visual analysis is a kind of image that shows the relationship between knowledge development process and structure based on knowledge domain [3].

### 2.2 Date Sources

The literature data in this paper are from China National Knowledge Infrastructure. The main search condition settings are shown in Table 1.

**Table 1.** Data acquisition methods

| Search subjects | Search contents and results |
|---|---|
| Acquisition time | 2020May 14, 2004 |
| Retrieval time range | 2000–2020 |
| Database | Core journals and CSSCI |
| Literature type | Periodical |
| Search keywords | Cold chain logistics |
| Search results | 760strip |

Finally, a total of 760 articles were manually checked and screened, 49 papers with low relevance to the theme, such as conferences, news and scholars' essays, were deleted. CiteSpace software is used to transform data format, reduce noise and remove duplicate, finally, 681 articles were obtained.

# 3 Analysis of Research Status

From the literature distribution, cooperation network and research dimensions, this paper analyzes the sample literature, comprehensively presents the research status of cold chain logistics.

## 3.1 Literature Distribution Analysis

The total number of papers published in cold chain logistics research journals, core papers and CSSCI papers in CNKI database from 2000 to 2020 were counted respectively. The descriptive statistical analysis was conducted by Excel software, the quantity distribution of the research literature was obtained, as shown in Fig. 1.

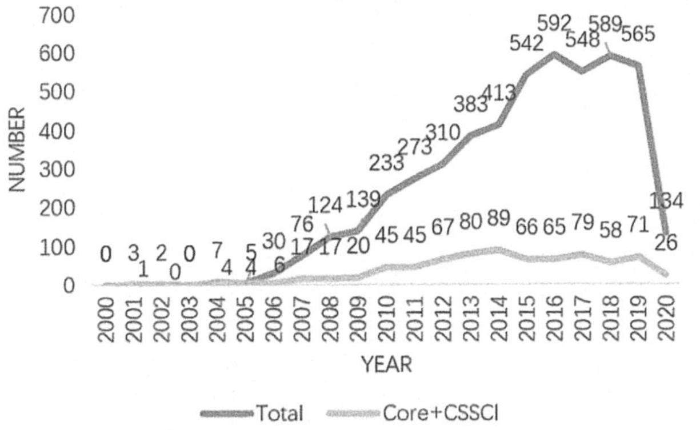


**Fig. 1.** Distribution of cold chain logistics research literature from 2000 to 2020

The total number of papers on cold chain logistics research was 4968 in Fig. 1, the number of papers published by core and CSSCI was 760. Therefore, the quality of research literature in the field of cold chain logistics needs to be further improved. From 2000 to 2006, domestic cold chain logistics research began to sprout, but due to the lag of social and economic level, the development state of cold chain logistics was scattered and disordered, the research results were less, with only 6 papers per year; from 2007 to 2009, the research was gradually warming up, but the progress was slow, with an average annual number of 113 articles, which was greatly improved compared with the previous stage; after 2010, the research ushered in an explosive growth, with an average annual number of 416 articles. Only part of the literature in 2020 has been collected, so the trend chart from 2019 to 2020 shows a downward trend.

In this paper, we use the quantitative visualization analysis function of CNKI to analyze the total number of papers published in cold chain logistics research related 40 journals. The top six journals are as follows: logistics engineering and management, logistics technology, China logistics and procurement, logistics technology and Application, China storage and transportation, logistics science and technology. The number of papers published in these six journals accounts for more than 50%.

### 3.2 Cooperation Network Analysis

The size of nodes in the visualization map represents the frequency, the connection between nodes represents the co-occurrence intensity [4]. Based on this, the author and the organization are taken as the node types to analyze the cooperation network, obtains the cooperation network graphs of authors and institutions of the sample literature, as shown in Fig. 2 and Fig. 3.

**Fig. 2.** Cooperation network of authorship of sample literature from 2000 to 2020

There are 91 nodes in Fig. 2 and 97 connections between nodes, the network density is 0.0237. The number of nodes is large, the area is small, and the node connections are relatively small. There are many research subjects in the field of cold chain logistics, but lack of communication and cooperation among them. From the perspective of cooperation network, the research subjects of cold chain logistics show the characteristics of small concentration and large dispersion, in other words, the cold chain logistics research has formed a small number of core research groups, but the connection strength between the groups is weak.

**Fig. 3.** Cooperation network of publishing institutions of sample literature from 2000 to 2020

It can be seen from Fig. 3 that there are 58 nodes and 14 connections between nodes, the network density is 0.0085. The number of nodes is large, the area is small, the number of nodes is relatively small, which indicates that the interagency cooperation between authors is relatively small. The institutions with more papers are China

Agricultural University (16 articles), Shanghai Maritime University (11 articles), Beijing Jiaotong University (9articles) and Beijing Wuzi University (6 articles).From the exchange and cooperation relationship between Beijing Business University and Beijing Jiaotong University, Hubei Institute of material circulation technology and Wuhan University, it can be seen that there is intra city cooperation among domestic cold chain logistics research institutions, and cross provincial cooperation research is less. From the perspective of the types of issuing institutions, the main types of issuing institutions are colleges and universities, which shows that the scientific research ability of colleges and universities is strong and occupies an important position in the field of cold chain logistics research. From the perspective of regional distribution, there are many research results of institutions located in the north of China, especially in Beijing, Tianjin and Hebei region.

To sum up, there is less connection between scholars and institutions of cold chain logistics. With the rapid development of China's logistics industry, to strengthen the relationship between scholars and institutions is an urgent problem.

### 3.3 Dimension Analysis

Five dimensions with more research results are summarized through manual verification of cold chain logistics research literature, which are "cold chain logistics system", "cold chain logistics distribution", "cold chain logistics cost", "cold chain logistics development" and "cold chain logistics enterprise evaluation". In order to add coherence to the status analysis, we then analyzed the current situation according to the dimension difference and list the representative academic achievements.

Cold chain logistics system aims as the construction and optimization of cold chain logistics system with the main idea of green and low-carbon. For example, Z (2016) used information entropy and the state parameters of food cold chain logistics system to establish the collaborative model of food cold chain logistics system [5]. L, W&Y (2018) constructed the carbon footprint model of cold chain logistics system [6].

Cold chain logistics distribution includes green cold chain logistics distribution path optimization and distribution center location. The former is to reduce costs, increase efficiency and environmental protection. Based on a certain perspective, the path optimization model is constructed. With the help of emerging technologies and intelligent algorithms (such as ant colony algorithm, simulated annealing algorithm, genetic algorithm, tabu search algorithm), and compared with other models, the paper puts forward countermeasures and suggestions. The latter is usually based on travel time reliability, safety reliability and cargo damage cost. Single or multiple distribution center location method is adopted, which comprehensively considers the business environment, land use requirements, natural conditions, transportation conditions and other factors, the location model is established and optimized. For example, Z &Y analyzed the research on solving time variable based on simulated annealing algorithm and average speed method for product distribution vehicle scheduling problem under time-varying road network environment [7].

Cold chain logistics cost includes theoretical analysis method and dynamic modeling and simulation method. Theoretical analysis is a pure theoretical discussion and qualitative analysis. For example, Y (2020) analyzed the composition of sales cost of fresh agricultural products in the e-commerce environment from two dimensions of initial investment cost and logistics cost [8]. Dynamic modeling is a combination of theoretical analysis and modeling simulation, which is combined qualitative with quantitative analysis. For example, Z (2019) used ant colony algorithm to optimize litchi cold chain logistics distribution cost [9].

Cold chain logistics development includes cold chain logistics development status, mode, influencing factors and path research. Among them, the research on cold chain logistics mode includes four aspects: a) financial model. Such as X, H &Y (2017) built a cold chain logistics financial model based on option purchase and sale contract [10]. b) Operation mode. Wang (2017) proposed specific optimization strategies for food supply chain logistics operation mode according to the influencing factors and requirements of different logistics modes [11]. c) Distribution mode. Yang (2018) divided the main modes of food cold chain logistics distribution under O2O mode into self-operation, outsourcing, self-delivery and crowdsourcing mode [12]. d) Development mode. Z& L (2020) established a full chain modern agricultural products cold chain logistics development mode of "urban big database" and "personal micro library" connected through big data [13].

Cold chain logistics enterprise evaluation includes enterprise performance evaluation, service quality evaluation and customer satisfaction evaluation. Methods such as analytic hierarchy process, fuzzy comprehensive evaluation method, entropy method, data envelopment analysis, rank sum ratio method and grey correlation analysis are commonly used. For example, H (2019) established the customer satisfaction evaluation index system of cold chain logistics enterprises, and conducted empirical research through Structural Equation Model [14].

In conclusion, there are many dimensions of cold chain logistics research, and they are relatively separated and integrated. The overall research is moving towards cost reduction, efficiency improvement, green sustainability and improve the present situation. Research methods are diverse, focusing on the combination of qualitative and quantitative, theory and model algorithm.

## 4 Analysis of Research Hotspots

CiteSpace software was used to analyze the co-occurrence, emergence and clustering of keywords in the sample literature. The top 10 high-frequency keywords, the top 10 emergent words are as shown in Table 2 and Table 3.

According to Table 2 and Table 3, the centrality of "cold chain", "fresh agricultural products" and "food safety" ranked top three, which played important roles in the co-occurrence network. The emergence words have evolved as time goes on. Since 2004, most of them are refrigerated vehicles, heat preservation vehicles and trucks and other infrastructure. Since 2007, the emergence intensity of the cold chain has reached a maximum, the food cold chain and food safety emergence intensity are similar. Since 2016, the emergence intensity of fresh agricultural products ranked top two, with algorithms and optimization becoming hot spots. Keywords such as genetic algorithm, path optimization and carbon emissions have emerged. On this basis, this paper carries on the cluster analysis to the key words, the 10 clusters are divided into 5 groups: #0, #1 and

**Table 2.** Top 10 high frequency keywords of sample literatures from 2000 to 2020

| Serial number | Frequency/time | Centraity | key word | particular year |
|---|---|---|---|---|
| 1 | 402 | 0.78 | Cold chain logistics | 2004 |
| 2 | 72 | 0.43 | Agriculture products | 2010 |
| 3 | 45 | 0.18 | Cold chain | 2007 |
| 4 | 40 | 0.22 | Fresh agricultural products | 2009 |
| 5 | 35 | 0.25 | Cold chain logistics of Agricultural products | 2010 |
| 6 | 24 | 0.07 | Logistics | 2007 |
| 7 | 19 | 0.18 | Food Safety | 2008 |
| 8 | 14 | 0.07 | Genetic algorithm | 2016 |
| 9 | 12 | 0.02 | Electronic Commerce | 2014 |
| 10 | 11 | 0.19 | Refrigerated truck | 2004 |

**Table 3.** Top 10 prominent words in the sample literatures from 2000 to 2020

| Emergent words | Emergence intensity | start time | End time | 2000-2020 |
|---|---|---|---|---|
| Refrigerated truck | 3.2685 | 204 | 2005 | |
| Insulation car | 3.7564 | 2004 | 2012 | |
| Truck | 2.5731 | 2005 | 2012 | |
| Food cold chain | 2.7796 | 2007 | 2010 | |
| Cold chain | 7.0557 | 2007 | 2011 | |
| Food Safety | 2.8981 | 2013 | 2015 | |
| Fresh agricultural products | 5.6491 | 2016 | 2017 | |
| Genetic algorithm | 4.6879 | 2016 | 2020 | |
| Path optimization | 3.8242 | 2017 | 2020 | |
| Carbon emission | 2.6375 | 2018 | 2020 | |

#6; #2 and #3; #5 and #8; #4 and #9 are classified into marketization degree, cold chain logistics infrastructure, demand, informatization construction, logistics infrastructure, professionals.

## 5 Frontier Trend Analysis

The evolution path of cold chain logistics research is summarized as follows: from 2000 to 2010, the research trend shifted from facilities to cold chain logistics distribution mode to agricultural products cold chain logistics. From 2010 to 2015, the research focused on integrating emerging technologies into the cold chain logistics system, Such as RFID has the feature of traceability to ensure food safety [15]. From 2016 to 2020, the fourth phase of the industrial revolution is a period of intelligent manufacturing in which artificial intelligence, the Internet of things, and big data have a greater impact in the manufacturing process [16]. Current technologies not only increased the amount of data, but also showed us data is actually getting generated in various formats [17]. Scholars began to use genetic algorithm, system dynamics and other algorithm models to optimize the cold chain logistics distribution path, reduce the damage to the environment caused by economic development, and develop green cold chain logistics. In the same time, organizations need information security to reduce the risk of unauthorized information disclosure, use, modification and destruction [18]. Cold chain logistics enterprises can use information technology for data protection. In addition, there are data clustering, it aims to group similar data together from the collected data [19]. This method can be used to mine the potential data in the operation of cold chain logistics.

The future development trend of cold chain logistics can be summarized as follows: first, the scale of cold chain logistics market continues to expand. The frequent introduction of national policies has created a good policy environment for the development of cold chain logistics. On the one hand, it provides infrastructure and financial support for the cold chain logistics industry, rectifies the whole industry, and makes the development of cold chain logistics more normative; on the other hand, it makes the cold chain logistics field become the focus of attention from all walks of life. Scholars carry out scientific research from multiple perspectives, and the public's acceptance of premium generated by cold chain logistics is increasing. In addition, with the rapid development of e-commerce technology in China, the number of online orders increases. Under the dual influence of policy and technology, the demand of cold chain logistics will continue to rise. Second, the competition pattern of cold chain logistics is diversified. With the rapid development of domestic cold chain logistics industry, competitors from all walks of life have entered the cold chain market. For example, JD.com Logistics cooperated with China International Cargo Airlines to build a global cold chain logistics system. These giant enterprises increase the number of cold chain logistics enterprises, and promote the competition pattern of cold chain logistics industry begins to show diversification. Third, the cold chain logistics system is more information-based and intelligent. With the improvement of consumers' requirements for food freshness and the enhancement of the government's supervision and inspection of cold chain logistics process, cloud computing, big data, blockchain and other technologies will be more applied to the cold chain process, and the informatization and intelligence degree of the cold chain logistics system will gradually improve.

## 6 Conclusion

This paper makes an in-depth analysis of the current situation, hotspot and evolution trend of cold chain logistics research and draws the following conclusions:

First, regional distribution of research scholars and research institutions is not balanced, and the cooperation and exchange are not close. Since 2004, the total number of papers published in cold chain logistics research has been on the rise, but the number of papers published in core journals and CSSCI accounts for a small proportion of the total. In the knowledge map of cooperation network, the nodes are less connected, most nodes are distributed independently, only a few cooperation and communication teams are formed, and most of them are distributed in Beijing, Tianjin and Hebei region.

Second, the hot spot of cold chain logistics research evolves with time. In 2004, at the initial stage of cold chain logistics development, the research focus was mainly on refrigerated vehicles, insulated vehicles and freight cars and other infrastructure. In 2010, the research focus changed to the application of cutting-edge technology in cold chain logistics system. In 2020, algorithm and model optimization will become a hot spot. The key words cluster into 10 categories, which are summarized into 5 hot directions: marketization degree of cold chain logistics, infrastructure, information construction, demand and logistics talents.

Third, the future development trend of cold chain logistics industry is: the market scale continues to expand, the competition pattern is diversified, and the cold chain logistics system is more information-based and intelligent. Driven by policy and technology, the demand for cold chain logistics is increasing gradually, the application of emerging technologies is enhanced, and the market prospect is good, which attracts competitors from all walks of life to join the cold chain logistics industry.

Fourth, the research of cold chain logistics tends to combine "intelligent cold chain logistics" with "green sustainable development". Scholars use algorithms and models to optimize the distribution path, explore the improvement of transportation vehicles centered on energy saving and environmental protection, and realize green and low-carbon while promoting the development of the industry.

## References

1. Wang, Y., Zhao, H.: Market-oriented research on cold chain logistics of edible fungi. Edible Fungi China **38**(4), 45–47 (2019). (in Chinese)
2. Chen, Y., Chen, C., Liu, Z., Hu, Z., Wang, X.: The methodology function of CiteSpace mapping knowledge domains. Stud. Sci. Sci. **33**(2), 242–253 (2015)
3. Fu, Y., Xia, S., Li, C.: Analysis on mapping knowledge domains of academic research on cold chain logistics in China. Railway Transp. Econ. **38**(4), 45–47 (2019)
4. Liu, Y., Wang, Z., Qian, K.: The evolution and emerging trends of China's green supply chain research: visual analysis of CiteSpace based on CSSCI literatures. Sci. Technol. Manage. Res. **39**(21), 200–207 (2019)
5. Zhang, B.: Research of food cold chain logistics system on the synergy management. Food Res. Dev. **37**(21), 215–218 (2016)
6. Liu, G., et al.: Carbon footprint of cold chain system and a case analysis. J. Refrigera. **39**(4), 19–25 (2018)

7. Zhang, J., Yang, Z.: Research on multi temperature cold chain distribution route optimization under time varying road network environment. J. Chongqing Norm. Univ. (Nat. Sci.) **37**(1), 119–126 (2020)
8. Yue, J.: Construction of agricultural products cold chain logistics and analysis of e-commerce sales cost composition. Agric. Econ. (3), 140–142 (2020)
9. Zeng, Z., Zou, C., Wei, J., Lu, H., Lv, E., Ruan, Q.: Optimization of distribution cost model of cold chain logistics for litchi based on ant colony algorithm. Packag. Eng. **40**(11), 58–65 (2019)
10. Xin, C., Hou, J., Yang, X., Jin, Y., Zhu, W.: Study on cold-chain logistics finance mode based on option. Railway Transp. Econ. **39**(8), 15–19 (2017)
11. Wang, F.: Study on operation mode and mechanism of food supply chain logistics. J. Commer. Econ. (9), 100–101 (2017). (in Chinese)
12. Yang, S.: Research on food cold chain logistics distribution under O2O model. Food Ind. **39**(10), 231–234 (2018)
13. Zhang, X., Li, H.: Research on development mode of modern agricultural cold chain logistics based on large data. Sci. Technol. Manage. Res. **40**(7), 234–240 (2020)
14. He, S.: Research on customer satisfaction of cold chain logistics enterprises based on structural equation model. Soc. Sci. **6**, 31–37 (2019). (in Chinese)
15. Ting, P.-H.: An efficient and guaranteed cold-chain logistics for temperature-sensitive foods: applications of RFID and sensor networks. Int. J. Inf. Eng. Electron. Bus. **5**(6), 1–5 (2013)
16. Süzen, A.A.: A risk-assessment of cyber attacks and defense strategies in industry 4.0 ecosystem. Int. J. Comput. Netw. Inf. Secur. **12**(1), 1–12 (2020)
17. Rida, Q.: A roadmap towards big data opportunities, emerging issues and hadoop as a solution. Int. J. Educ. Manage. Eng. **10**(4), 8–17 (2020)
18. Yadav, P., Dhall, S.: Comparative analysis of steganography technique for information security. Int. J. Math. Sci. Comput. **6**(4), 42–69 (2020)
19. Fahim, A.: Finding the number of clusters in data and better initial centers for K-means algorithm. Int. J. Intell. Syst. Appl. **12**(6), 1–20 (2020)

# Channel Strategy of Supply Chain Extended Warranty Service Considering Retailers' Fairness

Jiali Pan[1], Xingjian Zhou[2], and Yong Wang[3,4](✉)

[1] School of Management, Guangzhou College of Commerce, Guangzhou 511363, China
[2] Postdoctoral Research Station in Management Science and Engineering, Nanchang University, Nanchang 330031, China
[3] School of Transportation, Wuhan University of Technology, Wuhan 430063, China
[4] School of Logistics, Wuhan Technology and Business University, Wuhan 430065, China

**Abstract.** With the rapid development of e-commerce, more and more manufacturers are building multiple channels to expand the sales of their products. Under this trend, different channel strategies are very prone to channel conflicts. After channel conflicts occur, the cooperation between enterprises may break or even end the cooperative relationship. This article provides reasonable channel strategies for supply chain members under different channel structures, maintains the stability of channels and supply chains, and provides support for enterprises in channel strategy decisions. The results show that manufacturers can reasonably adjust the wholesale price to maintain the stability of the channel and supply chain. The addition of channels for manufacturers increases their own interests and damages the interests of other members. With the increase of channels, the utility of traditional retailers will decrease with the increase of traditional retailer's fairness concern.

**Keywords:** Fairness concern · Online and offline · Channel strategy · Extended warranty service · Supply chain

## 1 Instruction

As more and more companies adopt multi-channel distribution, channel conflicts between manufacturers and retailers have become more and more obvious. Channel conflicts are mainly due to the double marginal effect when the goals of the members of the supply chain are inconsistent, or when the goals of the members of the supply chain are the same but there is no coordinated action [1]. In order to alleviate channel conflicts, the supply chain members of the strong side can make reasonable pricing by considering the fair concerns of the weak side and distribute the benefits to avoid channel conflicts caused by price competition. With the ever-changing global economic environment and the emergence of new business models, the main way of competition between enterprises has become a competition for overall price + value-added services [2]. Consumers not only pay attention to the quality and price of the product itself, but also pay attention


Z. Hu et al. (Eds.): ICAILE 2021, LNDECT 82, pp. 317–326, 2021.
https://doi.org/10.1007/978-3-030-80475-6_31

to the services that companies provide to consumers who have product failures. Companies have introduced a paid extended warranty service strategy for this. Consumers only need to pay a certain fee, and after the product passes the warranty period, they will continue to provide extended warranty services for a period of time. The introduction of this service not only reduces the cost of users using the product during the warranty period, but also becomes a new source of corporate profit growth.

## 2 Literature Review

In response to the problem of channel conflicts, the existing business and academic circles have begun pay attention to the "channel strategy", some scholars have conducted related research. Tsan, Chen, and Chung discovered that the trade-off between delay and order cost advantages, compare different franchise contracts, study different scenarios, find solutions to avoid channel conflicts and cannibalism between franchisees and brand owners [3]. Dan, Xu, and Liu discovered that approximately 42% of top industrial suppliers sell directly to consumers through direct channels [4]. According to Kabadayi, conducted an empirical study on the supply chain to increase direct channels or indirect channels on the basis of the original channels [5]. The results show that in the face of high external and internal environmental uncertainties, choosing to add additional direct channels is the most beneficial to corporate decision-making. Tsao and Su discovered that the dual-channel supply chain includes the manufacturer's direct channels and retail channels to determine the best price and shelf life [6]. Chiang, Chhajed, and Hess discovered that manufacturers can profit from dual channels, but for retailers may cause loss of profits [7]. From data prepared by Dumrongsiri et al., manufacturers using dual channels are more advantageous than single channels when wholesale prices, demand changes and customer valuations are low, and retailers have higher marginal costs [8].

A large number of experimental results show that the members in the game are willing to give up part of their interests to achieve a fairer result due to fair concerns. Therefore, fair concern behavior affects and acts on the main body of supply chain decision-making to a certain extent. Cui, Raju, and Zhang discovered that suppliers can coordinate the supply chain under members with fair preferences by adjusting wholesale prices [9]. Loch and Wu discovered that under normal circumstances, members in a weak position will pay more attention to their own profits, and find a way to coordinate a "fairness" by comparing with the other party's earnings [10]. Tan and Yi discovered that fairness concerns about the delivery time strategy of online direct sales and product pricing information under dual channels [11]. He and Li discovered that the influence of retailers with fairness concerns on the decision-making of dual-channel supply chain networks [12]. Chen believes that online retailers can provide special services to meet the needs of online customers and allow customers to enjoy the online shopping process [13]. However, the existing extended warranty service channel strategy research has not yet considered the simultaneous online and offline operations.

In real life, there are multiple online and offline channels coexisting, which is very critical for the practical application of enterprises. Therefore, this article intends to build an extended warranty service model under three different channel structures, taking into account the fairness of traditional retailers in the supply chain, and obtaining optimal pricing and reasonable profit distribution for extended warranty services through optimal decisions. The calculation example analyzes the influence of the traditional retailer's fairness concern coefficient on the supply chain decision-making. Conducive to alleviating channel conflicts, improving the satisfaction of traditional retailers, and providing reasonable channel strategies for supply chain members [14].

# 3 Model Description and Model Assumption

## 3.1 Model Description

During the channel building process, the manufacturer occupies the channel control position; the traditional retailer belongs to the subordinate position. The channel structure model composed of manufacturers and traditional retailers, this article will be divided into offline distribution channel structure model, offline distribution online direct sales channel structure model, offline distribution, direct sales and online direct sales channel structure model, specific channels Structural model, as shown in Fig. 1.

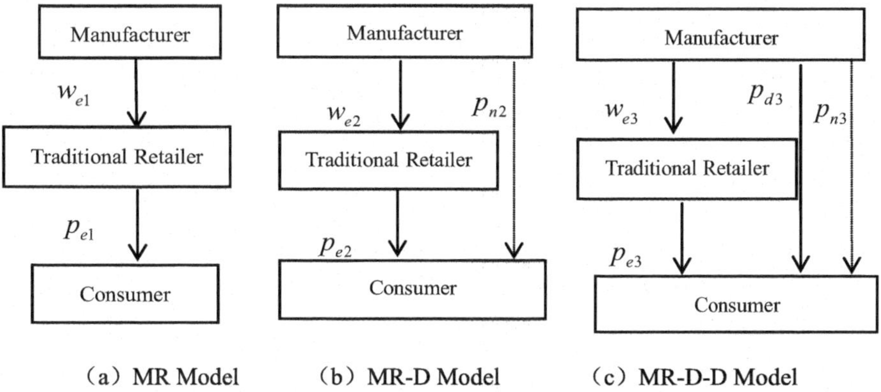


**Fig. 1.** Structure diagram of the extended warranty service channel

## 3.2 Model Assumption

This article considers that manufacturers only produce one extended warranty service product. The variable names and their meanings involved in this issue are shown in Table 1.

In order to facilitate the solution of the model, this article does not consider the production cost of the extended warranty service manufacturer for the time being. Through practical verification, this approach does not affect the model's conclusion. Suppose that traditional retailers have fair concerns about preference, use the profits of manufacturers in the system as a reference point to measure their own utility. Most of this fairness concern exists in the game relationship between principal and agent, which

**Table 1.** Variable names and their meanings

| Variable name | Meaning |
|---|---|
| $a$ | Potential market demand for this product $a > 0$ |
| $k$ | Coefficient of price competition between different channels $0 < k < 1$ |
| $p_e$ | Retail price of offline distribution extended warranty service |
| $p_{d3}$ | Offline direct sale extended warranty service price |
| $p_n$ | Online distribution extended warranty service price |
| $w_e$ | Wholesale price of offline distribution extended warranty service |
| $\theta$ | Proportion of offline channel market demand $0 < \theta < 1$ |
| $\alpha$ | Proportion of offline distribution market demand $0 < \alpha < 1$ |
| $\Pi_m$ | Manufacturer profit |
| $\Pi_r$ | Traditional retailer profits |
| $U_r$ | Traditional retailer utility |
| $\lambda$ | Fair concern factor $0 < \lambda < 1$ |

can be called vertical fairness concern or distribution fairness concern [15]. Then the traditional retailer utility function can be expressed as:

$$U_r = \Pi_r - \lambda(\Pi_m - \Pi_r) \tag{1}$$

# 4 Channel Model of Extended Warranty Service Under Fair Concerns

## 4.1 Traditional Channel Structure Model (MR Model)

Under the traditional channel structure, the manufacturer first decides the wholesale price of the extended warranty service, then the traditional retailer with fair concerns acts to determine the retail price of the extended warranty service, from which the manufacturers profit function can be obtained as:

$$\Pi_{m1} = w_{e1}(a - p_{e1}) \tag{2}$$

The retailer's profit function can be expressed as:

$$\Pi_{r1} = (p_{e1} - w_{e1})(a - p_{e1}) \tag{3}$$

The retailer's utility function can be expressed as:

$$U_{r1} = \Pi_{r1} - \lambda(\Pi_{m1} - \Pi_{r1}) \tag{4}$$

### 4.2 The Channels Structure Model of Offline Distribution and Online Direct Sales (MR-D Model)

Under the structure of offline distribution and online direct sales channels, the manufacturer first determines the direct sales price of the extended warranty service and the wholesale price of the extended warranty service distribution channels, and then the traditional retailer with a fair concern behavior determines the retail price of the extended warranty service, from which the manufacturers profit function can be obtained as:

$$\Pi_{m2} = w_{e2}(\theta a - p_{e2} + kp_{n2}) + p_{n2}((1-\theta)a - p_{n2} + kp_{e2}) \tag{5}$$

The retailer's profit function can be expressed as:

$$\Pi_{r2} = (p_{e2} - w_{e2})(\theta a - p_{e2} + kp_{n2}) \tag{6}$$

The retailer's utility function can be expressed as (Table 2):

$$U_{r2} = \Pi_{r2} - \lambda(\Pi_{m2} - \Pi_{r2}) \tag{7}$$

**Table 2.** MR model, MR-D model and MR-D-D model solution results.

| | MR model | MR-D model | MR-D-D model |
|---|---|---|---|
| Offline wholesale price | $w_{e1}^* = \frac{a(1+\lambda)}{2(1+2\lambda)}$ | $w_{e2}^* = \frac{\begin{pmatrix} k(1-\theta)(1+2\lambda) \\ +\theta(1+\lambda)+\lambda ck^2 \end{pmatrix} a}{2(1-k^2)(1+2\lambda)}$ | $w_{e3}^* = \frac{\begin{pmatrix} \theta(k\alpha(2+3\lambda)-\alpha(1+\lambda)) \\ -k(1+2\lambda)-2\theta\alpha k^2\lambda \end{pmatrix} a}{2(1+k)(2k-1)(1+2\lambda)}$ |
| Offline retail price | $p_{e1}^* = \frac{3a}{4}$ | $p_{e2}^* = \frac{\left(-\theta k^2+2k(1-\theta)+3\theta\right)a}{4(1-k^2)}$ | $p_{e3}^* = \frac{\left(2\theta\alpha k^2+5k\theta\alpha-2k-3\theta\alpha\right)a}{4(1+k)(2k-1)}$ |
| Online direct selling price | / | $p_{n2}^* = \frac{(k\theta+1-\theta)a}{2(1-k^2)}$ | $p_{n3}^* = \frac{(k-1+\theta(1-2k))a}{2(1+k)(2k-1)}$ |
| Offline direct selling price | / | / | $p_{d3}^* = \frac{((1-\alpha)(2k\theta-\theta)-k)a}{2(1+k)(2k-1)}$ |
| Manufacturers profit | $\Pi_{m1}^* = \frac{a^2(1+\lambda)}{8(1+2\lambda)}$ | $\Pi_{m2}^* = \frac{\left(\theta^2k^2(3\lambda+1)+A\right)a^2}{8(1-k^2)(1+2\lambda)}$ | $\Pi_{m3}^* = \frac{-a^2\left(kC+2k^2\theta^2\alpha^2(3\lambda+1)+C_1\right)}{8(1+2\lambda)(1+k)(2k-1)}$ |
| Retailer profit | $\Pi_{r1}^* = \frac{a^2(1+4\lambda)}{16(1+2\lambda)}$ | $\Pi_{r2}^* = \frac{\theta^2a^2(1+4\lambda)}{16(1+2\lambda)}$ | $\Pi_{r3}^* = \frac{a^2\theta^2\alpha^2(1+4\lambda)}{16(1+2\lambda)}$ |

*(continued)*

**Table 2.** (*continued*)

| | MR model | MR-D model | MR-D-D model |
|---|---|---|---|
| Retailer utility | $U_{r1}^* = \frac{a^2(1+4\lambda)}{16(1+2\lambda)}$ | $U_{r2}^* = \frac{a^2(8k\lambda\theta(1-\theta)+B)}{16(1-k^2)}$ | $U_{r3}^* = \frac{a^2\left(k^2\theta^2\alpha^2(10\lambda+2)+kD+D_1\right)}{16(1+k)(2k-1)}$ |

Note:
$A = (1+2\lambda)(2(1-2\theta)+4k\theta(1-\theta)) + \theta^2(3+5\lambda), B = \theta^2\left((1-3\lambda)-k^2(5\lambda+1)\right) - 4\lambda(1-2\theta)$

$$C = \left(-\theta^2\alpha^2(7+13\lambda) - 8\theta^2(1-\alpha)(1+2\lambda) - 2(1+2\lambda)(1+4\theta)\right),$$
$$C_1 = \theta^2\alpha^2(3+5\lambda) + 2(1+2\lambda)\left(1-2\theta+2\theta^2(1-\alpha)\right)$$

$$D = -11\theta^2\alpha^2\lambda + 16\lambda\theta(1-\theta+\theta\alpha) - 4\lambda + \theta^2\alpha^2,$$
$$D_1 = -\theta^2\alpha^2(1+3\lambda) - 8\theta\lambda(1-\theta+\theta\alpha) + 4\lambda$$

### 4.3 Structure Model of Offline Distribution, Direct Sales and Online Direct Sales Channels (MR-D-D Model)

Under the structure of offline distribution, direct sales and online direct sales channels, manufacturers first determine the direct selling price of extended warranty services and the wholesale price of extended distribution channels, and then traditional retailers with fair concern behaviors decide the retail sales price of extended warranty services. Manufacturers profit function can be expressed as:

$$\Pi_{m3} = w_{e3}(\alpha\theta a - p_{e3} + k(p_{d3}+p_{n3})) + p_{d3}((1-\alpha)\theta a - p_{d3} + k(p_{e3}+p_{n3})) \\ + p_{n3}((1-\theta)a - p_{n3} + k(p_{e3}+p_{d3})) \tag{8}$$

The retailer's profit function can be expressed as:

$$\Pi_{r3} = (p_{e3} - w_{e3})(\alpha\theta a - p_{e3} + k(p_{d3}+p_{n3})) \tag{9}$$

The retailer's utility function can be expressed as:

$$U_{r3} = \Pi_{r3} - \lambda(\Pi_{m3} - \Pi_{r3}) \tag{10}$$

## 5 Results and Discussion

In order to verify the previous research results of the supply chain extended service channel, and at the same time get more management inspiration, we can carry out the analysis example shown in Fig. 2, 3, 4, 5 and 6.

The numerical simulation experiment here can take the following basic data: $a = 500$, $k = 0.4$, $\theta = 0.5$, $\alpha = 0.6$

### 5.1 An Analysis of Examples of Extended Warranty Price and Retailers' Fair Concern

The wholesale price of extended warranty service decreases with the increase of the traditional retailer's fair concern coefficient, while the extended service retail price does not change with the change of the traditional retailer's fair concern coefficient, as shown in Fig. 2 and Fig. 3. This is because with the increase in the fairness factor of traditional retailers, the traditional retailers' unfair perception of profit distribution and channel negotiation capabilities have also increased.

In real life, in order to maintain the stability of channels and supply chains, manufacturers adjust wholesale prices and make concessions by reducing the wholesale prices of traditional channels. However, traditional retailers and manufacturers are in a rational state when selling extended warranty services, and do not consider the fairness concern coefficient and still sell them at the best price.

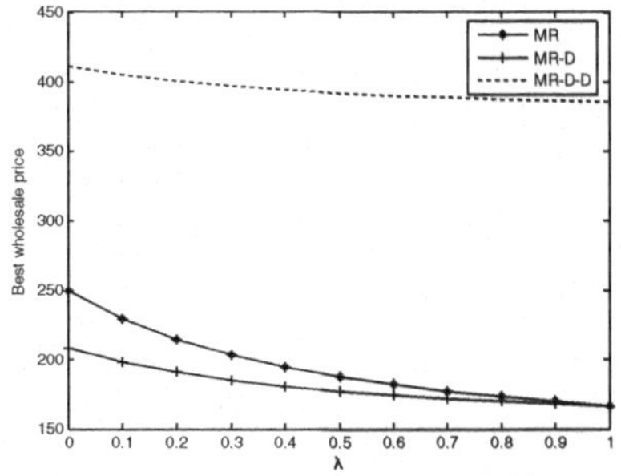


**Fig. 2.** The influence of the traditional retailer's fairness concern coefficient on wholesale prices;

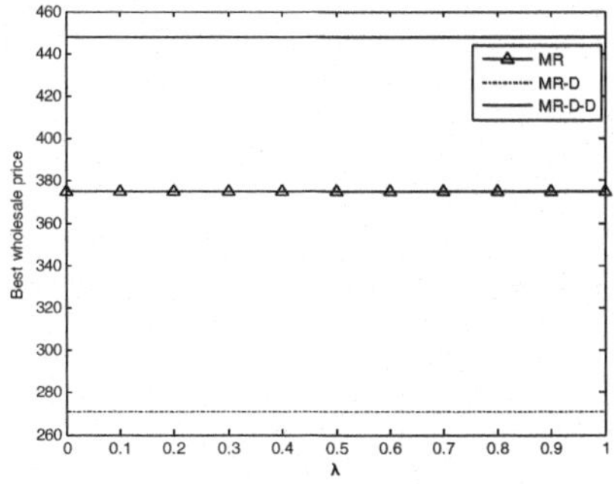


**Fig. 3.** The influence of the traditional retailer's fair concern coefficient on offline retail

### 5.2 Analysis of Supply Chain Members' Profit and Retailers' Fair Concern Coefficient

The manufacturer's profit decreases with the increase of the traditional retailer's fairness factor, while the traditional retailer's profit increases with the increase of the traditional retailer's fairness factor, as shown in Fig. 4, Fig. 5 and Fig. 6.

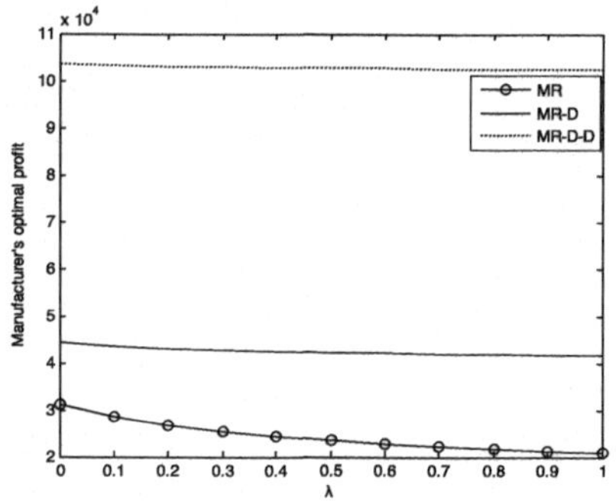


**Fig. 4.** Manufacturer's optimal profit.

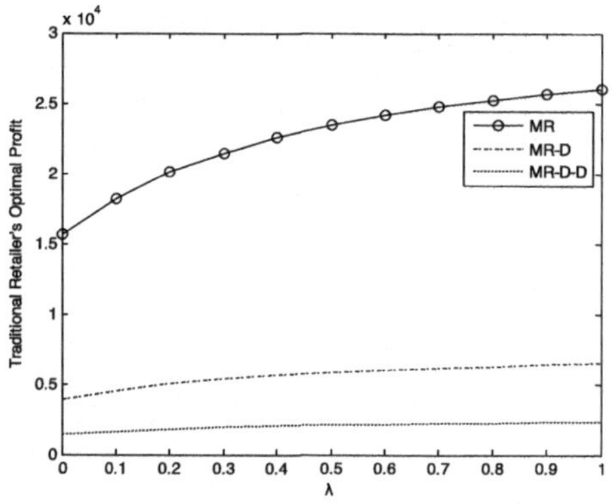


**Fig. 5.** Traditional retailer's optimal profit

In order to maintain the stability of the channel, manufacturers give traditional retailers by way of concession, causing loss of their own profits. At the same time, traditional retailers should try to avoid excessive focus on fairness concerns and irrational competition with manufacturers. The fair attention behavior of traditional retailers enhances the ability to negotiate prices with manufacturers, thereby increasing their own profits.

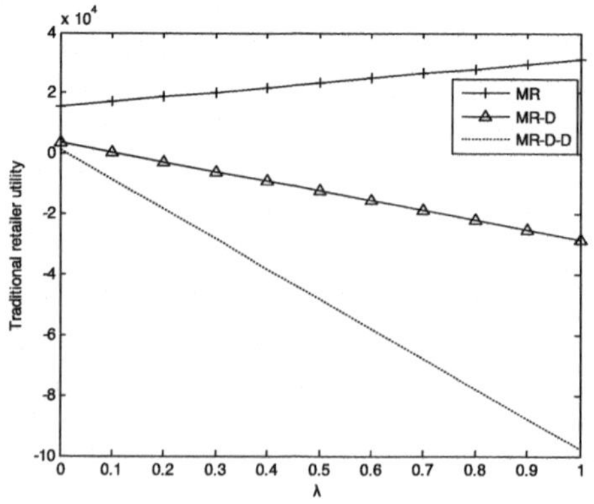


**Fig. 6.** Utility of traditional retailers

## 6 Conclusion

As a dominant manufacturer, in order to ensure the stable delivery of their products downstream, they should actively formulate relevant measures to ease channel conflicts. The irrational behaviors of traditional retailers have caused channel instability and damage to all aspects of the supply chain's profits. The supply chain profits can be reasonably distributed through reasonable negotiation.

As for traditional retailers, they should not pay too much attention to fairness and irrational competition with manufacturers. They can use their own characteristics to achieve differentiated sales, improve the purchase satisfaction and purchase effectiveness of downstream customers, and make use of their advantages to make up for deficiencies.

This article has limited rational analysis of the manufacturer's channel strategy selection. There are still many things that need to be considered in depth. The follow-up work can be studied from the retailer's channel strategy.

**Acknowledgment.** This project is supported by Key Projects of CAST (China Association of Science and Technology) Project (2018CASTQNJL33); Fundamental Research Funds for the Central Universities (2019-JL-008); MOE Project of Humanities and Social Sciences 14YJCZH154); WTBU Academic Team (XSTD2015004); National Natural Science Foundation of China (72062019); Postdoctoral Research Project of Jiangxi Province (2019KY13).

## References

1. Zhao, Y., Zhu, W.: Research on online and offline channel conflicts and price games of retail clothing brands. Shanghai Text. Technol. **45**(10), 61–63 (2017). (in Chinese)
2. Cheng, Q., Tang, S.: Manufacturer's pricing strategy for supply chain with service level-dependent demand. Int. J. Eng. Manuf. **4**(4), 1–13 (2014)
3. Choi, T.M., Chen, Y., Chung, S.H.: Online-offline fashion franchising supply chains without channel conflicts: choices on postponement and contracts. Int. J. Prod. Econ. **215**(09), 174–184 (2016)
4. Dan, B., Xu, G., Liu, C.: Pricing policies in a dual-channel supply chain with retail services. Int. J. Prod. Econ. **139**(1), 312–320 (2012)
5. Kabadayi, S.: Adding direct or independent channels to multiple channel mix. Direct Mark. Int. J. **2**(2), 66–80 (2013)
6. Tsao, Y.C., Su, P.Y.: A dual-channel supply chain model under price and warranty competition. Int. J. Innov. Comput. Inf. Control Ijicic **8**(3B), 2125–2135 (2012)
7. Chiang, W.Y.K., Chhajed, D., Hess, J.D.: Direct marketing, indirect profits: a strategic analysis of dual-channel supply chain design. Manage. Sci. **49**(1), 1–20 (2003)
8. Dumrongsiri, A., Fan, M., Jain, A., et al.: A supply chain model with direct and retail channels. Eur. J. Oper. Res. **187**(3), 691–718 (2008)
9. Cui, T.H., Raju, J.S., Zhang, Z.J.: Fairness and channel coordination. Manage. Sci. **53**(8), 1303–1314 (2007)
10. Loch, C.H., Wu, Y.: Social preferences and supply chain performance: an experimental study. Manage. Sci. **54**(11), 1835–1849 (2008)
11. Tan, C., Yi, W.: Dual-channel supply chain pricing and online direct sales delivery time strategy. J. Syst. Eng. **34**(05), 683–699 (2019). (in Chinese)

12. He, L., Li, Z.: Decision analysis of cooperative advertising in dual-channel supply chain considering retailers' fairness concerns. Ind. Eng. Manage. **22**(04), 107–114 (2017). (in Chinese)
13. Chen, X., Xiao, K.: An analysis on price dispersion in online retail market based on the different of the product levels. Int. J. Eng. Manuf. **2**(4), 54–58 (2016)
14. Ting, P.-H.: An efficient and guaranteed cold-chain logistics for temperature-sensitive foods: applications of RFID and sensor networks. Int. J. Inf. Eng. Electron. Bus. **5**(6), 1–5 (2013)
15. Jiang, L., Zhang, Y., Jian, J.: Supply chain decision model of competition between two manufacturers under different reference points of fairness concerns. Comput. Integr. Manuf. Syst. **25**(01), 224–234 (2019). (in Chinese)

# Strategic Choice of Smart Logistics Business Model Based on Supply Chain Perspective

Na Liu and Xiaojun Zhang(✉)

School of Econoimics and Business Foreign Languages, Wuhan Technology and Business University, Wuhan 430065, China

**Abstract.** As a new business model, smart logistics has the development advantages of intelligence, informatization, networking and integration compared with traditional logistics mode. It solves the low efficiency problem of traditional logistics mode. However, due to the lack of coordination between logistics enterprises and logistics operation processes, there are still high cost operation problems in China's smart logistics business model. Therefore, it is of great significance to explore new logistics development mode. From the perspective of supply chain, this paper puts forward the construction scheme of supply chain smart logistics mode based on logistics information service platform, analyzing the advantages and characteristics of the typical supply chain smart logistics mode crowdsourcing logistics mode, and according to the actual situation of China's logistics industry, putting forward specific strategies for the transformation of supply chain smart logistics mode.

**Keywords:** Smart logistics · Supply chain · Crowdsourcing logistics

## 1 Introduction

With the rapid development of economic globalization and communication information technology, logistics industry is in constant reform and development. As a comprehensive service industry, logistics industry not only supports the steady development of China's national economy, but also determines the quality of China's economic development [1]. The government attaches great importance to the logistics industry and encourages business model innovation of logistics industry [2]. Therefore, the logistics industry should seize the development opportunities of the times, constantly explore the combination with modern Internet technology, realize the intelligent and scientific development of traditional logistics industry through big data, cloud computing and artificial intelligence technology, promote the transformation and upgrading of traditional logistics industry, and innovate the intelligent logistics business mode [3].

Compared with the traditional logistics mode, the smart logistics mode has the following characteristics: Firstly, the development of information technology. All operational links of logistics enterprises can be controlled artificially such as collection, transmission, summary, identification and tracking; Secondly, the development of intelligent.


Z. Hu et al. (Eds.): ICAILE 2021, LNDECT 82, pp. 327–335, 2021.
https://doi.org/10.1007/978-3-030-80475-6_32

Logistics system have self perception and learning ability using big data, cloud computing and artificial intelligence, making self judgment on surrounding environment in time, and solving problems in the process of logistics business operation independently; Thirdly, the development of network. In the field of logistics, logistics enterprises are closely connected with each other through intelligent equipment, and logistics information is processed by data, so that logistics enterprises can process logistics information anytime and anywhere, and further excavate logistics information and resources; Fourthly, the development of collaborative. In a logistics supply chain system, logistics enterprises within the supply chain are closely around the core enterprises to carry out logistics business, and different logistics enterprises need to coordinate and operate synchronously jointly to complete the logistics service objectives [4]. The biggest difference between smart logistics and traditional logistics is that smart logistics integrates internet technology, intelligent sensing technology, cloud computing, big data and other technologies to form a more efficient new logistics format [5].

Although the smart logistics business model shown in Fig. 1 can solve the problem of low operation efficiency of traditional logistics industry, due to the lack of coordination of various operation links and the lack of organization of upstream and downstream industries, the existing smart logistics business model still has the problem of high operating cost. It is necessary to further build a more sound and collaborative intelligent logistics provider from the perspective of industrial chain business model [6].

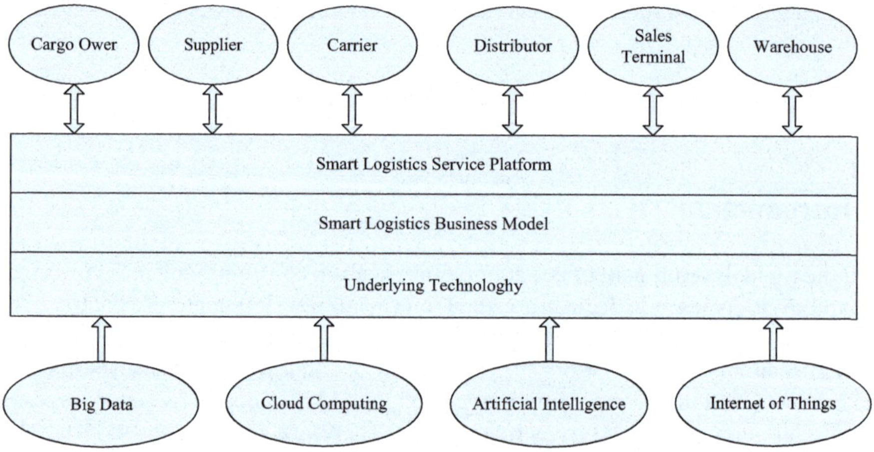


**Fig. 1.** Smart logistics business model

## 2 Characteristics and Difficulties of Smart Logistics Supply Chain System

### 2.1 Characteristics of Smart Logistics Supply Chain

Smart logistics supply chain can solve the problem of high cost in China's smart logistics industry [7]. By strengthening the connection between the upstream and downstream links of the supply chain, integrating information data such as service flow, information flow and material flow, and directly connecting with consumers, it can achieve the goal of improving the operation efficiency of smart logistics, reducing costs and improving the economic benefits of logistics enterprises. Smart logistics supply chain has strong management capabilities, which makes the advantages of smart logistics business model further enlarged, with stronger network, complexity, circulation and information characteristics [8]. Specifically speaking:

Firstly, the network characteristics of the smart logistics supply chain are more obvious, especially the internet technology strengthens the tightness between upstream and downstream enterprises and logistics operation links, and they are not purely subordinate relationship between each other, and some operational links are also intersecting, so the network structure is more tight.

Secondly, the smart logistics supply chain is more complex. Since the smart logistics business model is integrated into the supply chain management, the chain of the whole smart logistics business model is longer. Logistics enterprises need to consider not only their own development, but also the development of upstream and downstream enterprises. Therefore, in order to avoid unnecessary operation links and eliminate additional costs, the smart logistics supply chain architecture needs to simplify the business operation process as much as possible, and eliminate unnecessary intermediate procedures.

Thirdly, the smart logistics supply chain has stronger collaborative management ability, which is mainly reflected in inventory management. Supply chain management technology can make the inventory of logistics enterprises copy along the chain and produce bullwhip effect, so as to reduce the horizontal enterprise inventory and vertical enterprise inventory in the supply chain system [9].

Fourthly, the smart logistics supply chain structure breaks the information gap between enterprises and eliminates the information island problem to a certain extent. The supply chain system can provide a common information sharing platform for all enterprises in the whole logistics industry chain, so that a large number of logistics basic data can be collected and stored in a centralized way. This will enable the core logistics enterprises to master more information and data, better coordinate the various operation links and make the most scientific and correct decisions, so as to achieve the goal of reducing cost and increasing efficiency of smart logistics.

## 2.2 Problems of Smart Logistics Supply Chain Development

Although the development of smart logistics supply chain will bring many logistics industry innovation, help logistics enterprises to accurately predict the logistics market and formulate demand plans, it can have a deeper understanding of the development trend of smart logistics, and then obtain competitive advantages [10]. However, the logistics enterprises in transition will also face many problems in the process of developing smart logistics and supply chain development, such as cost budget, how to obtain and use information data, realize data value and make scientific decision. Specifically speaking:

The information data in the logistics market is very large, how can logistics enterprises transform a large number of information data into effective decision-making is helpful to the development of logistics enterprises through reasonable selection. At present, internet technologies such as big data and cloud computing have been relatively mature, and logistics enterprises have been able to obtain information data in the logistics market by using these technologies. However, in order to bring the value of information data into full play, it is necessary for logistics enterprises to establish a set of relatively perfect workflow when using big data and cloud computing technology, so as to avoid the logistics enterprises blindly using technical analysis, which will bring large analysis cost to logistics enterprises [11].

Although the development of smart logistics supply chain can improve the service ability of each link of the logistics industry, it can meet the market demand of different consumers through accurate prediction ability. However, as the logistics supply chain is a comprehensive service mode involving more than ten industries, how to coordinate the operation ability of upstream and downstream enterprises and different operation links undoubtedly poses a great challenge to the logistics core enterprises.

In the process of the transformation of logistics enterprises to smart logistics and supply chain development mode, there are still technical and human development difficulties. For example, the management system of traditional logistics enterprises for logistics operation is relatively backward, and how to upgrade the whole supply chain enterprises in a unified way remains to be solved. At the same time, the development of smart logistics can not do without the support and protection of talents. The traditional development mode of logistics in China belongs to labor-intensive industry, and its talent base is obviously insufficient. Therefore, it is urgent to establish the corresponding personnel training mechanism as soon as possible.

## 3 The Mode of Smart Logistics Supply Chain

### 3.1 The Core Foundation of Smart Logistics Supply Chain Development

The essence of smart logistics supply chain mode is the upgrading of traditional supply chain logistics, which is the advanced stage of logistics industry development. The smart logistics supply chain should take the logistics management system as the core to integrate production, processing, sales and distribution [12]. In the process of connecting the subsystems, it is necessary to play the intelligent role of the system to connect the enterprises in the supply chain in the form of automation and intelligence, so as to realize the exchange of resources and information. Therefore, the key point of smart logistics supply chain mode is to establish a service platform with information sharing and resource sharing, and manage the whole logistics supply chain service through the platform resource integration ability. Meanwhile, in order to reflect the intelligent characteristics of logistics industry in the internet era, it is necessary to establish different levels of information service platform to connect with social resources departments and provide integrated services for logistics enterprises in the supply chain system [13].

Smart logistics supply chain information service platform should include three subsystems, including management system, information system and service system. Among them, the management system is the embodiment of logistics enterprise intelligence, including distribution goods tracking, transportation vehicle scheduling, etc.; the information system is the operation center, all the information data in the supply chain system will be completed through the information system, including information interaction, information query, information management and transaction, etc.; service system is the key and core of the smart logistics supply chain mode, which is different from other logistics modes, and is an important manifestation of the integration of smart logistics supply chain mode. In the past, the financing, policy analysis, commodity transfer and other business links that need to be independently completed by logistics enterprises can be completed through information and information service platform.

### 3.2 Analysis on the Typical Model of Smart Logistics Supply Chain - Crowdsourcing Logistics Model

The emergence of crowdsourcing logistics business model is closely related to the rapid development of China's logistics industry and the lack of logistics distribution resources. On the one hand, the rapid development of the logistics industry provides the necessary premise and foundation for the innovation of the new logistics mode; on the other hand, the disadvantages and defects of the traditional logistics industry distribution mode based on human resources are constantly emerging with the increase of human resource cost. Therefore, the essence of crowdsourcing logistics mode is to solve the problem of the shortage of labor resources and the rising labor cost in the logistics industry. Through the internet channel, the work required by the professional distribution personnel of logistics enterprises is allocated to other social human resources in the form of paid and voluntary, and the goods are delivered to customers [14]. Therefore, the initial business model of crowdsourcing logistics is a complete operation process consisting of crowdsourcing logistics platform, crowdsourcing logistics carrier, crowdsourcing logistics developer and final customers, which is composed of information flow, logistics and capital flow. Then, the crowdsourcing logistics platform will spread the distribution service on the platform to find distribution personnel. After receiving the order, the crowdsourcing logistics platform will deliver products to customers. In terms of capital delivery, when the final customer receives the product, it will notify the crowdsourcing logistics initiator, and then pay the funds to the crowdsourcing platform and distribution personne [15]. At the same time, after the completion of distribution services, the distribution personnel will also inform the platform to complete a complete crowdsourcing logistics operation process.

The crowdsourcing logistics business model in Fig. 2 is the main operation mode in the initial stage of crowdsourcing logistics mode, which is suitable for the distribution business of small and non-important products. With the continuous development of crowdsourcing logistics, more and more businesses and internet e-commerce platforms join or participate in the crowdsourcing logistics industry. The business scope of crowdsourcing logistics mode is becoming wider and wider, and the products and services distributed are more and more. In this regard, many crowdsourcing logistics platforms and other supplier platforms introduce insurance institutions into the crowdsourcing logistics industry chain to strengthen the safety and quality assurance of distribution products. In the new business model of crowdsourcing logistics, the profit model of crowdsourcing logistics platform is to charge franchise fee, advertising fee, distribution order extraction fee and service fee from other platforms.

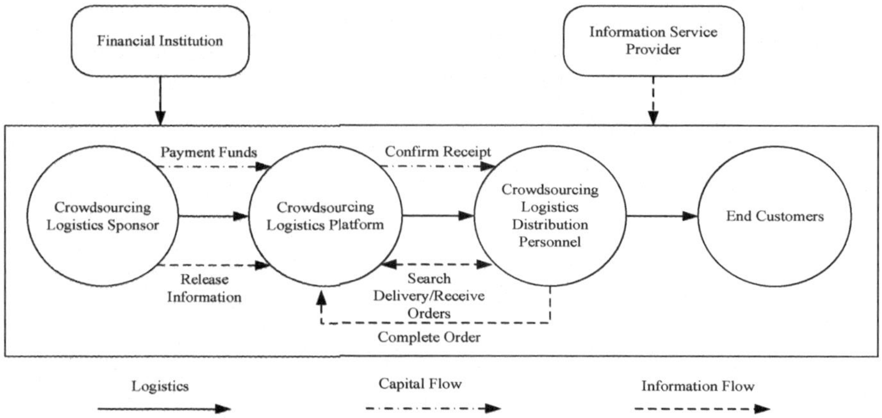


**Fig. 2.** Crowdsourcing logistics business model

# 4 Strategic Choice of Promoting Healthy Development of Crowdsourcing Logistics in China

At present, crowdsourcing logistics is developing rapidly in China, and its business model has been basically formed. Through o2o (online and offline) integration, the logistics cost has been greatly reduced, the logistics industry boundary has been extended, and the social capital integration function has been effectively realized. However, the crowdsourcing logistics mode, as a new business form, is still in disorder. It is necessary for relevant regulatory authorities to strengthen risk supervision and business compliance supervision, promote the healthy and orderly development of crowdsourcing logistics mode through supply chain integration and information sharing [16].

## 4.1 Introduce Laws and Regulations for Crowdsourcing Logistics Mode and Improve the Corresponding Regulatory Mechanism

The development of any new things can not do without the protection of laws and policies. Therefore, it is necessary to speed up the introduction of laws and regulations for the crowdsourcing logistics mode, so that the crowdsourcing logistics mode has laws to abide by. It is suggested that relevant legislative departments should modify the existing laws and regulations of the logistics industry according to the problems existing in the developing crowdsourcing logistics mode, so as to meet the development needs of the emerging business model. Through the formulation of laws and regulations, protect the crowdsourcing logistics enterprises, the interests of carriers and the rights and interests of consumers [17]. Firstly, in terms of mechanism construction, establish blacklist system, and take corresponding punishment measures for illegal platforms and individuals. According to the violation situation, it can be punished in the form of warning, rectification, fine and industrial withdrawal. Secondly, establish a deposit system to fully protect the interests of consumers. The practice of the deposit system can refer to the financial industry, and deposit into the accounts of third-party enterprises in proportion to the cost of delivering products. Once consumers' interests are damaged, they should be compensated in time. Thirdly, strengthen industry self-discipline and establish industry standards. It is suggested that the logistics industry association should issue policies to guide the crowdsourcing logistics enterprises to establish business operation standards to ensure the quality of logistics services in accordance with their own business conditions.

## 4.2 Pay Attention to the Cultivation of Industrial Talents and Enhance the Professional Ability of Service Personnel

At present, the problem of uneven level of talent quality in crowdsourcing logistics industry is becoming increasingly serious, and the quality of logistics service is declining day by day. In addition, most of the service personnel of crowdsourcing logistics come from the part-time society. Their own business ability is insufficient, and they are lack of experience in dealing with emergencies. Therefore, it is necessary to strengthen the training of logistics personnel. Firstly, establish a professional training mechanism and process, develop a targeted training plan, and choose online and offline training methods according to the characteristics of crowdsourcing logistics business. Some skills need to be taught offline and face-to-face, while examinations and assessments are optional to choose online mode. Secondly, we should establish a talent training pool. Relying on modern internet technologies such as big data and cloud computing to integrate resources, centralize service personnel engaged in related business, carry out regular professional training, and cultivate a batch of crowdsourcing logistics distribution talents with professional quality [18]. Thirdly, establish a reasonable reward and punishment system. Crowdfunding logistics enterprises need reasonable incentive measures to ensure the enthusiasm and attitude of service personnel, so it is necessary to establish a targeted reward and punishment mechanism.

## 4.3 Give Full Play to the Technological Advantages of Crowdsourcing Logistics Enterprises and Explore New Profit Mechanism Mode

As an important driving force of economic development, science and technology is also the engine of crowdsourcing logistics enterprises. Make full use of technologies such as big data, cloud computing, internet of things and artificial intelligence will help to improve the service quality and efficiency of crowdsourcing logistics industry. First of all, it is suggested that the crowdsourcing logistics enterprises should establish the industrial information database, and focus on the collection and analysis of industrial data to provide the optimal processing scheme for business operation. Secondly, in terms of order allocation, according to the fit degree of product distribution path, the orders are combined scientifically and distributed to the minimum number of distributors to achieve the optimal allocation of resources. Thirdly, in terms of profit model, in addition to the traditional profit-making methods such as advertising and business, we can also rely on the principle of complementary resources and collaborative development to actively carry out strategic cooperation with other supply platforms, internet technology enterprises, logistics and transportation enterprises. In order to ensure the long-term and efficient development of crowdsourcing logistics mode, we should give full play to the concept of sharing economy, realize resource sharing and collaborative operation in a wider range of cross industry forms, so as to ensure the long-term and efficient development of crowdsourcing logistics mode.

## 5 Conclusion

Smart logistics, as a new mode of business combination of internet technology and traditional logistics industry, has advantages of automation, intelligence and network development compared with traditional logistics, effectively solving the problem of high cost and low efficiency in traditional logistics operation. From the perspective of supply chain, crowdsourcing logistics mode has the advantages of human resources and logistics cost, and has become a new mode of smart logistics development. In order to promote the development of crowdsourcing logistics mode, we should establish coordination mechanism, issue laws and regulations, strengthen personnel training, promote enterprise transformation, so as to improve the logistics adaptability and evolution ability and realize further development.

**Acknowledgment.** This project is supported by school Projects of Wuhan Technology and Business University (2019ZX01).

## References

1. He, J.: Analysis of the aggregation and innovation mechanism of intelligent logistics industry development. Logist. Eng. Manage. (02), 60–62 (2020). (in Chinese)
2. Ting, P.-H.: An efficient and guaranteed cold-chain logistics for temperature-sensitive foods: applications of RFID and sensor networks. Int. J. Inf. Eng. Electron. Bus. **5**(6), 1–5 (2013)
3. Pan, Z.: Research on the application of blockchain in the development of intelligent logistics. Prices Monthly (05), 45–48 (2019). (in Chinese)
4. Cheng, Q., Tang, S.: Manufacturer's pricing strategy for supply chain with service level-dependent demand. Int. J. Eng. Manuf. **4**(4), 1–13 (2014)
5. Zhen, Q.: Development mode, problems and countermeasures of smart logistics in China. Logist. Manage. (12), 89–92 (2018). (in Chinese)
6. Dietaenbacher, E., Los, B., Stehrer, R., et al.: The construction of world input-output tables in the WIOD project. Econ. Syst. Res. **25**(01), 71–98 (2013)
7. Li, F., Tu, Y., Guo, X.: Who is the free rider in the drop-shipping supply chain? Int. J. Inf. Eng. Electron. Bus. **6**(17), 44–51 (2011)
8. Pallis, A.A., Vitsounis, T.K., Langen, P.W.D.: Port economics, policy and management: review of an emerging research field. Transp. Rev. (11), 115–161 (2013)
9. Gomm, M.L.: Supply chain finance: applying finance theory to supply chain management to enhance finance in supply chains. Int. J. Logist. Res. Appl. **13**(02), 133–142 (2012)
10. Puro, L., Teich, J.E., Wallenius, H., Wallenius, J.: Borrower decision aid for people to people lending, decision support systems. J. ACM **49**, 52–60 (2010)
11. Wen, J., He, L., Zhu, F.: Warm robotics control and communications: imminent challenges for next generation smart logistics. IEEE Commun. Mag. **56**(07), 102–107 (2018)
12. Jiang, L., Wang, Y., Liu, D.: Logistics cost sharing in supply chains involving a third-party logistics provider. Cent. Eur. J. Oper. Res. **24**(1), 207–230 (2014). https://doi.org/10.1007/s10100-014-0348-5

13. Xing, J.: An intelligent logistics tracking system based onwireless sensor network. Int. J. Online Eng. (01), 17–28 (2018)
14. More, D., Basu, P.: Challenges of supply chain finance: a detailed study and a hierarchical model based on the experiences of an India firm. Bus. Process. Manag. J. **19**(4), 624–647 (2013)
15. Mcfarcfarlane, D., Giannikas, V.: Intelligent logistics: involving the customer. Comput. Ind. **81**(06), 105–115 (2016)
16. Liu, C., He, B.: Effect of information sharing and process coordination on logistics outsourcing. Ind. Manage. Data Syst. (07), 41–63 (2015)
17. Ramirez, H., Maschke, B., Sbarbaro, D.: Modelling and control of multienergy systems: an irreversible port-Hamiltonian approach. Eur. J. Control **19**(06), 513–520 (2013)
18. Wang, Z., Wei, S.: Measures of participationin global value chains and global business cycles. J. Natl. Bureau Econ. Res. (05), 75–90 (2017)

# Investment Income Analysis of Cold Chain Transportation Platform Based on the Construction of Third Party Logistics

Xudong Deng[1,2(✉)], Ning Gao[1], and Ao Liu[1,2]

[1] Hengda School of Management, Wuhan University of Science and Technology, Wuhan 430081, China
[2] Service Science and Engineering Research Center, Wuhan University of Science and Technology, Wuhan 430081, China

**Abstract.** Due to the particularity of the cold chain problem, it is a better solution to solve the transportation problem of fruits and vegetables products by using the logistics sharing transportation platform. The biggest function of the platform is to integrate resources. This paper first studies the operation mode of the cold chain shared transportation platform, and then analyzes the pricing and profit problems of the platform. Through determining the investment and charging mode of the platform, the modeling and simulation help the platform to determine the level at which the best profit of the platform will reach the maximum. Finally, the development of logistics sharing transportation platform is analyzed and prospected.

**Keywords:** Shared transportation platform · Cold chain express · Platform profit

## 1 Instruction

With the improvement of people's living standard in recent years. People's demand for fresh fruits and vegetables is getting higher and higher. Jozsef Vancza, etc. Studied the function and function of information platform in production logistics are analyzed. Mainly through the integration of information to solve the problem of information asymmetry. In order to make better prediction results [1]. Valentin Robu, etc. Studied a logistics platform for automatic bidding is designed to help the warehouse to distribute and transport goods [2]. Brigita GAJŠEK, etc. Studied a logistics platform based on logistics hub is studied. To help establish regional logistics center [3]. Domenico Gattuso, etc. Studied the focuses on the receiving area of a logistics platform operating in the agri-food sector [4]. With the acceleration of urbanization and the continuous improvement of people's living standards, consumer demand for multi-category, differentiated and safe fresh fruits and vegetables is increasing [5]. Sales of fresh fruit and vegetable products based on e-commerce platform are developing very rapidly. However, because of synergy, the cost of cold chain transportation and infrastructure costs remain high [6]. Logistics information platform is the inevitable product of logistics information, in improving the utilization rate of information and equipment facilities, optimizing the management of


Z. Hu et al. (Eds.): ICAILE 2021, LNDECT 82, pp. 336–344, 2021.
https://doi.org/10.1007/978-3-030-80475-6_33

transportation organizations, improving the load rate, improving the quality of service, energy conservation and emission reduction have a very great role in promoting [7]. Because of the characteristics of cold chain express: First, Professional means of transport are required. Second, Perishability. Third, Demand dispersion. Therefore, we can use the shared transportation platform to solve the problem of high cost of cold chain express transportation [8]. There are two modes of construction of general transportation platforms: First, E-commerce platform self-built logistics and transportation platform. Transportation and resource sharing through the development of platforms. Second, Build a shared transport platform through third-party transport companies. For all merchants and shipping companies. Zhao Xu, etc. studied the ways in which e-commerce platforms build their own logistics and transportation systems [9]. Liu Weiqi, etc. have studied the pricing of bilateral platforms under different merger structures [10]. Wang Zhihong, etc. have studied the pricing strategy of multi-attributed freight-sharing transportation platforms [11]. Zhao Yanfei etc. have studied the optimal profit of duopoly competition platform which provides value-added services for buyers [12]. Guowei Dou, etc. have studied the investment strategy in the case of bilateral platform is studied, and the two-way charging and pricing of platform are studied [13]. Jean-Charles Rochet have studied the competitive model of bilateral platforms was studied. The above scholars are based on bilateral platforms [14].

Previous research is aimed at ordinary types of goods, domestic and foreign scholars on cold chain express research is less. In this paper, cold chain express delivery as the research object, put forward the logistics-based sharing transport platform based on the mode of transport. Analyze the best profits through the computing platform's business strategy and investment strategy analysis platform.

## 2 Model Description and Problem Posing

First of all, we need to analyze the characteristics of cold chain express. Cold Chain Express has: First, Demand is dispersed. Second, A single order is smaller in volume. Third, Transport costs remain high [15]. Based on the characteristics of cold chain express delivery, this paper puts forward a transportation model oriented towards sharing transportation platform and shared cold storage. First of all to achieve information sharing, based on this plus the overall planning of the provincial cold storage as the core to coordinate the fruit and vegetable cold chain from order to transport to the hands of customers the entire process. See Fig. 1. The platform is the center to integrate information and resources. This can greatly reduce the number of transfers [16]. Realize the common transportation and distribution of goods.

Because the platform is an information hub connecting multiple parties, its profit structure is more complex. It is not only necessary to charge a certain registration fee and transaction fee for the users who enter the platform. Also consider their own investment direction and service level. Logistics enterprises should strengthen the platform quality construction, improve the matching efficiency of e-commerce search platform, improve the attractiveness of the platform and increase the user stickiness [17]. Attract more customers to settle in the platform and promote more transaction volume, so as to form a virtuous circle. If the relationship between the investment level and the transaction

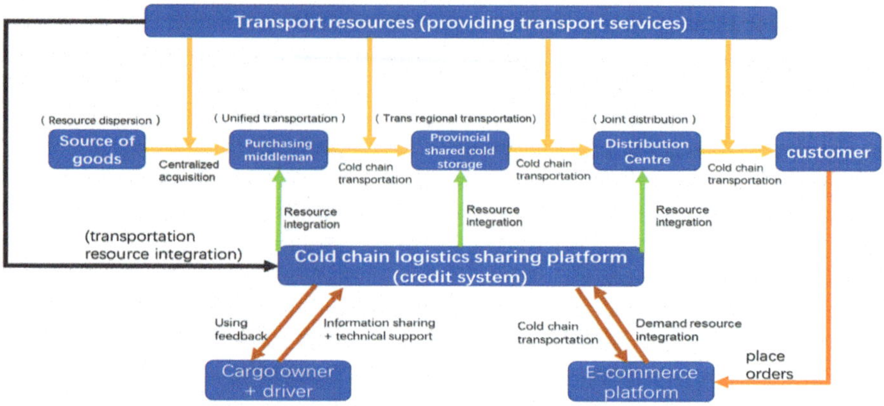


**Fig. 1.** Transportation mode under shared transportation platform

price is unbalanced, the platform profit may decrease or even become negative [18]. Therefore, this paper will study the profit of the platform under the influence of various factors. It provides some guidance methods for the operation of the real platform.

## 3 Based on the Platform Profit Model of Transportation Pricing

A shared transport platform does not determine the transaction price, but the platform needs to determine its own charging method based on the transaction price [19]. In general, the benefits of the platform come from two parts. Part of the registration fee for the user's presence. Charges for this portion are in the form of a one-time fee or an annual fee. The other part is to charge according to the transaction price and volume of the goods. Therefore, the platform on the one hand to invest advertising and training costs to attract more customers to stay on the platform, on the other hand, to develop a reasonable proportion of charges for commodity transactions (Table 1).

In the light of the above, it is available to:

$$R_1 = \beta PQ \tag{1}$$

$$R_2 = dn_t \tag{2}$$

$$C_1 = e \tag{3}$$

$$C_2 = \frac{\alpha s^2}{2} \tag{4}$$

$$C_3 = f \tag{5}$$

$$C_4 = gn_t \tag{6}$$

**Table 1.** Description of the basic parameter symbols

| The basic symbol | Description |
|---|---|
| P | Average of the closing price over a fixed period |
| d | Personal registration fees received during the unit cycle |
| β | A fixed percentage of the transaction amount is paid to the platform after the successful transaction |
| e | Platform cycle time maintenance costs |
| f | The level of advertising investment |
| s | The level of expenditure on supporting value-added services |
| v | Market capacity |
| g | Level of training expenses |
| $n_t$ | The number of new customers per issue |
| W | Fixed vehicle and personal insurance premiums are deducted for each transaction |
| Q | The average volume per cycle |
| $\rho_1$ | The rate of damage to the goods |
| $\rho_2$ | Probability of default |
| $\alpha$ | Service cost parameters |
| $\theta$ | Penalty factors for breach of contract, etc |
| h | A fixed constant related to the size of the market |
| k | Customer Growth Potential Index |
| π | The elasticity coefficient between the intersection price and the volume |
| μ | The elasticity factor between the level of service and the volume of transactions |
| δ | The elasticity factor between the level of advertising input and the volume of transactions |
| $\sigma$ | The extent to which new customers affect the volume of transactions |

$$C_5 = wQ \tag{7}$$

$$C_6 = \rho_1 PQ + \rho_2 PQ\theta \tag{8}$$

$$R_t = R_1 + R_2 - C_1 - C_2 - C_3 - C_4 - C_5 - C_6 \tag{9}$$

$$R_t = \beta PQ + dn_t - e - \frac{\alpha s^2}{2} - f - gn_t - wQ - \rho_1 PQ - \rho_2 PQ\theta \tag{10}$$

$$n_t = \frac{vg}{1 + e^{h-kt}} \tag{11}$$

$$s = \frac{\mu\beta P}{\alpha} \tag{12}$$

$$Q = -\pi P + \mu s + \delta\sqrt{f} + \sigma n_t \tag{13}$$

$$st. \quad 4\pi - \delta^2 > 0 \tag{14}$$

$$\mathrm{A}\pi > \mu^2 \tag{15}$$

$$\Pi > max\{\mu,\ \delta\} \tag{16}$$

$$v > 4P \tag{17}$$

$$\Pi,\ \mu,\ \delta,\ h,\ \alpha,\ k > 0 \tag{18}$$

$$\beta,\ \rho_1,\ \rho_2,\ \sigma,\ \theta \in (0,\ 1) \tag{19}$$

$R_1$ is transaction income, $R_2$ is registration fee income, $C_1$ is platform maintenance fixed expenses, $C_2$ is spending on value-added services for the platform, $C_3$ is the level of expenditure for advertising. $C_4$ is the costs level of expenditure for training costs, $C_5$ is fixed insurance expenses, $C_6$ is Transportation and cargo loss costs. Constraints make the calculation of the model meaningful. So the platform benefits $R_t$ is:

$$R_t = \beta P\left(-\pi P + \frac{\mu^2\beta P}{\alpha}\delta\sqrt{f} + \sigma\frac{vg}{1+e^{h-kt}}\right) + d\frac{vg}{1+e^{h-kt}} - e - \frac{\alpha\left(\frac{\mu\beta P}{\alpha}\right)}{2} - f$$
$$-\frac{vg^2}{1+e^{h-kt}} - (w + \rho_1 P + \rho_2 P\theta)\left(-\pi P + \frac{\mu^2\beta P}{\alpha}\delta\sqrt{f} + \sigma\frac{vg}{1+e^{h-kt}}\right) \tag{20}$$

This paper sets the platform service level to be related to the average transaction price and volume of the platform, indicating that the higher the transaction price and volume, the higher the service level that the platform is willing to pay. However, the trading price and volume of the platform are related to both sides of the transaction, so the average transaction price P and the volume Q of the unit cycle are not related to the platform. So what the platform wants to maximize its revenue within its capabilities is the level of investment in advertising and training. So look for a first-order conductor separately.

$$R'_{tf} = \frac{\delta}{2\sqrt{f}}\beta P - 1 - (w + \rho_1 P + \rho_2 P\theta)\frac{\delta}{2\sqrt{f}} \tag{21}$$

$$R'_{tg} = \frac{v\sigma}{1+e^{h-kt}}\beta P + \frac{vd}{1+e^{h-kt}} - \frac{2vg}{1+e^{h-kt}} - (w + \rho_1 P + \rho_2 P\theta)\left(\frac{v\sigma}{1+e^{h-kt}}\right) \tag{22}$$

When seeking the maximum profit of the platform. Make $R'_{t(\mathrm{f})} = 0$, $R'_{t(\mathrm{g})} = 0$:

$$f = \frac{\sigma^2(\beta P - W - \rho_1 P - \rho_2 P\theta)^2}{4} \tag{23}$$

$$g = \frac{2\sigma\sqrt{f} + d\delta}{2\delta} \tag{24}$$

Put the formula (23), (24) in (22). Because of the formula for the final result at the time of generation is too complicated. So only put (24) into the (22). And the formula (23) as a stand-alone argument. Available:

$$R_t = \beta P\left(-\pi P + \frac{\mu^2\beta P}{\alpha} + \delta\sqrt{f} + \frac{v\sigma\left(2\sigma\sqrt{f}+d\delta\right)}{2\delta\left(1+e^{h-kt}\right)}\right) + \frac{vd\left(2\sigma\sqrt{f}+d\delta\right)}{2\delta\left(1+e^{h-kt}\right)} - e - \frac{(\mu\beta P)^2}{2\alpha}$$
$$-m - (w + \rho_1 P + \rho_2 P\theta)\left(-\pi P + \frac{\mu^2\beta P}{\alpha} + \delta\sqrt{f} + \frac{v\sigma\left(2\sigma\sqrt{f}+d\delta\right)}{2\delta\left(1+e^{h-kt}\right)}\right) \tag{25}$$

$$st. \quad m = f + \frac{v\frac{2\sigma\sqrt{f}+d\delta}{2\delta}^2}{1 + e^{h-kt}} \tag{26}$$

The ultimate best profit function of the platform available from above. And $m$ is an intermediate variable. Under the fixed transaction average price, the platform's advertising investment level and training input level reach certain conditions when the platform's profit income is the largest. And there is a function relationship between the level of advertising input and the level of training input.

## 4 Simulation

This paper mainly studies the cold chain transportation mode with the cold chain sharing freight platform as the core. According to the content of the article, the platform profit model and distribution utility model are mainly assigned to simulate. Suppose there is a cold chain platform-based enterprises to cold chain express transport.

As an operations manager outside of third-party transportation, assume that the properties of the goods shipped are consistent. Indicates that the average closing price is fixed over a fixed period of time. Assign the parameters: n = 20, $\beta = 0.5$, p = 30, w = 3, $\rho_1 = 0.01$, $\rho_2 = 0.02$, $\theta = 0.1$, $\delta = 2$, h = 2, k = 1, t = 1, e = 150, $\pi = 5$, $\mu = 3$, $\alpha = 2$, v = 150, d = 1000. The tool used in this paper is MATLAB2018a, which calculates the best profit for the cold chain logistics shared transportation platform $R_t = 238109.94$. We analyze the sensitivity of the transaction price p during the period. Represented by an image:

Shown as Fig. 2 and Fig. 3, the investment status and profits of the platform will change as the trading price changes during the cycle. Because training investment g is positively related to advertising investment f, it can be considered that the price impact is similar. Because the closing price is negatively related to volume Q, the best profit of the

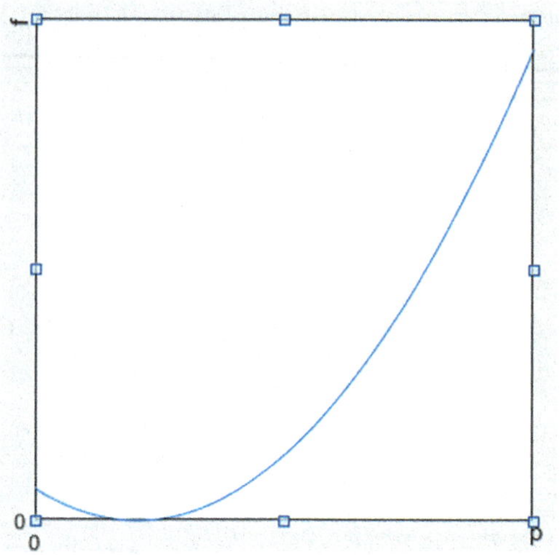


**Fig. 2.** Relationship between transaction pricing and platform investment

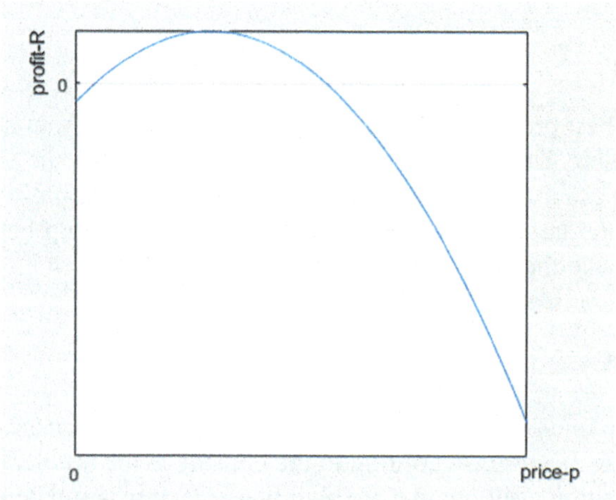


**Fig. 3.** Relationship between transaction pricing and platform profit

platform shows a trend of growth and then decline when the trading price keeps getting bigger. It's in line with reality. Therefore, the platform needs to control the transaction price and volume at the corresponding level in order to get the best profit. In addition, the platform needs to increase investment in order to get more new customers into the platform.

## 5 Conclusion

In summary, the cold chain relies on logistics to share the transport platform to achieve the cold chain express transport. For small batches of multi-batch goods using a shared cold storage method for centralized transshipment and distribution. The loading rate of cold chain logistics vehicles is improved. This avoids the waste of resources. This paper does a detailed analysis of the operation mode of cold chain express delivery. The feasibility of successful operation of the platform is reflected by calculating the profit of the cold chain sharing platform.

1) Building a sharing mechanism of cold chain transportation. we are still unable to achieve the low cost of cold chain express delivery. The best solution is for the courier companies to come together and share their resources. Each company bears part of the cost of infrastructure construction, and jointly realizes the transportation of cold chain express. How to maximize corporate profits and distribute them fairly is a difficult problem to solve.
2) Perfect transportation and distribution system. In the real life of the variety of goods are diverse, price, quality, shelf life, corruption time, etc. are not the same. Some fruits and vegetables can not be transported together and other issues are not taken into account in this article. Relying on supermarkets, retail stores on the basis of cold storage to build cold chain distribution network is also a way.
3) Building a government led credit system. That's to avoid the platform alone problem. Turn resources to your side for greater profit. Therefore, the construction of the platform must be under strict control and under the supervision of government enterprises. Only in this way can the interests and interests of consumers and other companies be guaranteed. Build a credit evaluation mechanism. Any individual users or enterprise users who join the platform should monitor the fair operation of the platform.

Although this paper makes some assumptions about the model of the platform, but the construction of the platform involves many links, the value system of the platform is more complex, so this paper cannot fully reveal the full content of the value of the platform, need to be further expanded.

## References


1. Váncza, J., Egri, P., Karnok, D.: Planning in concert: a logistics platform for production networks. Int. J. Comput. Integr. Manuf. **23**(4), 297–307 (2010)
2. Robu, V., Noot, H., Poutré, H.A., Schijndel, W.J.V.: A multi-agent platform for auction-based allocation of loads in transportation logistics. Expert Syst. Appl. **38**(4), 3483–3491 (2010)
3. Gajšek, B., Kovač, J., Hazen, B.T.: An organizational framework for logistic platform and its subtypes in a search for more logistically attractive regions. Organizacija **51**(1), 20–34 (2018)
4. Gattuso, D., Cassone, G.C., Pellicanò, D.S.: A micro-simulation model for performance evaluation of a logistics platform. Transp. Res. Procedia **3**(3), 574–583 (2014)
5. Xicai, Z., Hailing, L.: Research on the development mode of agricultural products modern cold chain logistics based on big data. Sci. Technol. Manag. Res. **40**(07), 234–240 (2020)
6. Ting, P.-H.: An efficient and guaranteed cold-chain logistics for temperature-sensitive foods: applications of RFID and sensor networks. Int. J. Inf. Eng. Electron. Bus. **5**(6), 1–5 (2013)
7. Weihua, G., Tingting, Z., Yuwei, Z.: On RFID application in the information system of rail logistics center. Int. J. Educ. Manag. Eng. **3**(2), 52–58 (2013)

8. Qingying, Z., Zhimin, C.: HACCP and the risk assessment of cold-chain. Int. J. Wirel. Microw. Technol. **1**(1), 67–71 (2011)
9. Zhao, X., Yong, W., Weijun, H.: Socialized competition mechanism of self built logistics system of e-commerce platform. Comput. Integr. Manuf. Syst. **26**(08), 2288–2298 (2020)
10. Weiqi, L., Zhang, S.: Analysis of pricing under the strategy of bilateral platform merger. China Manag. Sci. **25**(05), 17–24 (2017)
11. Wang, Z., Fu, C.: Study on pricing strategy of freight sharing platform under different user ownership behaviors. J. Manag. **16**(07), 1081–1087 (2019)
12. Yanfei, Z., Yong, W.: Value added services and pricing competition strategy of duopoly bilateral platform considering partial multi ownership of seller users. Forecast **39**(04), 76–82 (2020)
13. Dou, G., He, P., Xu, X.: One-side value-added service investment and pricing strategies for a two-sided platform. Int. J. Prod. Res. **54**(13), 3808–3821 (2016)
14. Rochet, J.C., Tirole, J.: Platform competition in two-sided markets. Eur. Econ. Assoc. **1**(4), 990–1029 (2003)
15. Behzadi, G., Sundarakani, B., Mardaneh, E.: Robust optimization model for the cold food chain logistics problem under uncertainty. Int. J. Logist. Econ. Globalization **5**(3), 167–179 (2013)
16. Qiuxiong, L.: Research on cold chain logistics of fresh agricultural products under online shopping mode. Logist. Eng. Manag. **42**(05), 109–111 (2020)
17. Fatao, W.: Network effect and competitive strategy selection of e-commerce multilateral platform from the perspective of evolution. China Circ. Econ. **33**(11), 54–64 (2019)
18. Chunlin, L., Xiaobing, M., Xin, T.: Research on demand information sharing strategy in network platform sales mode. China Manag. Sci. **25**(08), 149–157 (2017)
19. Yuyan, W., Liang, S.: Research on e-supply chain advertising strategy based on third-party e-commerce platform sales. Manage. Rev. **31**(10), 124–132 (2019)

# Simulation and Optimization of M-commerce Small Commodities Logistics System Based on Arena

Xin Li[1], Yong Wang[2(✉)], Yuan Chen[1], Hong Jiang[1], and Xuanzhi Ye[1]

[1] School of Logistics, Wuhan Technology and Business University, Wuhan 430065, China
[2] School of Transportation, Wuhan University of Technology, Wuhan 430063, China
wang-yong@whut.edu.cn

**Abstract.** In the traditional physical retail operation mode, due to the limitation of time and space, the system is inadequate for the tasks it has to perform. It cannot meet the more diversified needs of consumers, which mainly leads to many problems, such as many links, high cost, receiving place is not centralized, multi-layer transmission of resources, information cannot be shared in time and so on. Therefore, we should make full use of many advantages of the M-commerce platform transaction. Based on it, this paper analyzes the M- commerce small commodity logistics system, use arena simulation software to compare two models, and establishes the evaluation method and index system of small commodity M-commerce. Combined with the relevant theoretical knowledge of queuing theory, adjusting the setting parameters, real-time monitoring, comparing the simulation output results of the traditional model and the optimized model. Finally, summarize the experimental data, draw a conclusion, and put forward a feasible optimization scheme for the logistics system of M-commerce small commodities logistics System.

**Keywords:** Small commodities logistics · M-commerce · Simulation · Arena software · Feasible optimization

## 1 Instruction

When people begin to get used to and rely on the convenience brought by the Internet, online shopping has gradually become the main form of consumption in the current market. And the Internet M-commerce industry is also giving the traditional market a strong voice [1].

Especially in recent years, M-commerce industry has gradually become one of the core competitiveness of the future development of domestic major industry market. When M-commerce in professional markets is blooming everywhere, all industries and markets have seized the competition pattern of M-commerce business opportunities in order to share the "big cake" of the Internet online market. Small commodity market is no exception.

In people's daily life, such small commodities as small general merchandise, hardware, daily necessities and cultural goods, which can be called "just needed", have a


Z. Hu et al. (Eds.): ICAILE 2021, LNDECT 82, pp. 345–354, 2021.
https://doi.org/10.1007/978-3-030-80475-6_34

stable and rising market demand. Especially with the improvement of people's material living standard, the demand for small commodities used in daily life is also increasing day by day, which is enough to make the small commodity industry grow rapidly both in scale and in enterprise efficiency.

In the traditional physical retail operation mode, due to the limitation of time and space, products and services are limited, which cannot meet the more diversified needs of consumers. Although in the industry, many enterprises have already laid out the Internet M-commerce platform. However, in most of the platforms, they still operate in the mode of one-way information output and lack of analysis and mining of user data, which not only cannot really meet the rising consumer demand, but also cannot effectively solve many problems in the industry.

In order to solve these problems, we should fully learn from the experience, continuously integrate and optimize the upstream and downstream supply chain ports of the industry with advanced Internet technology, and then connect to many high-quality manufacturers and operators in the industry with the platform as the center, so as to break the barriers between suppliers, platforms and end users.

By improving the efficiency of the supply chain and integrating the supply chain ports, the small commodity platform city users and businesses directly connect, breaking the information barrier among people, goods and factories, and providing users with more choices and more high-quality, cost-effective small business products with the mode of direct supply by manufacturers [2].

## 2 Simulation of Arena

### 2.1 The Summary of Simulation Based on Arena

Arena is a computer simulation software. Developed by system modeling company in 1993, it is a general simulation software based on SIMAN simulation language [3]. It can simulate the discrete continuous hybrid system, and has the function of visual interaction. Users can set the parameters according to the needs, establish the simulation model, and analyze and process the actual system. The software has built-in VBA (Visual Basic for application) tools, which can be integrated with other software. It is widely used in manufacturing, logistics and supply chain, service, medical, military and other fields. For large or complex models, such as the logistics system model in this paper, hierarchical modeling is particularly important. Arena provides a hierarchical modeling tool (sub-model) to make the logistics system model from macro to micro divided into several levels, and through the port to connect, effectively establish the model, analyze the data, improve the modeling efficiency and data analysis [4]. The logistics system studied in this paper is based on discrete event simulation, using arena simulation software to simulate the discrete event system [5, 6]. The steps of the Arena simulation experiment include: drawing the workflow diagram of the system, determining the arrival model, service model and queuing model (they constitute the simulation model of discrete event system), compiling the running program describing the system activities and executing the program on the compute [7].

## 2.2 The Logic Simulation Model

Discrete event simulation describes a process with a set of unique, specific events in time. These flexible, activity-based models can be effectively used to simulate almost any process. The M-commerce small commodities logistics System is essentially a discrete system based on queuing model, with arrival, waiting and processing as the process [8, 9]. Arena simulation model is used to simulate the process, as shown in the figure below (Fig. 1).

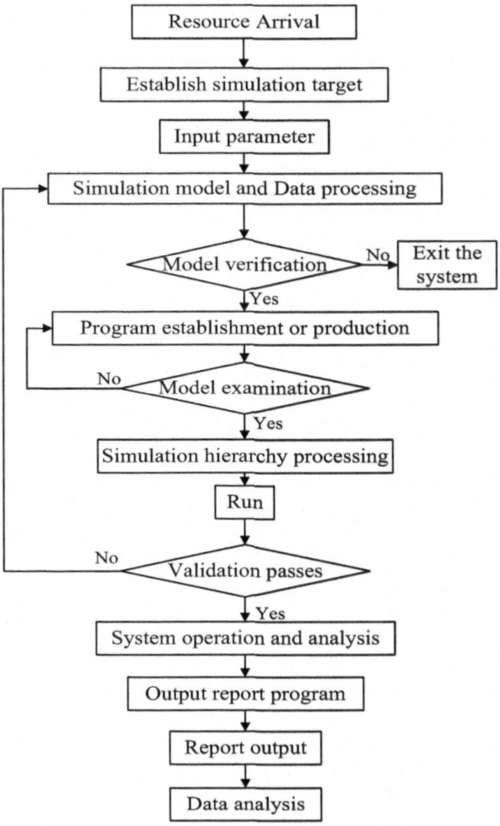


**Fig. 1.** Arena model simulation flow chart

In the MSCLS (M-commerce Small Commodities Logistics System), Entities can enter and leave Arena system 24 h a day, fully simulating mobile customers' entering and leaving mobile Internet logistics system for small commodity transactions and services. This mode can effectively combine the upstream and downstream, and the upstream small commodity enterprises form an integrated system to the end users, that is, the integrated process of customer entry, demand response, processing center, mobile M-commerce logistics system and after-sales maintenance service [10] (Fig. 2).

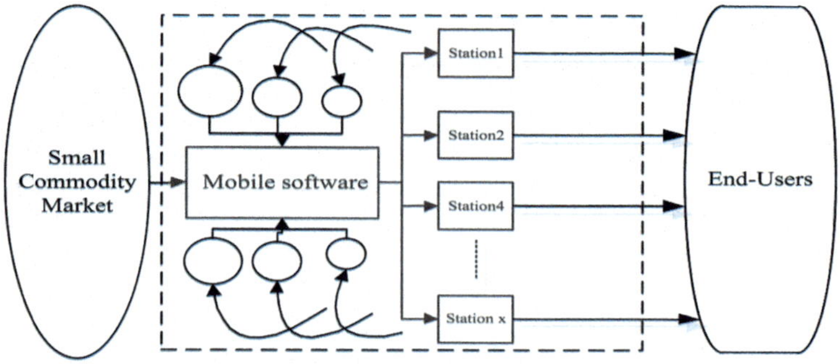


**Fig. 2.** Model of M-commerce small commodities logistics system

# 3 Simulation Model and Parameter Setting

## 3.1 Related Methods and Evaluation System

The process from customers entering the system to demand operation follows the Franc (M/N/S) Queuing System [11].

M: The entity arrival system follows the law of Poisson distribution process;
N: The service time follows exponential distribution;
S: The queuing process of entering the system follows the principle of first in first out.

The following formula is executed when the system is in a stable state:

$$p_n(s) = \begin{cases} \frac{(sp)^n P_0(s)}{n!}, n \leq s \\ \frac{p^n s^s}{s!} P_0(s) = p^{n-s} p_s(s), n > s \end{cases} \tag{1}$$

$$p_0(s) = \frac{1}{\left[\sum_{n=0}^{s-1} \frac{(ps)^n}{n!} + \frac{(ps)^s}{s!}\left(\frac{1}{1-p}\right)\right]} \tag{2}$$

$s$–Number of servers;
$n$–Number of customers in and out;
$p_n(s)$–The queue length follows an instantaneous distribution;
$\rho$–Traffic intensity of the corresponding queueing system;
If definite $\rho < 1$, the system is stable.

Using the analytical calculation method of queuing theory, the measurement index can be obtained from the above results. This simulation will use the following five commonly used evaluation indicators to test the effectiveness of the system operation. They are automatically generated by the system after the simulation runs.

$$L_q = \sum_{k=s+1}^{\infty} (k - s) P_n(s) = \frac{(\rho s)^s}{(1 - \rho)^2 s!} p_n(s) \tag{3}$$

$L_q$–The average queue length, which is the expected number of customers waiting for service at any time in steady state.

$$L_s = L_q + \rho s \tag{4}$$

$L_s$–The average queue team, which represents the expected value of the number of customers at any time in the steady state of the system.

$$W_s = \frac{L_s}{\lambda} \tag{5}$$

$W_s$–The average of sojourn time, which is the expected residence time of customers entering the system at any time when the system is stable.

$$W_q = \frac{L_q}{\lambda} \tag{6}$$

$W_q$–The average of waiting time, which is the expected value of customer waiting time in steady state.

$\mu$–Resource utilization, It refers to the use of different system resources.

$$\mu = \lim_{T\to\infty} \int_0^T \frac{\beta(t)\mathrm{d}t}{T} \tag{7}$$

$\beta(t)$ is the running state,the resource is '1' when it is busy and '0' when it is idle.

The first four are commonly used steady-state performance evaluation indicators, which are average queue length, average waiting time, average staying time and resource utilization. Compared with average waiting queue, average queue length can more accurately express the congestion degree of logistics distribution center system. These indicators can comprehensively and reasonably reflect the operation ability and actual effect of mobile M-commerce logistics system, as well as the comparative performance analysis of the two systems studied in this paper [12].

## 3.2 Simulation Model Design

Some assumptions have been made as follows:

1) In a certain system environment, check whether all processors are running before simulation.
2) From the customer entering the system to leaving the system, even if the process ends, there is no need to take into account the same customer repeatedly entering and leaving the system.
3) Ignore the running loss and vacancy rate of system server.
4) Ignore the cost fluctuation of small commodities.

The simulation model–MSCLS:

The mode–MSCLS which is composed of six sub-models, it follows the working time of M-commerce platform. Various resources can be used 24 h a day, and users can access the system and respond to demands. To reduce the "multi-level" of the process, we can rely on the M-commerce platform to solve the problem through offline logistics integration, and carry out more effective logistics integrated management and resource allocation through logistics information technology statistics and instructions (Fig. 3).

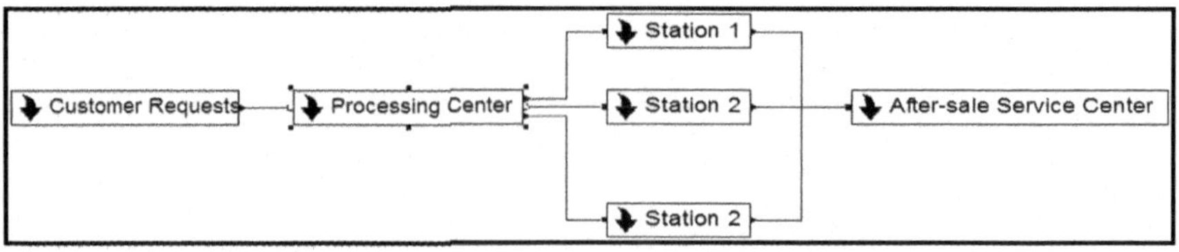


**Fig. 3.** Framework of the mode -MSCLS

### 3.3 Arena Simulation Data Input

Simulation is based on the logical analysis and the process of M-commerce, through the quantitative data of the logistics system level, through the adjustment of different parameters, to discover the influencing factors of the service process in M-commerce logistics system, find out the problems, and point out the optimization design, so as to improve the performance and efficiency of the whole logistics system (Table 1).

**Table 1.** System time and parameter allocation of model -MSLCS

| Name | Time Definition | Function Expression |
|---|---|---|
| Demand response | Response time | −0.001 + Expo (5) |
| Arrival time | Entities arrival time | Tria (1.5, 3.5, 5.5) |
| Browse products | Browsing time | Tria (0.5, 1, 1.5) |
| Order for goods | Payment time | Uniform (1, 3, 1.8) |
| Order processing | Processing time | Tria (1, 3, 5) |
| Order delivery | Delivery time | Tria (2, 4, 6) |
| Handling | Handling time | Tria (1, 3, 5) |
| Returns and services | After-sales service time | Tria (1, 3, 5) |

### 3.4 Analysis of Simulation Output Results

After running arena simulation software, system will automatically generate a TXT document, the established model outputs statistical data, and the results are obtained in the form of graph and table reports. The data statistics of the entities entering the system will be carried out, sorted and analyzed (Fig. 4, Table 2).

From the output performance of the above five indicators, MSLCS's various performance indicators are better than the traditional small commodity purchase process, so its operation mode is reasonable. Mobile commerce small commodity logistics system has certain practical significance for promoting the development of small commodity logistics in China [13].

| Scheduled Utilization | Average | Half Width | Minimum Average | Maximum Average |
|---|---|---|---|---|
| Resource 1 | 0.6580 | .17 | 0.5007 | 0.8724 |
| Resource 2 | 0.5581 | .13 | 0.4252 | 0.7138 |
| Resource 3 | 0.1249 | .06 | 0.04641785 | 0.1668 |
| Resource 4 | 0.3365 | .09 | 0.2646 | 0.4551 |

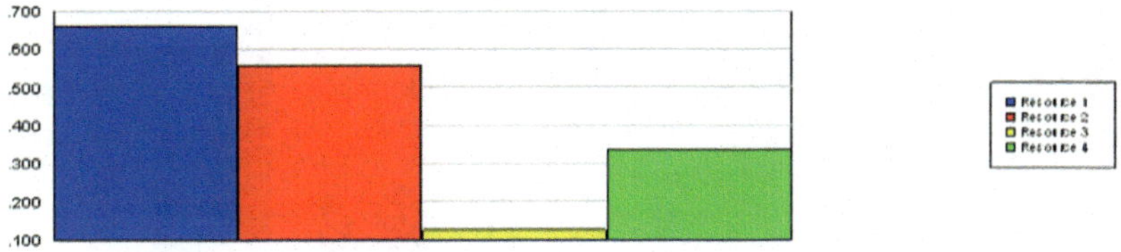


**Fig. 4.** Average resource utilization of model- MSLCS

**Table 2.** Output data of performance indicators

| Performance indicators | Model of MSLCS |
|---|---|
| Average waiting time(hours) | 4.5 |
| Average sojourn time(hours) | 13.5 |
| WIP | 3.0 |

## 4 Design of Optimization Parameter

According to the above data output, four groups of optimal regulation schemes are designed. Through the allocation of system modules, processing time of process and pass rate and reasonable parameter adjustment, the optimized system has corresponding improvement and performance improvement in each subsystem and processor.

The specific parameter adjustment scheme is as follows:

1) The purchase time is as follows: Tria (1, 2, 3), other parameters remain unchanged.
2) The protocol service time is as follows: Tria (1, 4, 8), other parameters remain unchanged.
3) The operation time of transportation and distribution is as follows: Tria (1, 2, 3), other parameters remain unchanged.
4) After sales service time is as follows: Tria (1, 4, 8), other parameters remain unchanged (Fig. 5, Table 3).

The output results show that: the average resource utilization rate of model-1 is 78%, and the system load is serious, which will lead to slow response of customers and too many queues in the system. The results show that the average resource utilization of model-4 is reasonable, and the data analysis shows that the system is not abnormally busy or excessively idle. Through the comparison of the output index data, the upper limit value of changing the exponential distribution can be detected. The average residence time and average waiting time of model-3 system are shorter than those of model-1 and model-2 distribution, the queue is shorter, and the import entity speed is faster. It can be

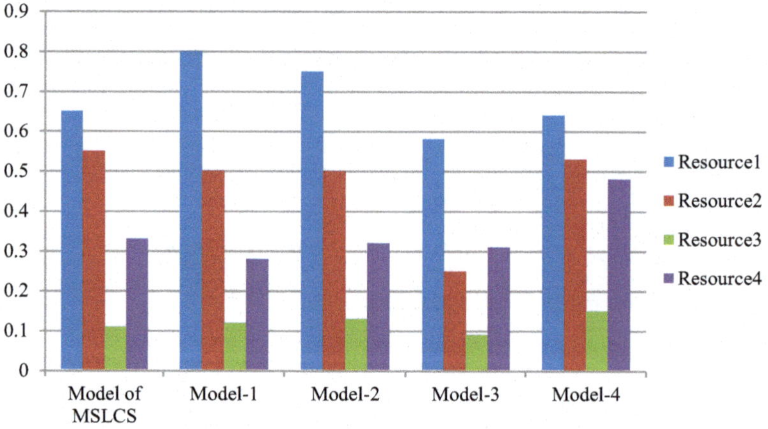


**Fig. 5.** Average resource utilization index under different models

**Table 3.** Comparison of performance indicators under different parameters

| Performance indicators | Model of MSCLS | Model-1 | Model-2 | Model-3 | Model-4 |
|---|---|---|---|---|---|
| The average waiting time(hours) | 4.5 | 5.5 | 5.7 | 1.8 | 4.9 |
| The average sojourn time(hours) | 13.5 | 15.5 | 15.6 | 10.1 | 14.7 |
| WIP | 3.0 | 3.1 | 3.4 | 1.8 | 3.3 |

concluded that the system has short residence time, short queuing time, high processing efficiency and high transmission speed in the process of customer waiting quick [14].

According to the five performance indicators of these outputs, the optimal results can be obtained by changing the parameters and resource parameters of the logistics system in each stage of operation through various indicators of different modes. The model can clearly reflect the demand of mobile M-commerce platform for small commodity logistics system, including user request, processing center, payment platform service, logistics distribution and other modules. Optimizing the indicators and parameters or the service time of each stage in the system can get different results, so that the logistics system is in a stable and efficient state [15].

## 5 Conclusion

In this paper, through the simulation operation and result output, combined with the model and data analysis, the following conclusions are drawn as follows:

1) From the above core comparison chart and simulation data analysis, it can be concluded that after customers place an order, many dimensions of order processing

and circulation are optimized and improved compared with some problems existing in the traditional system, reaching the "layer reduction" effect mentioned above.

2) Moreover, comparing the average utilization rate of resources of the two systems, the operation of the new operation system is higher than that of the traditional operation system, the utilization rate of resources has been significantly improved, the purpose of cost saving and effective allocation of resources has been achieved, and the service quality has also been greatly improved.
3) Then, the average stay time and average waiting time of the system, the timeliness of the operation of the new system are more dominant, and the system response ability and service level are higher.
4) Finally, in the average queue length of the system, the new system reduces the queue length, reflecting the advantages of M-commerce platform integration technology and the convenience of system operation.

**Acknowledgment.** This research is supported by the Logistics School of Wuhan Technology and Business University and School of Transportation of Wuhan University of Technology.

## References

1. Gong, W., Stump, R.L., Maddox, L.M.: Factors influencing consumers' online shopping in China. J. Asia Bus Stud. **7**(3), 214–230 (2013)
2. Fadzly, M.K., et al.: Analysis on flexible manufacturing system layout using arena simulation software. AIP Conf. Proc. **1885**(1), 1–10 (2017)
3. Guneri, A.F., Seker, S.: The use of arena simulation programming for decision making in a workshop study. Comput. Appl. Eng. Educ. **16**(1), 1–11 (2008)
4. Haibo, J.: An optimal algorithm for real-time Ethernet buffer queue based on queuing theory. J. Dalian Univ. Tech. **52**(1), 95–99 (2012). (in Chinese)
5. Ting, P.-H.: An efficient and guaranteed cold-chain logistics for temperature-sensitive foods: applications of RFID and sensor networks. Int. J. Inf. Eng. Electron. Bus. **5**(6), 1–5 (2013)
6. Weihua, G., Tingting, Z., Yuwei, Z.: On RFID application in the information system of rail logistics center. Int. J. Educ. Manag. Eng. **3**(2), 52–58 (2013)
7. Samosir, A.S., Gusmedi, H., Purwiyanti, S., et al.: Modeling and simulation of fuzzy logic based maximum power point tracking (MPPT) for PV application. Int. J. Electr. Comput. Eng. **8**(3), 13–15 (2018)
8. Randhawa, J.S., Ahuja, I.S.: An approach for justification of success 5S program in manufacturing organisations using fuzzy-based simulation model. Int. J. Prod. Qual. Manag. **25**(3), 331–348 (2018)
9. Jairak, R., Praneetpolgrang, P., Chirawichitchai, N.: A roadmap for establishing trust management strategy in M-commerce services using quality based assessment. Int. J. Inf. Eng. Electr. Bus. **6**(5), 1–9 (2014)
10. Yong, X., Jian, L., Baomei, M., et al.: Service mechanism and pricing based on fairness preference of customers in queuing system. J. Syst. Sci. Inf. **6**(06), 3–16 (2018)
11. Tahar, A.B., Jean-Marie, A.: The fluid limit of the multiclass processor sharing queue. Queung Syst. **71**(4), 347–404 (2012)
12. Micallef, A., Apap, M., Spiteri-Staines, C., et al.: Single-phase microgrid with seamless transition capabilities between modes of operation. IEEE Trans. Smart Grid **6**(6), 2736–2745 (2015)

13. Changxi, M., Ruichun, H.: Green wave traffic control system optimization based on adaptive genetic-artificial fish swarm algorithm. Neural Comput. Appl. **31**(7), 2073–2083 (2019)
14. Nicholas, J., Boster, A., Wu, N., et al.: Comparison of disease-modifying therapies for the management of multiple sclerosis: analysis of healthcare resource utilization and relapse rates from us insurance claims data. Pharmacoeconomics **2**(1), 31–41 (2018)
15. Faulhammer, E., Zellnitz, S., Wutscher, T., et al.: Performance indicators for carrier-based DPIs: Carrier surface properties for capsule filling and API properties for in vitro aerosolisation. Int. J. Pharm. **536**(1), 326–335 (2018)

# Role of Government and Regulatory System of Medical Waste Recycling and Transportation Based on Game Theory

Hengyu Zhang[1] and Qiguang Du[1,2](✉)

[1] School of Economics and Management, Tianjin Agricultural University, Tianjin 300384, People's Republic of China
[2] College of Management and Economics, Tianjin University, Tianjin 300072, People's Republic of China

**Abstract.** At the beginning of 2020, the COVID-19 broke out in China, and the use of medical supplies surged in various regions of China, resulting in an increase in the amount of medical waste. Against this background, the illegal industrial chain behind the medical waste has attracted more and more attention. This paper analyses the reasons for the emergence of the medical waste industry chain and the role of government supervision in the medical waste recycling system, and establishes the game relationship between manufacturers who can use medical waste as raw materials under the role of government supervision to choose raw materials. Finally, the setting of the government reward and penalty amount in the four numerical intervals and the corresponding game decisions of the manufacturers are obtained, to solve the problem of medical waste flow downstream of the industrial chain by changing the total yield of manufacturers.

**Keywords:** Medical waste · Dynamic game · Expected revenue

## 1 Introduction

With the continuous improvement of the living conditions of residents in China, medical ability, as the main measure of residents' living conditions, has been rapidly developed. Compared with the rapid development of medical capacity, however, there are still many problems to be solved in the field of medical waste recycling, such as privately transporting waste and recycling supervision [1]. In this case, the level of medical waste treatment does not match the growing demand for medical services. Secretly transporting medical waste often occurs, mainly due to insufficient supervision and a lack of responsibility of personnel in the industry.

Under the influence of the COVID-19, more attention has been paid to the recovery process of medical waste. This paper analyses the reasons for the phenomenon of waste outflow in the process of recycling, and establishes a game model for downstream enterprises that may use medical waste as raw materials under different levels of government supervision [2]. By improving the cost of secretly transportation of terminal


Z. Hu et al. (Eds.): ICAILE 2021, LNDECT 82, pp. 355–364, 2021.
https://doi.org/10.1007/978-3-030-80475-6_35

employees in the industrial chain, the responsibilities and self-regulatory consciousness of employees throughout the industrial chain are improved. The reliability of ensuring the internally recycling system, thus decrease the difficulty of medical waste recycling supervision [3].

# 2 Causes and Present Situation of Dark Industrial Chain of Medical Waste

The dark industrial chain of medical waste refers to that the employees involved in the recycling process send the waste to the non-designated recycling company driven by interests. The recycling and reprocessing of medical waste is mainly plastic products, which, at the same time, have high recycling value. As the non-designated recycling companies do not have the ability to disinfect and sterilize waste, the recycled medical waste is put on the market without adequate treatment, posing potential hazards to consumer groups.

## 2.1 Cause Analysis

### 2.1.1 The Temptation of High Profits

According to the dark medical waste industrial chain revealed by news, taking the recycled medical waste as raw materials, it can reap three to five times as much revenue as the normal purchase of raw materials without considering the cost of sterilization and disinfection. For the recycling enterprises, the recovery of medical waste can generate four to five times the income compared with general waste[4].

According to media in some areas, a ton of medical waste can earn 4000 yuan after treatment, which is four times of the same amount of ordinary waste. It is really profitable. On March 15, a program on CCTV (China Central Television) exposed the dark industrial chain of medical waste recovery, with a profit of about 1000 yuan per ton of crushed materials [5].

### 2.1.2 There is a Big Gap Between the Revenue and Expenditure of Regular Recovery Channels

Medical waste, a kind of hazardous goods, is usually disposed by incineration. According to the data of Shanghai General Administration of Health, the thermal effect of centralized disposal of medical waste in Shanghai is 80%, and the incinerator can produce about 50000 GJ heat per year at a price of 30 yuan /GJ. It will generate 1.5 million yuan per year on average if the heat can be fully utilized. However, the fact is that the thermal efficiency is always less than 100%. And the annual cost of waste treatment is about 3.2 million yuan [6]. It can be seen that the medical waste incineration costs a lot with low revenue.

### 2.1.3 Poor Self-regulation of Medical Institutions

There are some medical institutions which do not comply with relevant regulations to dispose of medical waste. For example, they do not disinfect the medical waste as required, and the relevant management departments of hospitals do not directly participate in the supervision and inspection of medical waste treatment. They hire temporary workers without professional training to deal with medical waste. As a result, the medical waste from the hospital is treated like the general garbage. As a matter of fact, few medical clinics have the ability to separate and recycle the medical waste. Some medical institutions even treat a large number of disposable infusion bottles (bags) as general waste by the hospital logistics, either as domestic waste or sold as waste products [7].

#### 2.1.4 The Collection System of Medical Waste Needs to Be Improved

At present, the medical waste recovery system adopted by most hospitals in China is composed of clinicians and nurses, staff of logistics department or property management department and staff outside the hospital. The logistics personnel collect the waste from each department every day and put it in the temporary storage place, and then send it to the recycling company for treatment [8]. There are some problems in this process, such as the lack of sense of responsibility, uneven cultural level, and complicated handover process, inaccurate measurement of medical waste quantity. As a result, it is difficult to trace the source of medical waste even if there are illegal products appearing in the market.

### 2.2 Present Situation of the Dark Industrial Chain

With the advent of the novel Coronavirus pandemic, the demand for medical supplies is increasing day by day, which generates more and more medical waste correspondingly. According to the data from the *Report on the Disposal of Medical Waste and Wastewater and Environmental Detection* issued by the Ministry of Ecology and Environment, in February, when the epidemic was serious, the daily maximum output of medical waste reached 2,719.1 tons, and the actual treatment of medical waste was 2 749.8 tons; The daily output of medical waste reached the peak of 3 135.7 tons in March, and the actual treatment of medical waste was 3,157.7 tons.

This paper believes that to manage and supervise the reproduction enterprises, namely, the downstream of the dark industrial chain of medical waste, will be more effective to solve the problems mentioned above. It is going to evaluate the behavior of manufacturers using medical waste as raw materials under the supervision of the government, so as to get appropriate reward and punishment coefficient.

## 3 Construction of Game Theory Between Manufacturers

Based on the government's supervision of medical waste recovery, this paper sets up a reward and punishment mechanism on whether the manufacturer takes recycled waste as raw material for reproduction. For manufacturers, if they use recycled waste as raw materials in the reproduction process, they will be punished. On the contrary, they will be rewarded if they choose proper channels to buy raw materials. With the increasing amount of medical waste, this system will play a greater role, which can effectively limit the outflow of medical waste to avoid the formation of dark industrial chain, so as to ensure that all waste are properly treated and recycled.

### 3.1 Mixed Game Model

According to the above, this paper constructs a duopoly game model under the supervision of the government, including Channel A (a regular plastic raw material supplier), Channel B (taking illegal recycling of medical waste as raw material), and two oligarchic plastic manufacturers who purchase raw materials from the suppliers according to their production plans [9]. By comparing manufacturers' choices of supply channels in different situations, this paper analyzes the possible gains of manufacturers, determines the degree of government rewards and punishments. It makes an in-depth study to find out the equilibrium between the profit maximization and the source of raw materials in the process of enterprise operation [10] (Fig. 1).

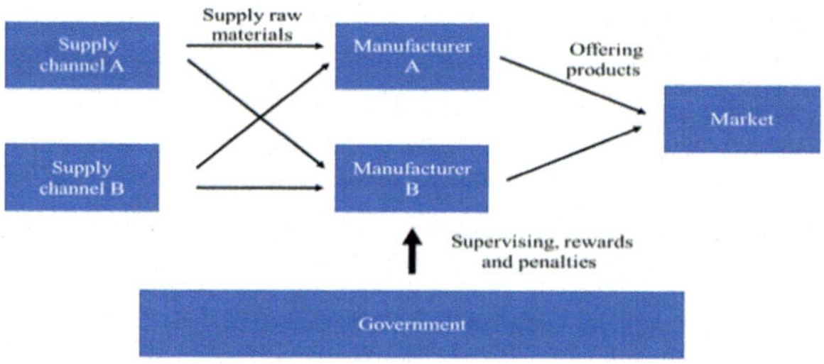


**Fig. 1.** Choices of manufacturers under the supervision of government

The symbols and assumptions used later are as follows:

$\omega$: The total cost incurred by the manufacturer in purchasing regular raw materials from suppliers;

$P$: Unit price of products sold by manufacturers;

$C$: The production cost of manufactures;

$\theta$: The ratio of the price of medical wastes as raw materials to the price of purchasing regular raw materials, $0 < \theta < 1$;

$\pi$: The profit made by the manufacturer in selling the products;

$F$: Reward / punishment for medical waste set by government policy.

There are the following assumptions:

**Hypothesis 1:** The production capacity of manufacturer A is the same as that of manufacturer B. Manufacturers buy raw materials from different supply channels according to their needs. Their products and the unit production cost $\omega$ are the same. This paper does not take the competitive relationship between suppliers into account.

**Hypothesis 2:** Based on the hypothesis of rational man, manufacturers tend to choose the supply channel with low cost of raw materials, that is, recycling medical waste, in the absence of government supervision [11].

**Hypothesis 3:** The government has ability to distinguish the raw materials of plastic products in the market. In addition, it is able to trace the source of products.

Based on the above assumptions, manufacturer A and manufacturer B get raw materials from different supply channels (A or B). Both manufacturers may choose to use recycled medical waste as raw materials in the production process. When government finds out which manufacturer producing the plastic products with medical waste as raw materials, it will serve as the basis for the government to implement punishment and reward. Affected by government policies, the manufacturer will measure the risks and possible benefits generated when choosing medical waste as raw materials for production [12]. In this model, there will be four situations: manufacturer A and B purchase regular raw materials; manufacturer A purchases regular raw materials while manufacturer B chooses waste as raw materials; manufacturer B purchases regular raw materials while manufacturer A chooses waste as raw materials; manufacturer A and B choose waste as raw materials.

## 3.2 Both Manufacturer A and Manufacturer B Choose to Purchase Regular Raw Materials

According to hypothesis 1, manufacturer A and manufacturer B share the market equally under the condition of the same production capacity. In other words, they will get the same profit.

The relationship between price and output is expressed as: $P = a - b(Q_1 + Q_2)$, the profit is expresses as:

$$\pi_1 = (P - \omega - C)Q_1 = (a - bQ_1 - bQ_2 - \omega - c)Q_1 \tag{1}$$

$$\pi_2 = (P - \omega - C)Q_2 = (a - bQ_1 - bQ_2 - \omega - c)Q_2 \tag{2}$$

According to the formulas (1) and (2), we can get the first derivative of the profit function. Because the second derivative is less than 0, so we can let $\frac{\partial \pi_1}{\partial Q_1} = 0$, $\frac{\partial \pi_2}{\partial Q_2} = 0$ in order to get the maximum profits of manufacturer A and manufacturer B, then:

$$\pi_1 = \pi_2 = \frac{(a - \omega - c)^2}{9b} \tag{3}$$

$$\pi = \pi_1 + \pi_2 = \frac{2(a - \omega - c)^2}{9b} \tag{4}$$

## 3.3 One Purchases the Materials from Regular Channels, and the Other Uses Waste as Raw Materials

According to hypothesis 1, manufacturers with the same production capacity will generate different profits when using different raw materials. In this case, the profit functions of manufacturer A and manufacturer B are as follows:

$$\pi_1' = (P - \omega - C)Q_1 = (a - bQ_1 - bQ_2 - \omega - c)Q_1. \tag{5}$$

$$\pi_2' = (P - \theta\omega - C)Q_2 = (a - bQ_1 - bQ_2 - \theta\omega - c)Q_2 \tag{6}$$

According to the formulas (5) and (6), Then we have

$$\pi_1' = \frac{(a - 2\omega - \theta\omega - c)^2}{9b} \tag{7}$$

$$\pi_2' = \frac{(a + \omega - 2\theta\omega - c)^2}{9b} \tag{8}$$

$$\pi' = \pi_1' + \pi_2' = \frac{(a - 2\omega - \theta\omega - c)^2 + (a + \omega - 2\theta\omega - c)^2}{9b} \tag{9}$$

Because of $0 < \theta < 1$, it can be known that $\pi_1' < \pi_2'$. In this case, the manufacturer who chooses medical waste as raw materials can get more profits.

### 3.4 Both Manufacturer A and Manufacturer B Choose Waste as Raw Materials

According to hypothesis 1, manufacturer A and manufacturer B share the market equally under the condition of the same production capacity. In other words, they will get the same profit. In this case, the profit functions of manufacturer A and manufacturer B are as follows:

$$\pi_1'' = (P - \theta\omega - C)Q_1 = (a - bQ_1 - bQ_2 - \theta\omega - c)Q_1 \tag{10}$$

$$\pi_2'' = (P - \theta\omega - C)Q_2 = (a - bQ_1 - bQ_2 - \theta\omega - c)Q_2 \tag{11}$$

According to the formulas (10) and (11), the output, price, profit and total market profit of these two manufacturers are:

$$\pi_1'' = \pi_2'' = \frac{(a - \theta\omega - c)^2}{9b} \tag{12}$$

$$\pi'' = \pi_1'' + \pi_2'' = \frac{2(a - \theta\omega - c)^2}{9b} \tag{13}$$

### 3.5 Game Matrix

According to the above results, the Game Matrix between manufacturer A(MA) and manufacturer B(MB) is established as in Table 1:

Because of $(a - \omega - c)^2 < (a + \omega - 2\theta\omega - c)^2$, $(a - 2\omega - \theta\omega - c)^2 < (a - \theta\omega - c)^2$ and $(a - 2\omega - \theta\omega - c)^2 < (a - \omega - c)^2$, it can be easily known that $\pi_1 < \pi_2'$, $\pi_1' < \pi_1''$ and $\pi_1' < \pi_1$.

It can be seen that the Nash equilibrium of the matrix is that both manufacturer A and manufacturer B choose to use recycled medical waste as raw materials. In order to make the values in the matrix comparable, $\pi_1'$, $\pi_2'$, $\pi_1''$ are expressed by the difference with $\pi_1$.Then: $A = \pi_1 - \pi_1'$; $C = \pi_1'' - \pi_1'$; $B = \pi_2' - \pi_1$ (Table 2).

$$A = \frac{(a - \omega - c)^2 - (a - 2\omega + \theta\omega - c)^2}{9b} = \frac{(2a - 2c - 3\omega + \theta\omega)(\omega - \theta\omega)}{9b} \tag{14}$$

**Table 1.** The matrix games between manufacturer A and B

| | | MB | |
|---|---|---|---|
| | | Regular channel | Recycling waste |
| MA | Regular channel | $(\pi_1, \pi_1)$ | $\left(\pi_1^{'}, \pi_2^{'}\right)$ |
| | Recycling waste | $\left(\pi_2^{'}, \pi_1^{'}\right)$ | $\left(\pi_1^{''}, \pi_1^{''}\right)$ |

$$C = \frac{(a - \theta\omega - c)^2 - (a - \omega - c)^2}{9b} = \frac{(2a - 2c - 3\omega + \theta\omega)(\omega - \theta\omega)}{9b} \tag{15}$$

$$B = \frac{(a + \omega + 2\theta\omega - c)^2 - (a - \omega - c)^2}{9b} = \frac{2(2a - 2c - 2\theta\omega)(\omega - \theta\omega)}{9b} \tag{16}$$

**Table 2.** The matrix games between manufacturer A and B combined by B, C, A and $\pi_1$

| | | MB | |
|---|---|---|---|
| | | Regular channel | Recycling waste |
| MA | Regular channel | $(\pi_1, \pi_1)$ | $(\pi_1 - A, \pi_1 + B)$ |
| | Recycling waste | $(\pi_1 + B, \pi_1 - A)$ | $(\pi_1 + C, \pi_1 + C)$ |

## 4 Game Model Under the Role of Government Supervision

In order to change the optimal choice of the decision-makers in the upper matrix, this paper introduces the government supervision mechanism based on hypothesis three. The government will impose penalties on manufacturers who use medical wastes as raw materials, and reward manufacturers who still choose to purchase raw materials from regular channels despite such behaviors in the market. The amount of reward and punishment is F.

In order to make manufacturer A and manufacturer B tend to purchase raw materials from regular channels, according to the above matrix, when one chooses to purchase raw materials from regular channel, the total profit of the other who chooses recycled waste as raw material needs to be less than that of regular channel (Table 3).

That is, $\pi_1 + B - F < \pi_1$, Then B $<$ F, the critical value is $F = B$.

When one chooses recycled medical waste as raw material, the total profit of the other who chooses recycled waste as raw material should be less than the total profit of purchasing through regular channels. That is, $\pi_1 + C - F < \pi_1 - A + F$, Then $\frac{A+C}{2} < F$, the critical value is $F = \frac{A+C}{2}$.

**Table 3.** The matrix games between manufacturer A and B with governmental influence

| | | MB | |
|---|---|---|---|
| | | Regular channel | Recycling waste |
| MA | Regular channel | $(\pi_1, \pi_1)$ | $(\pi_1 - A + F, \pi_1 + B - F)$ |
| | Recycling waste | $(\pi_1 + B - F, \pi_1 - A + F)$ | $(\pi_1 + C - F, \pi_1 + C - F)$ |

### 4.1 Model Analysis

According to the above matrix, we can get the following four situations by changing the critical value of Nash equilibrium in the matrix.

**Scenario 1:** when $0 < F < \frac{A+C}{2}$, the equilibrium point in the matrix does not change, and the best choice for manufacturer A and manufacturer B is to use medical waste as raw material.

**Scenario 2:** when $\frac{A+C}{2} < F < C$, the equilibrium point in the matrix changes. If the value F is in this range, there will be two equilibrium points in the matrix, namely, manufacturer A chooses to purchase through regular channels and manufacturer B uses waste as raw materials; Manufacturer A chooses to use waste as raw material, and manufacturer B chooses to purchase from regular channel. In order to determine the equilibrium point in this case, the probability of manufacturer A purchasing from suppliers is q, and the probability of using recycled waste as raw material is (1 – q); The probability of manufacturer B purchasing from suppliers is p, and the probability of using recycled waste as raw material is (1 – p).

**Scenario 3:** When C < F < B, the equilibrium point in the matrix changes. If the value F is in this range, there will be two equilibrium points in the matrix, namely, manufacturer A chooses to purchase through regular channels and manufacturer B uses waste as raw materials; Manufacturer A chooses to use waste as raw material, and manufacturer B chooses to purchase from regular channel.

**Scenario 4:** When F > B, the equilibrium point in the matrix changes. If the value F is in this range, the matrix reaches Nash equilibrium again. In this case, both manufacturer A and manufacturer B choose to purchase from regular channels.

According to the above four situations, with the increase of government rewards and punishments, manufacturers will have different choices on the sources of raw materials. For the multiple equilibrium points in the above scenarios 2 and 3, this paper makes a mixed game model to determine its probability [13] (Table 4).

### 4.2 Mixed Model Analysis

As for scenarios 2 and 3, there are multiple equilibrium points between manufacturer A and manufacturer B. When $\frac{A+C}{2} < F < B$, there is:

$$q\pi_1 + (1-q)(\pi_1 - A + F) = q(\pi_1 + B - F) + (1-q)(\pi_1 + C - F) \tag{15}$$

**Table 4.** The expected revenue in matrix games between manufacturer A and B

| | | M B | | |
|---|---|---|---|---|
| | | Regular channel | Recycling waste | |
| M A | Regular channel | $(\pi_1, \pi_1)$ | $(\pi_1 - A + F, \pi_1 + B - F)$ | $q$ |
| | Recycling waste | $(\pi_1 + B - F, \pi_1 - A + F)$ | $(\pi_1 + C - F, \pi_1 + C - F)$ | $(1 - q)$ |
| | | $p$ | $(1 - p)$ | |

$$p\pi_1 + (1 - p)(\pi_1 - A + F) = p(\pi_1 + B - F) + (1 - p)(\pi_1 + C - F) \tag{16}$$

Then: $q = \frac{A+C-2F}{A+C-B-F}$, $p = \frac{A+C-2F}{A+C-B-F}$. Therefore, within the range of $\frac{A+C}{2} < F < B$, when $q = \frac{A+C-2F}{A+C-B-F}$ and $p = \frac{A+C-2F}{A+C-B-F}$, the game equilibrium between manufacturer A and manufacturer B is $\left(\frac{A+C-2F}{A+C-B-F}, \frac{A+C-2F}{A+C-B-F}\right)$.

## 5 Conclusion

This paper introduces the phenomenon of waste loss existing in the medical waste recycling process, and analyzes the main causes of medical waste loss: profit-driven, insufficient practitioner responsibility and imperfect recycling system.

To solve the problems existing in the current medical waste treatment system, this paper analyzes the risks and benefits of enterprises that may use medical waste as raw materials to choose to privately transport medical waste, and then establishes a game model among manufacturers who may use recovered medical waste as raw materials. It aims to prevent manufacturers from using recycled medical waste as raw materials by adjusting government supervision. It concludes that the manufacturers will make different decisions when the amount of rewards and penalties set by the government is in different ranges. Then the purpose of preventing manufacturers from secretly transporting waste is achieved by adjusting the amount of prizes and penalties.

This paper integrates with new ideas to put forward theoretical solutions to some problems existing in the medical waste treatment system. However, in-depth research is still needed in concrete practice because of different professional fields.

**Acknowledgment.** This project is supported by National Natural Science Foundation of China (No 71472134).

## References

1. Qayyum, R.: A roadmap towards big data opportunities, emerging issues and hadoop as a solution. Int. J. Educ. Manage. Eng. **10**(4), 8–17 (2020)
2. Beltadze, G.N., Giorgobiani, J.A.: The stability of equilibrium situation in lexicographic strategic games. Int. J. Mod. Educ. Comput. Sci. **8**(12), 38–45 (2016)

3. Xia, D., Wu, F.: Towards more sustainability: a dynamic recycling framework of discarded products based on SD theory. Int. J. Intell. Syst. Appl. **3**(1), 43–50 (2011)
4. ZhenHua, Z.: Medical waste processing plant in the compound. Fang Yuan **19**, 46–47 (2018). (in Chinese)
5. Looking out. Why it is difficult to stop the illegal production of medical waste, such as needles and blood bags turning into basin and toys. People's Daily, 18 May 2019 (in Chinese)
6. Bin, L.: Economic analysis of shanghai medical waste treatment system. Environ. Sci. Manage. **11**, 6–9 (2018). (in Chinese)
7. Kargar, S., Pourmehdi, M., MahdiPaydar, M.: Reverse logistics network design for medical waste management in the epidemic outbreak of the novel coronavirus (COVID-19). Sci. Total Environ. **12**(1), 1–11 (2020)
8. Yanli, D., Yang, H., Yuhong, Z.: Study on innovation mode of medical waste classification management. Front. Med. Sci. Res. **1**(1), 74–81 (2019)
9. Wenhui, Y.: Research on Game Behavior of Duopoly Automobile Manufacturers Under the Intervention of Carbon Emission Reduction Policy. Tianjin University, Tianjin (2017).(in Chinese)
10. Beltadze, G.N.: Game theory - basis of higher education and teaching organization. I. J. Mod. Educ. Comput. Sci. (6), 41–49 (2016)
11. Clark, C.M.A.: Catholicism and economics: towards a "deeper reflection on the nature of the economy and its purposes." Am. J. Econ. Sociol. **78**(2), 409–441 (2019)
12. Beltadze, G.N.: The solution of scalar bimatrix games in preferred pure strategies. I. J. Mod. Educ. Comput. Sci. (3), 1–7 (2020)
13. Zeng, S., Wu, T.: Research on the standardized cold chain supervision of fresh food enterprises based on mixed strategy game. Value Eng. (20), 116–118 (2020) (in Chinese)

# Simulation and Analysis of Handling Technology Equipped with Single Box Gauge Crane

Jingjing Zheng[1], Qiang Zhou[1](✉), and Ting Wang[2]

[1] Wuhan University of Technology, Wuhan 430063, China
[2] China Railway Construction Heavy Industry Corporation Limited, Changsha 410100, China

**Abstract.** To solve the problem of high energy consumption and low efficiency in the current automated container terminal loading and unloading process, we go into a new single-box gauge container gantry crane. Combined with this kind of crane, we deeply study a new type of automated container terminal layout and loading and unloading process, analyze its technical characteristics, and establish a simulation test platform based on the layout of the terminal. Through simulation experiments, the technical indicators such as terminal capacity, energy consumption, loading and unloading efficiency, and land utilization rate are studied to evaluate the feasibility and existing problems of the new process system.

**Keywords:** Automatic container terminal · Single box gauge crane · Handling technology · Simulation experiments

## 1 Instruction

With the increase of container freight volume and the development of large-scale container ships, in order to reduce labor costs and improve competitiveness, container terminal loading and unloading is gradually developing towards automation [1]. At present, the automatic terminal mainly adopts the "quay crane +AGV + gantry crane" process, and the layout of the storage area is vertical to the shore line [2]. The gantry crane only transfers the containers to the transport vehicles at the intersections at both ends of the yard area [3]. This has changed the long travelling mechanism from the past auxiliary travelling mechanism to the main travelling mechanism, resulting in a longer running distance of the 250–300 ton orbital bridge and larger energy consumption. At the same time, the reversing storage mode is adopted at the junctions of both ends of the storage area, which not only leads to the low efficiency of the coupling work between the horizontal transportation vehicles and the field bridges, but also causes the coupling work to occupy too much land of storage area, which reduces the effective utilization rate of the land.

## 2 Related Literature

At present, domestic and foreign scholars' research on the loading and unloading technology of automated terminals mainly focuses on loading and unloading equipment and


Z. Hu et al. (Eds.): ICAILE 2021, LNDECT 82, pp. 365–377, 2021.
https://doi.org/10.1007/978-3-030-80475-6_36

horizontal transportation equipment. There is little research on the innovation of yard operation equipment, especially for small-span light container gantry crane loading and unloading technology. Jorg Wiese presented an approach based on a mixed integer linear model to find promising layout configurations for container terminals and established simulation modal to validate and evaluate the attained layout configuration [4]. Florian Jaehn consider the problem of scheduling two identical rail mounted gantry cranes (twin cranes) working within a single storage area at a seaport [5]. In terms of loading and unloading technology, Zhou Qiang introduced a new loading and unloading process for elevated rail transit container terminal, studied the interference problem between loading and unloading vessels and horizontal transportation of collection and distribution [6]. Lin Hao analyzed the development and main technical characteristics of foreign automatic terminals, and introduced a new type of full track automatic terminal handling technology scheme [7]. Luo Xunjie focused on technical parameters the efficiency, capability, energy consumption, cost and advanced technology are compared. The two horizontal transportation modes of automatic guided vehicle and automatic cross vehicle are compared [8]. Silva, CA describes the modelling of a container terminal using Petri nets with predicates, making it possible to perform a terminal optimization based on the study of long-term performance indicators under different conditions [9].

In the background of the rapid development of computer technology, simulation technology is increasingly used to solve the practical problems encountered in the process of terminal production and operation management. Mustafa Egemen Taner established simulation models for the container terminals to examine the effect of transporter dispatching rules and resource allocation strategies in terms of total annual handling amount. According to the results, terminal performance is significantly affected by terminal layout design under different transporter dispatching rules and allocation strategies [10]. The yard operation plays an important role in the daily running of a container terminal 3D simulation model was built by Yu Jingjing to simulate and visualize the yard operation in a container terminal [11]. Legato Pasquale presented a discrete-event simulation model to investigate the best yard layout in terms of block position, number and capacity in a straddle carrier-based transshipment hub bearing a perpendicular yard layout [12]. Hu Zequn established a cooperative model of the three, which is suitable for the actual berths, quay cranes, and trucks dispatching and distribution of the terminal, and uses algorithms to solve them [13].

Combined with a new invention patent for single-box gauge crane, this paper proposes a new automatic loading and unloading process for single-box gauge cranes. The single-box gauge crane with the minimum self-weight and cargo-weight ratio is the core of the yard operation, and design a container terminal layout which is vertical to the shore line. The system simulation technology is used to carry out comprehensive technical analysis and evaluation of the loading and unloading process.

## 3 Problem Definition

### 3.1 Parameters and Process of Single Box Gauge Crane

A coal transportation project is composed of four parts: mine area, long-distance belt conveyor, port area and sea lightering. At present, after coal mining, dump trucks are used to transport to the port stockyard for loading, and the annual coal transportation capacity is about 11 million tons. After the implementation of the logistics system, it is expected to achieve the goal of increasing the coal delivery capacity to 20 million tons.

The single-box gauge container crane is mainly composed of a bridge mechanism, a hoisting mechanism, and a large-car operating mechanism. The difference between the crane and the previous design is that the gauge and door-to-leg spacing are only slightly larger than the length of a 40-inch container, so the name is "single-box gauge". The equipment can realize the stacking requirements of the four-and five-layer container, and the vertical loading and unloading of the spreader and the container is realized by the rigid sliding rod, and the cantilever can extend at least one container width along the traveling direction of the cart. Table 1 shows the technical parameters of a single-box gauge crane and a conventional track-field bridge.

**Table 1.** Comparison of technical parameters of two types of gantry cranes

| Parameters | Single-box gauge gantry crane | Conventional gantry crane |
|---|---|---|
| Total Weight (t) | 73 | 254 |
| Long travel speed (m/min) | 240 | 60 |
| Trolley travel speed (m/min) | No move | 120 |
| Hoist speed (m/min) | 26 | 30 |
| Lifting height (m) | 16.5 | 24.1 |
| Span (m) | 13.83 | 28.5 |
| Operation width | One berth | 9 row box area |

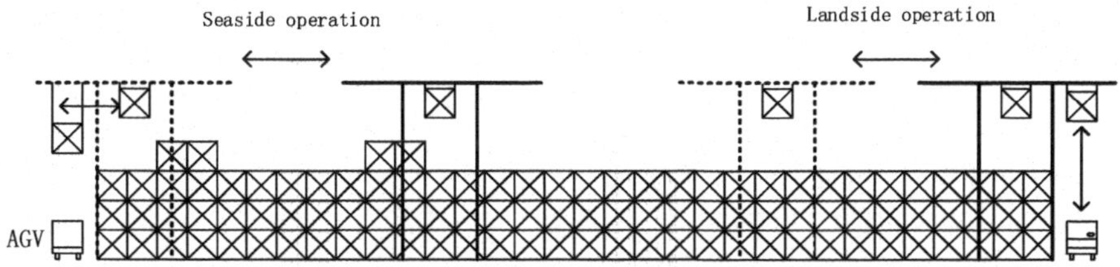


**Fig. 1.** Single-box gauge crane loading and unloading process

The single-box gauge crane is mainly responsible for the transshipment of containers in the container area and the handover with the vehicles at the two sides of the box area. Taking the operation of the aggregating containers and the distributing containers as an example, the working principle of the single-box gauge crane is illustrated, as shown in Fig. 1. In the aggregating container operation, the AGV transports container to the corresponding box area, and the seaside single-box gauge crane is commanded to run to the end of the box area. The spreader extends a certain distance along the cantilever, and vertically descends the container from the AGV to target storage area. In the operation of distributing container, the truck is operated to the corresponding box area. After receiving the command, the landside single-box gauge crane hoists the target container and runs to the end of the storage area. The spreader extends a certain distance along the cantilever and descends vertically. The container is placed on the vehicle which in the side of the storage area.

In order to facilitate the establishment of simulation system, the technical overview of each component is as follows, and mainly focuses on the determination of technical indicators and performance parameters, and the specific layout details need to be carried out in the following aspects.

### 3.2 Terminal Loading and Unloading System Composition

The container terminal of a seaport is 857m long and 784m deep. It is planned to build two 100, 000-ton container berths with a designed annual capacity of 1.2 million TEU. The terminal yard is 669m long and 480m wide. According to the container yard loading and unloading equipment, two technological schemes are put forward differently. Scheme one, quay crane + AGV + single box gauge crane technology; scheme 2, quay crane + AGV + rail field bridge typical process is proposed.

The terminal plane is designed based on the new process of quay crane + AGV + single box gauge crane. The container crane is used for loading and unloading operations. The container crane with 8 gauges of 35m is considered at the front of the dock. Four working channels are arranged under the arm. The container heavy-duty yard loading and unloading equipment is a single-box gauge crane with a gauge of 13.83 m, a one-bay container arranged parallel to the shoreline, and the lifting height meets the requirements of the stack 4 over 5; the container is arranged at a distance of 58 m from the dock shoreline. In the heavy box yard, 36 box areas are arranged perpendicular to the shore line, and a total of 72 single box gauge cranes are arranged.

The terminal plane is designed based on the typical process of the quay crane + AGV + rail-mounted container gantry crane. The loading and unloading operation are the same as that of the first scheme. The container heavy-duty yard loading and unloading equipment is a rail-type non-cantilever container gantry crane with a gauge of 32 m and 9 rows of containers arranged inside. The lifting height meets the requirement of stack 4 over 5; the container heavy container yard is arranged at the line 80 m, and 20 box areas are arranged perpendicular to the shore line. A total of 40 track field bridges are arranged, and 5 buffer parking spaces are arranged at the end of the yard. The other arrangements of the dock are the same as the plan one.

### 3.3 Comparison of Loading and Unloading Process

Compared with the typical process, the new process gantry crane has light weight, fast running speed and low running energy. The process characteristics are as follows:

1) Single-bay multi-row container box area. The typical process yard has a number of bays, each bay has multiple rows of containers, and the new process box is a bay + multi-row container, the number of rows determines the length of the box.

2) Proximity loading and unloading process. In the new process, the horizontal transport vehicles adopt the parallel berthing operation mode, which eliminates the need for cumbersome reversing operations, and converts the buffer parking spaces at the end of the yard into the storage area, effectively improving the utilization of the terminal area.
3) Under the new process, the width of the stacking box area is narrow, which will affect the vehicles entering and leaving the loading and unloading lanes in the adjacent sub-tank area. At the same time, the loading and unloading lanes at the two ends of the box area are adjacent to the traffic lanes, and only one horizontal transport can be stopped at each end of each box area.

The car is easy to cause traffic jams; in addition, when the vehicle enters the loading and unloading lane, it needs to change lanes, which is easy to affect the traffic flow of the traffic lane.

Through the terminal plane design and loading and unloading process analysis, compared with the typical process, the new technology has a high utilization rate of the dock area and low investment cost. However, under the influence of vehicle operation interference and traffic disturbance problem in the new technology, the new process energy consumption, the annual capacity and loading efficiency need to be studied in depth. The container terminal is a complex discrete event dynamic system. The performance of the terminal is affected by many random factors. It is often impossible to estimate correctly using conventional empirical formulas. It is planned to use system simulation technology to calculate.

# 4 Simulation Model

## 4.1 Modeling Purposes

There are several prominent problems in the single-box gauge crane process: 1 the width of the stacking box area is narrow, and there is interference in the adjacent sub-box area; 2 the queuing vehicles at both ends of the box area may have an impact on the main road traffic; 3 the yard box area There are many strips and many equipment configurations, and the overall energy consumption needs to be quantitatively analyzed. In addition, it is necessary to compare and analyze the new process and the classic process from the perspective of the overall operational capability of the terminal. Therefore, it is necessary to establish a simulation model of the two processes, design a simulation test, observe the operation process of the model, and deeply compare and analyze the feasibility and existing problems of the new process through the technical indicators such as terminal equipment energy consumption, annual throughput, loading and unloading efficiency.

## 4.2 Simulation Model Description

### 4.2.1 Model Range and Model Boundaries

A system model with dock loading and unloading operations as the core research object was established to simulate the process of container consignment, unloading, loading and unloading. The model includes anchorage, berth, yard, road network, entry and exit gates and loading and unloading operations equipment. The model has two boundaries: the dock gate is the land boundary and the anchor is the sea border.

#### 4.2.2 Ship Generation Module

Based on past historical data, the time-to-space interval of container ships arriving at the port is negatively distributed; the ship's carrying capacity (TEU) is subject to a truncated normal distribution, in which the export box is subject to TNORMAL (1400, 500, 900, 1500, 1) and the imported box is obeyed TNORMAL (1100, 300, 600, 1200, 1).

#### 4.2.3 Berth Operation Module

Judging the berthing conditions of the ship and controlling the berthing of the ship, arranging the appropriate number of shore container cranes, and cooperating with the AGV to carry out loading and unloading operations for the arriving ships.

#### 4.2.4 Road Segment Module

In the road segment model, the lanes are divided into sub-segments of equal length; the vehicle-following rules are simplified so that each sub-segment can only accommodate one vehicle; the lane-changing rule is that vehicles can be used when there is no car in the adjacent section and no vehicle in the next section of the adjacent section. Change lanes; first-come-first-served rules are used at intersections and turning areas, and if the front is congested, the rear vehicles are waiting in line. In addition, the speeds of different lanes are divided into different grades in the model. The speed of the vehicle on the outermost lane is 30 km/h, the speed of the vehicle on the loading and unloading lane is 15 km/h, and the speed of other lanes is between the two. The vehicle speed, driving direction and loading amount remain unchanged during one road.

#### 4.2.5 Yard Box Module

The storage box is stored in the yard box area, and the gantry crane operates the container according to the instructions. The container in the box area is 40 inches, and the box is not considered.

#### 4.2.6 Gate Module

There are 9 lanes at the entrance and exit of the terminal to manage the entry and exit of the outer card.

#### 4.2.7 Container Collection and Distribution Module

Driven and transported containers under the driving of the corresponding ship operation plan.

#### 4.2.8 Statistics Module

Through computer simulation test statistical energy consumption, annual throughput, ship time efficiency, ship loading and unloading operation time, average collection time of the external collection card, equipment utilization, etc., the energy consumption of the gantry crane in the statistical model is driven by the energy consumption of the lifting mechanism and the trolley The energy consumption of the mechanism and the energy consumption of the running mechanism of the vehicle are calculated according to the working range of 200m. In the statistical model, the heavy load, the no-load running distance and the lifting times of the crane are counted and combined. The data in the table calculates the total energy consumption of the crane.

### 4.3 Implementation of Simulation Model

Based on the layout of the wharf, a simulation model of the new process and the typical process is established on the Witness simulation platform. The model is mainly divided into 9 modules, namely the ship plan generation and management module, the container assembly module, the gate module, the terminal network module, the container box, berth operation module, container distribution module, AGV module, simulation data statistics module. The input and output among the nine modules constitute the simulation model of the whole container terminal.

## 5 Simulation Test

### 5.1 Interference Test in Adjacent Box Area of Yard

#### 5.1.1 Test Design

In combination with the characteristics of the new process, for the problem of simultaneous operation disturbance and interference of the operation traffic in the adjacent box area, the number of the compartments at the same time for the horizontal transportation vehicles is the change factor, according to no interval, one interval, two intervals, Interval three-box area design simulation test, each test condition is a peak operation state, and analyze the operation ability of different operation plans under peak conditions.

#### 5.1.2 Test Process and Data Analysis

Set the peak operation simulation time to 7 days, run the simulation model, observe the model running process, and find that under the four operation schemes, when there is no interval operation, the horizontal transport traffic is concentrated, and there is obviously a queuing phenomenon. When working in a box area, the vehicle operation is more dispersed and has fewer queuing phenomena. In addition, the average residence time of the AGV yard can reflect the road traffic conditions. The average residence time of the vehicle under peak operation is shown in Fig. 2.

The point connected to the abscissa axis indicates the box area. It is not the target box area for vehicle loading and unloading operations. According to the data of Fig. 2, in terms of the overall trend, under the same workload, the same transportation distance and the same transportation speed, the vehicle working time is relatively long when there is no compartment interval operation. Under this operation mode, the traffic flow is concentrated. There is a certain degree of work disturbance, which leads to a long queue waiting time during vehicle operation. When operating in a box area, the average storage time of the vehicle yard is relatively short. It can be seen that the traffic disturbance is less affected by this type of operation. The road traffic is in good condition. Therefore, in the face of the problem of box interference and traffic disturbance, the new technology terminal should adopt a strategy of spacing one box area.

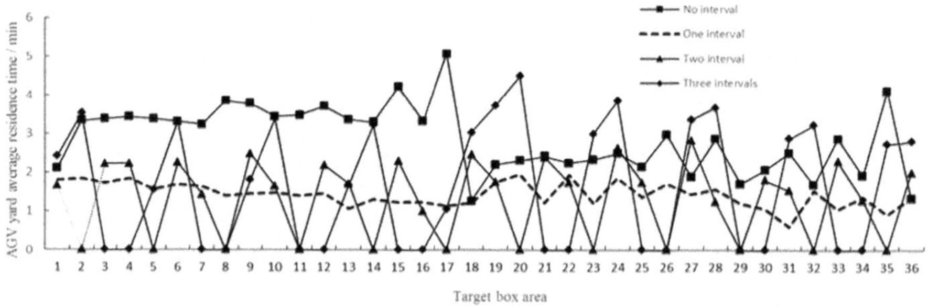


**Fig. 2.** Average residence time of AGV yard

## 5.2 Production Efficiency Test

### 5.2.1 Test Design

Based on the previous simulation analysis, the overall efficiency of the terminal is higher when the horizontal transport vehicles are operated in a box area under the new technology. Therefore, based on the change factor of ship arrival density, five kinds of working conditions are designed, respectively. The density obeys the negative exponential distribution of 1060 min, 1010 min, 960 min, 910 min, and 860 min. Considered with the effects of meteorological conditions and holidays, the simulation model runs for 330 days to study the production efficiency change of single box gauge crane technology in the process of ship arrival density change.

### 5.2.2 Test Process and Data Analysis

According to the five working conditions proposed in the previous section, the simulation experiment was carried out. By observing the running process of the model, it was found that the occupancy rate of the typical process terminal yard was higher than that of the new process under any working conditions, and the yard was always busy. The state and test statistics are shown in Table 2, Table 3 and Table 4.

*5.2.2.1 Energy Analysis* The total annual energy consumption of 72 single-box gauge cranes and 40 conventional orbital bridges is shown in Table 2. With the continuous intensive ship arrival, the total annual energy consumption of the single-box gauge crane gantry crane is reduced by about 20% compared to the typical process. It can be seen that the new process can make full use of the single-box gauge crane span compared with the typical process. The small span, light weight and high speed of the cart solves the problem of high-speed walking and high energy consumption of the bridge span of the long-span track field.

**Table 2.** Annual total energy consumption of single box gauge crane and conventional crane

| Arrival density (min) | Conventional crane (kwh) | Single box gauge crane (kwh) | Declining energy consumption (%) |
|---|---|---|---|
| NEGEXP (1060) | 3453111 | 2739257 | 21 |
| NEGEXP (1010) | 3540575 | 2891849 | 18 |
| NEGEXP (960) | 3623386 | 2910923 | 20 |
| NEGEXP (910) | 3656482 | 2942713 | 20 |
| NEGEXP (860) | 3662840 | 2955513 | 19 |

*5.2.2.2 Operation Ability Analysis* Statistical analysis of the average annual operating capacity of the two processes under five working conditions, the test data are shown in Table 3 and Table 4.

With the shortening of the ship arrival time interval, the daily occupancy rate of the import and export container yard and the equipment utilization rate increase, the yard operation enters a busy state, but under the five working conditions, the average utilization rate of the new process loading and unloading equipment is not high. The utilization rate of the single-box gauge crane does not exceed 20%. This is because the number of yards in the new process is large, the corresponding yard equipment is configured, and the loading and unloading operations are scattered, resulting in a decrease in the utilization rate of the gantry crane. By observing the running process of the model, it is found that the AGV under the quay crane has a queue waiting for operation, which will affect the loading and unloading operation of the box area to a certain extent.

In addition, under the new process, the loading and unloading operation time of the ship and the average residence time of the external collection card are significantly lower than the typical process, which reflects the rapid travel speed of the crane in the new process yard, high operating efficiency, good road traffic conditions, and overall service capacity of the terminal. Higher; as ships arrive in port, the ship-time efficiency of the new process is still higher than that of the typical process. It can be seen that the new process terminal has a strong ability to cope with peak operations. At the same time, in the same simulation time, the number of service ships and the annual container throughput of the terminal under the new technology are larger than the typical process, and the overall capacity of the terminal is higher. However, compared with the number of annual service ships of the two process terminals under the five working conditions, it is found that the growth rate from working condition 1 to working condition 2 is relatively large, but the

**Table 3.** Test data of terminal operation capacity under two loading and unloading processes

| Arrival density (min) | Number of service ships per year | | Annual throughput (10, 000 TEU) | | Ship time efficiency (move/h) | | Average working time of the ship (h) | | Average collection time of the outer card (min) | |
|---|---|---|---|---|---|---|---|---|---|---|
| | New process | Typical process | New process | Typical process | New process | Typical process | New process | Typical process | New process | Typical process |
| NEGEXP (1060) | 436 | 433 | 113.81 | 113.18 | 92.19 | 76.58 | 14.43 | 17.24 | 3.44 | 16.64 |
| NEGEXP (1010) | 460 | 444 | 118.85 | 115.98 | 92.56 | 76.39 | 14.19 | 17.25 | 3.46 | 16.56 |
| NEGEXP (960) | 463 | 454 | 119.52 | 118.81 | 89.8 | 75.47 | 14.93 | 17.51 | 3.53 | 16.54 |
| NEGEXP (910) | 468 | 457 | 121.51 | 119.47 | 93.31 | 74.74 | 14.33 | 17.66 | 3.52 | 16.39 |
| NEGEXP (860) | 470 | 456 | 122.86 | 118.66 | 93.11 | 74.87 | 13.79 | 17.52 | 3.56 | 16.05 |

**Table 4.** Test data of terminal operation capacity under two loading and unloading processes

| Arrival density (min) | Yard Average daily occupancy of import box (%) | | Yard Average daily occupancy of export box(%) | | Berth Utilization rate (%) | | Shore crane utilization (%) | | Gantry crane utilization (%) | |
|---|---|---|---|---|---|---|---|---|---|---|
| | New process | Typical process | New process | Typical process | New process | Typical process | New process | Typical process | New process | Typical process |
| NEGEXP (1060) | 53 | 50 | 30 | 78 | 44.50 | 46 | 41.81 | 47.60 | 11.51 | 49.09 |
| NEGEXP (1010) | 54 | 65 | 32 | 81 | 45 | 47 | 47.20 | 48.83 | 11.93 | 50.84 |
| NEGEXP (960) | 58 | 72 | 40 | 83 | 47 | 50 | 44.79 | 50.66 | 12.29 | 52.41 |
| NEGEXP (910) | 56 | 87 | 42 | 81 | 45 | 53 | 49.50 | 51.44 | 12.74 | 53.30 |
| NEGEXP (860) | 49 | 89 | 43 | 87 | 41 | 53 | 41.00 | 50.88 | 12.64 | 53.51 |

growth rate from working condition 2 to working condition is gradually decreasing, and the terminal gradually enters the operational bottleneck. Comprehensive analysis, the new loading and unloading technology of single-box gauge crane has certain feasibility, but there are also problems of interference in the box area and traffic disturbance. It is recommended to improve the operation management level to reduce the interference problem by optimizing the container storage strategy and equipment scheduling strategy.

## 6 Conclusion

In the typical process terminal with "quay crane + AGV + rail gantry crane" as the core, the long travelling mechanism of gantry crane becomes the main working mechanism, which has the characteristics of Heavy weight, long stroke and high energy consumption. And the single-box gauge crane in the process, the cranes of the yard are small in weight, fast in operation, low in energy consumption, and the initial investment cost of the terminal is low, and the land utilization rate is high utilization.

With the large-scale development of container ships, the shore container cranes have also become larger. However, the size of horizontal transport vehicles and containers handed over to container gantry cranes has not changed. Through the simulation analysis of the typical process and the single-box gauge crane process, it is found that the small equipment such as the single-box gauge crane can also meet the loading and unloading requirements, and the working efficiency is high. It can be seen that the miniaturized yard equipment is also worthy of further study.

In order to cope with the interference of the box interval and the problem of traffic disturbance caused by the narrow width of the yard box area of the single-box gauge crane process and no operation buffer area, the container of the same voyage should adopt the distributed storage strategy at the yard. At the same time, higher requirements are put forward for the formulation and management level of the terminal operation plan, that is, the information and intelligent level of the terminal needs to be improved.

## References

1. Kumar, V.K.N., Srinivasan, B.: Implementation and performance effect on electronic procurement and its ship management companies. Int. J. Inf. Eng. Electr. Bus. **5**(5), 10–16 (2013)
2. Chengduan, W.: Research and implementation on the flexible feasibility of AGV routing. Int. J. Educ. Manage. Eng. **2**(9), 22–28 (2012)
3. Meng, Y., Yanwei, Z.: Multi-agent-based fuzzy dispatching for trucks at container terminal. Int. J. Intell. Syst. Appl. **2**(2), 41–47 (2010)
4. Wiese, J., Suhl, L., Kliewer, N.: Mathematical programming and simulation based layout planning of container terminals. Int. J. Simul. Process Model. **5**(4), 313–323 (2009)
5. Jaehn, F., Kress, D.: Scheduling cooperative gantry cranes with seaside and landside jobs. Discret. Appl. Math. **242**, 53–68 (2017)
6. Qiang, Z., Qian, H., Jiancheng, L., Xiaocui, W.: New handling technology and performance evaluation of elevated tail transit container terminal. Port Waterw. Eng. **10**, 67–70 (2014)
7. Hao, L., Qinhua, T.: Loading and unloading technological schemes for new type automatic container terminals. Port Waterw. Eng. **01**, 158–163 (2011)

8. Luo, X.: Comparison of horizon transportation system of full automatic container terminal. Port Waterw. Eng. (9), 76–82 (2016)
9. Silva, C.A., Soares, C.G., Signoret, J.P.: Intermodal terminal cargo handling simulation using Petri nets with predicates. Proc. Inst. Mech. Eng., Part M: J. Eng. Marit. Environ. **229**(4), 323–339 (2015)
10. Taner, M.E., Kulak, O., Koyuncuoğlu, M.U.: Layout analysis affecting strategic decisions in artificial container terminals. Comput. Ind. Eng. **75**, 1–12 (2014)
11. Yu, J., Chen, L., Tang, G.: 3D simulation modeling of yard operation in a container terminal. In: Proceedings - 30th European Conference on Modelling and Simulation, ECMS, 87–92 (2016)
12. Pasquale, L., Mary, M.R.: A simulation model for designing straddle carrier-based container terminals. In: Proceedings - 2017 Winter Simulation Conference, WSC, 3138–3149 (2017)
13. Hu, Z.: Ningbo container terminal "berth-quayside bridge-container truck" optimization of work schedule. Sci. J. Intell. Syst. Res. **2**(10) (2020)

# Author Index

**A**
Alienin, Oleg, 3, 131

**B**
Bai, Qiang, 141
Berezko, Oleksandr, 171

**C**
Cao, Jingjing, 54
Cao, Lu, 76
Chen, Gang, 307
Chen, Gang'ao, 254
Chen, Qinxian, 209
Chen, Yuan, 345
Chu, Jie, 54

**D**
Deng, Xudong, 336
Doms, Volodymyr, 131
Dong, Mingwang, 219
Du, Qiguang, 355

**F**
Fang, Liang, 190
Fedushko, Solomia, 23

**G**
Gao, Ning, 336
Gordienko, Nikita, 3
Gordienko, Yuri, 3, 44, 131
Gu, Yong, 219, 277
Guo, MengMeng, 307

**H**
He, Zhiling, 298

**I**
Izonin, Ivan, 23

**J**
Jian, Huiling, 161
Jiang, Hong, 345
Jiang, Xiaocheng, 161

**K**
Kang, Di, 307
Kochura, Yuriy, 131
Kostiukevych, Kostiantyn, 3
Koziy, Dmytro, 23

**L**
Li, Bingbing, 265
Li, Hao, 254
Li, Liwei, 238
Li, Xin, 345
Li, Yanjia, 254
Li, Yu, 277
Lin, Yun, 238
Liu, Ao, 336
Liu, Hanzhe, 180
Liu, Na, 327
Liu, Ping, 85
Liu, Yanhui, 190
Liu, Yong, 96
Liu, ZhiPing, 141
Liu, Zhiping, 228


Z. Hu et al. (Eds.): ICAILE 2021, LNDECT 82, pp. 379–380, 2021.
https://doi.org/10.1007/978-3-030-80475-6

Liu, Zili, 265
Lu, Liuqing, 238

M
Ma, Wei, 85
Meng, Ruiceng, 190

O
Olga, Shurchkova, 54

P
Pan, Jiali, 317
Pan, Yanfang, 238
Petoukhov, Sergey V., 66, 288

Q
Qiu, Yi, 219

R
Rokovyi, Oleksandr, 3, 131

S
Shan, Qi, 200
Shilinh, Anna, 171
Statkevych, Roman, 44
Stirenko, Sergii, 3, 44, 131
Sun, Qunqun, 200

T
Tao, Dexin, 228
Tkachenko, Roman, 23
Tolokonnikov, Georgy K., 66

V
Vovk, Olena, 23

W
Wang, Chuanlei, 161
Wang, Haiyuan, 254
Wang, Lulu, 265
Wang, Meng, 32, 76, 107
Wang, Qi, 141
Wang, Ruiyu, 298
Wang, Ting, 365
Wang, Yong, 13, 317, 345
Wang, Zhengguo, 118
Wei, XueJiang, 107
Wei, Xuejiang, 32
Wen, Fuan, 96
Wu, Jingjing, 152

X
Xia, Jinling, 152
Xia, Qing, 32
Xie, Yumeng, 13
Xing, Saipeng, 209
Xing, YaHui, 141

Y
Yang, YanFang, 141
Yang, Zanxing, 254
Ye, Xuanzhi, 345
Yin, Fei, 54
Yu, Meng, 298
Yu, Mingzhu, 180

Z
Zeng, Kai, 96
Zhang, Hengyu, 355
Zhang, Peilin, 13
Zhang, Xiaojun, 327
Zhang, XiaoShu, 141
Zhang, Yi, 85
Zhang, Ying, 228
Zhang, Yuyan, 96
Zhang, Zhuoni, 118
Zheng, Jingjing, 365
Zheng, Yangyang, 277
Zheng, Zhong, 238
Zhezhnych, Pavlo, 171
Zhou, Qiang, 365
Zhou, Xiaoguang, 190
Zhou, Xingjian, 317
Zhu, Haojie, 228
Zub, Khrystyna, 23, 171

Printed by Printforce, the Netherlands